T0183827

Lecture Notes of the Institute for Computer Sciences, Social Informatics and Telecommunications Engineering 271

More information about this series at http://www.springer.com/series/8197

Bo Li · Mao Yang ·
Hui Yuan · Zhongjiang Yan (Eds.)

IoT as a Service

4th EAI International Conference, IoTaaS 2018
Xi'an, China, November 17–18, 2018
Proceedings

 Springer

Editors
Bo Li
Northwestern Polytechnical University
Xi'an, China

Mao Yang
Northwestern Polytechnical University
Xi'an, China

Hui Yuan
Shandong University
Jinan, Qinghai, China

Zhongjiang Yan
Northwestern Polytechnical University
Xi'an, Shaanxi, China

ISSN 1867-8211   ISSN 1867-822X   (electronic)
Lecture Notes of the Institute for Computer Sciences, Social Informatics
and Telecommunications Engineering
ISBN 978-3-030-14656-6   ISBN 978-3-030-14657-3   (eBook)
https://doi.org/10.1007/978-3-030-14657-3

Library of Congress Control Number: 2019932788

This Springer imprint is published by the registered company Springer Nature Switzerland AG
The registered company address is: Gewerbestrasse 11, 6330 Cham, Switzerland

Preface

IoTaaS is endorsed by the European Alliance for Innovation, a leading community-based organization devoted to the advancement of innovation in the field of ICT. The 4th International Conference on IoT as a Service (IoTaaS) aimed to contribute to the discussion on the challenges posed by the Internet of Things (IoT). The IoTaaS conference aims to bring together researchers and practitioners interested in IoT from academia and industry. IoTaaS attendees present novel ideas, exchange points of view, and foster collaborations. In 2018, the city of Xi'an, a famous historical and cultural city of China, hosted IoTaaS 2018.

IoTaaS 2018 consisted of two technical tracks and three workshops: Networking and Communications Technologies for IoT, IoT as a Service, International Workshop on Edge Computing for 5G/IoT, International Workshop on Green Communications for Internet of Things, and International Workshop on Space-Based Internet of Things. IoTaaS has become one of the major events in these areas in the Asia-Pacific region. It has been successful in encouraging interactions among participants, exchanging novel ideas, and disseminating knowledge.

Following the great success of the past IoTaaS 2014–2017 events, IoTaaS 2018 received more than 80 submitted papers, out of which 58 papers were selected for presentation. The Technical Program Committee (TPC) did an outstanding job in organizing a diverse technical program consisting of 12 symposia that covered a broad range of research areas in IoT technologies. Under the excellent leadership of the TPC co-chairs, Prof. Song Xiao and Prof. Yong Li, the TPC members handled the reviews of papers, with more than three reviews per paper on average.

The technical program featured three outstanding keynote speakers, who presented their vision of IoT in theory and practice: Prof. Yi-Bing Lin, Chiao Tung University (NCTU), Taiwan; Prof. Der-Jiunn Deng, Changhua University of Education, China; Prof. Pinyi Ren, Xi'an Jiaotong University, China.

We would like to thank the TPC co-chairs, TPC members, all the reviewers, the workshop co-chairs, the Web chairs, the publication chair, the local chairs, and all the members of the Organizing Committee, for their assistance and efforts to make the conference succeed. The continuing sponsorship by EAI and Springer is gratefully acknowledged. We also express our appreciation to the conference keynote speakers, tutorial speakers, paper presenters, and authors.

February 2019   Bo Li

Conference Organization

Steering Committee

Imrich Chlamtac   University of Trento, Italy

Organizing Committee

General Chair

Bo Li   Northwestern Polytechnical University, China

General Co-chair

Chilian Chen   China Electronics Technology Group Corporation (CETC), China

TPC Chair and Co-chairs

Song Xiao   Xidian University, China
Yong Li   Tsinghua University, China

Sponsorship and Exhibit Chair

Xiaoya Zuo   Northwestern Polytechnical University, China

Local Chair

Zhongjiang Yan   Northwestern Polytechnical University, China

Workshops Chair

Yong Li   Tsinghua University, China

Publicity and Social Media Chair

Hongwei Zhao   Northwestern Polytechnical University, China

Publications Chair

Hui Yuan   Shandong University, China

Web Chair

Mao Yang   Northwestern Polytechnical University, China

Technical Program Committee

Jinglun Shi	South China University of Technology, China
Jialiang Lu	Shanghai Jiao Tong University, China
Zhenyu Xiao	Beijing University of Aeronautics and Astronautics, China
Xiang Chen	Sun Yat-sen University, China
Shaohui Mei	Northwestern Polytechnical University, China
Zhong Shen	Xidian University, China
Qinghe Du	Xi'an Jiaotong University, China
Jianchao Du	Xidian University, China
Yong Niu	Beijing Jiaotong University, China
Yongqiang Hei	Xidian University, China
Guifang Li	Northwestern Polytechnical University, China
Zhengchuan Chen	Singapore University of Technology and Design, Singapore
Xiaoyan Pang	Northwestern Polytechnical University, China
Yongqian Du	Northwestern Polytechnical University, China
Mingwu Yao	Xidian University, China
He Guan	Northwestern Polytechnical University, China
Min Zhu	Xidian University, China
Zhongjin Liu	Coordination Center of China, China
Shimin Li	Winona State University, USA
Rongfei Fan	Beijing Institute of Technology, China
Zhou Zhang	China Electronic System Engineering Company, China
Xiaoya Zuo	Northwestern Polytechnical University, China
Ruonan Zhang	Northwestern Polytechnical University, China
Jie Zheng	Northwest University, China
Jingling Li	China Academy of Space Technology, Xi'an, China
Yiming Liu	China Academy of Electronics and Information Technology, China

Contents

IoT as a Service

Networking Technology for IoT

Multiple Access and Communication Technologies for IoT

Workshop on Edge Computing for 5G/IoT

Workshop on Green Communications for Internet of Things

Workshop on Space-Based Internet of Things

IoT as a Service

FrameTalk: Human and Picture Frame Interaction Through the IoT Technology

Wen-Shu Lai[2], Yi-Bing Lin[1], Chung-Yun Hsiao[1(✉)], Li-Kuan Chen[1], Chao-Fan Wu[2], and Shu-Min Lin[2]

[1] Department of Computer Science, National Chiao Tung University, Hsinchu 30010, Taiwan, R.O.C.
liny@cs.nctu.edu.tw, phoebe.cyhsiao@gmail.com, clikuan.cs05g@g2.nctu.edu.tw
[2] Institute of Applied Arts, National Chiao Tung University, Hsinchu 30010, Taiwan, R.O.C.
wndylai@gmail.com, dannylinsam@gmail.com, 294889468@qq.com

Abstract. The concept of frame plays a significant role in art design. However, traditional picture frames do not allow interaction with human. To provide another dimension of artwork demonstration of picture frames, we utilize the Internet of Things (IoT) technology to create human and picture frame interaction. Specifically, an application-layer IoT device management platform called IoTtalk is used to quickly establish connections and meaningful interactions between IoT devices. The frame and human interaction is achieved by implementing an output device of IoTtalk called FrameTalk, which provides animation in a digital frame to be controlled by the input devices (typically a smartphone). The FrameTalk device can be displayed in specific frame hardware or the screen of any computing device. We describe the functional block diagram of the FrameTalk device and its default applications. Then we show two FrameTalk application examples. The first example is CalligraphyTalk that allows the audience to interact with the frame to write the poems. The second example is PortraitGuess that allows the audience to guess historical figures through the frame.

Keywords: IoTtalk · Interactive visual design · Cyber physical interaction

1 Introduction

Picture frame has been used to demonstrate artworks since ancient times. The importance of the frame was emphasized by a quote of American novelist Chuck Palahniuk: "It's funny how the beauty of art has so much more to do with the frame than the artwork itself." Famous Canadian director David Cronenberg used "frame" to encourage the photographers: "That's rule number one for a photographer, isn't it? Fill your frame?" Although the concept of frame plays a significant role in art design, traditional picture frames do not allow interaction with human. In recent years, some commercial digital frames can be controlled to display the static image through mobile apps [1] or hand gestures [2]. Some can display dynamic graphics [3, 4]. To provide

© ICST Institute for Computer Sciences, Social Informatics and Telecommunications Engineering 2019
Published by Springer Nature Switzerland AG 2019. All Rights Reserved
B. Li et al. (Eds.): IoTaaS 2018, LNICST 271, pp. 3–11, 2019.
https://doi.org/10.1007/978-3-030-14657-3_1

another dimension of artwork demonstration of picture frames, this paper utilizes the Internet of Things (IoT) technology to create human and picture frame interaction. Specifically, an application-layer IoT device management platform called IoTtalk [5–7] is used to quickly establish connections and meaningful interactions between IoT devices without concerning the lower-layer IoT protocols. Figure 1 illustrates the simplified IoTtalk network architecture.

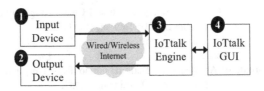

Fig. 1. IoTtalk network architecture.

In this architecture, the IoT devices (Fig. 1(1) and (2)) are connected to the IoTtalk engine (Fig. 1(3)) in the Internet through wireline or wireless technologies. An IoT device is called an input device if it is a group of sensors (such as color sensor and temperature sensor) or controllers (such as switches and buttons). The input device generates and sends data to the IoTtalk engine for processing (see path (1) → (3) in Fig. 1). Similarly, an IoT device is called an output device if it is a group of actuators (e.g., robot arm). The output device receives data from the IoTtalk engine (see path (3) → (2) in Fig. 1) to drive its actuators. In IoTtalk, every IoT device is characterized by its functionalities or "device features". A device feature (DF) is a specific input or output "capability" of the IoT device. For input devices such as a smartphone, the input device features (IDFs) are sensors (e.g., the microphone, the Gyroscope, the GPS, and the camera) or controllers (e.g., the keypad). For output devices such as an electric fan, the output device features (ODFs) are actuators including motor rotation, motor speed and so on. A network application defining the interaction between the IoT devices is automatically generated and executed at the IoTtalk engine. When the values of the IDFs are updated, an IoT device informs the network application to take some actions, and the network application sends the result to the ODF of the same or another IoT device to affect the actuators of that output device. With this view, the IoT devices interact with each other through their device features. Details of IoTtalk operations can be found in [5–7].

The frame and human interaction in IoTtalk is achieved by implementing an output device called FrameTalk, which provides animation in a digital frame to be controlled by the input devices. The FrameTalk device is cyber (an animation program) that can be displayed in specific frame hardware or the screen of any computing device. This paper is organized as follows. Section 2 describes the functional block diagram of the FrameTalk device and its default applications. Sections 3 and 4 show two FrameTalk applications and examples for interaction. Section 5 concludes our work with future research directions.

2 The FrameTalk Device

A FrameTalk device is implemented as a Java Processing program [8] with the functional block diagram described as follows. The message handler (MsgHandler; Fig. 2(1)) receives instructions from the IoTtalk engine through the ODFs of the FrameTalk device. There are two default ODFs. The *Mode* ODF determines which frame application to be displayed. A default application is Weather (Fig. 2(2)) to be discussed later. Another default ODF *Name-O* receives the folder name that corresponds to a painting file folder stored in the database (Fig. 2(3)).

A FrameTalk device is equipped with a QR code scanning mechanism (Fig. 2(4)). When the device detects that a person may want to interact with the frame, a QR code pops up at the right corner at the bottom of the frame display area (Fig. 2(5)). The person uses a smartphone to scan the QR code to enter the webpage of the FrameTalk application. In this way, the smartphone can interact with the frame through its browser without installing any mobile app. We will elaborate more on the mobile app in Sect. 4.

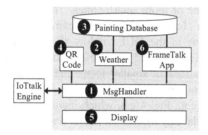

Fig. 2. Functional block diagram of a FrameTalk device.

Fig. 3. The frame automatically reflects the weather at this moment: (a) sunny with scattered clouds, (b) rainy, (c) cloudy.

The default display mode is weather-painting, where the frame executes the Weather module to automatically display a painting corresponding to the current weather condition [9]. For example, the frame displays "Impression, Sunrise" of Monet when it is sunrise with scattered clouds outdoors (Fig. 3(a)). The frame shows "The Umbrellas" by Renoir when it is raining (Fig. 3(b)), and shows "Autumn Landscape with a Flock of Turkeys" by Millet when it is cloudy (Fig. 3(c)). These paintings are stored in the database in Fig. 2(3). The weather information is provided by any input device that can detect or predict the local weather condition. The weather conditions can be 0 (clear sky), 1 (few clouds), 2 (scattered clouds), 3 (broken clouds), 4 (shower rain), 5 (rain), 6 (thunderstorm), 7 (snow), and 8 (mist).

Besides the weather application, we can develop new applications as software modules (Fig. 2(6)) to be integrated in the FrameTalk device. Switching among the applications is controlled by the *Mode* ODF. We will describe the development of two FrameTalk applications in the subsequent sections.

3   CalligraphyTalk: A Cyber Calligrapher that Talks

CalligraphyTalk is a FrameTalk application including the following input devices: Weather, Intel-Cam and one or more smartphones. The output device is a FrameTalk device called Calligrapher. IoTtalk implements interaction of these IoT devices by configuring the connections in the GUI window (Fig. 1(4)). In this GUI, an input device is represented by an icon placed at the left-hand side of the window (e.g., Fig. 4 (a), (c) and (e)), which consists of smaller icons that represent IDFs (e.g., Fig. 4(b), (d) and (f)). Similarly, an output device is represented by an icon placed at the right-hand side of the window (e.g., Fig. 4(g)), which includes ODF icons (e.g., Fig. 4(h)–(l)). For the purpose of readability, if both an IDF and an ODF have the same name, the GUI will append the IDF name with "-*I*" and the ODF name with "-*O*". For example, the configuration in Fig. 4 includes the *Name-I* IDF and the *Name-O* ODF. Figure 5 illustrates the functional block diagrams of the input and the output devices, and the connections correspond to the configuration in Fig. 4.

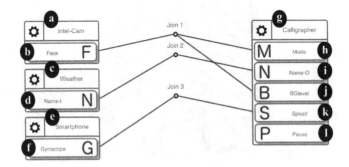

Fig. 4. The IoTtalk GUI for a CalligraphyTalk application.

Besides the weather-painting mode described in Sect. 2, CalligraphyTalk includes the interactive-calligraphy mode. Switching between the two modes is determined by

Intel-Cam (Fig. 4(a)). The hardware of Intel-Cam is an Intel RealSense Camera (model F200 or SR300), which can detect human facial expressions. The detection functions are provided by Intel RealSense SDK [10] as shown in the functional block diagram (Fig. 5(1)). When N persons stand in front of the frame, their faces are detected, and the *Face* IDF of Intel-Cam sends out the value N (the interactive-calligraphy mode). If no face is detected, it outputs value 0 (the weather-painting mode). In the weather-painting mode, the Weather input device (Fig. 4(c)) uses IP Geolocation API and OpenWeatherMap API (Fig. 5(2)) [9] to obtain the real-time weather condition (clear sky, raining and so on) of the frame's location, and output the corresponding folder name of the painting files through the *Name-I* IDF (Fig. 4(d)). The painting files are stored in the calligraphy database (Fig. 5(4)).

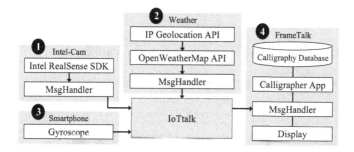

Fig. 5. The functional block diagrams of the CalligraphyTalk application.

The interactive-calligraphy mode is implemented as a FrameTalk application module (Fig. 2(6)), which allows the audiences to interact with the frame to write poems with different speeds. Besides the default ODFs (*Mode* and *Name-O*), Callig- rapher implements three ODFs including *BGlevel* (Fig. 4(j)), *Speed* (Fig. 4(k)) and *Pause* (Fig. 4(l)). Through Join 1, the *Mode* ODF is connected to the *Face* IDF of Intel-Cam to determine the display mode of Calligrapher. Through Join 2 the *Name-O* ODF is connected to the *Name-I* IDF of Weather to decide the "weather condition". The usage of the folder name in the weather-painting mode is described in Sect. 2. In the interactive-calligraphy mode, the folder name is used to determine the background animation pattern. For example, the calligraphy background pattern for a sunny day is shown in Fig. 6. The background pattern for a rainy day is shown in Fig. 7.

The *BGlevel* ODF gives the degree to emphasize the effect of the background pattern. There are three levels (from 0 to 2) as illustrated in Figs. 6 and 7. In the example in Fig. 4, the *BGlevel* ODF is connected to the *Face* IDF of Intel-Cam through Join 1. Therefore, if more people stand in front of the frame, the background effect will be more significant.

The *Speed* ODF (Fig. 4(k)) determines the writing speed of the poem. In our example, the ODF is connected to the *Gyroscope* IDF of a smartphone through Join 3. With the QR code scanning mechanism described in Sect. 2, the smartphone can connect the Gyroscope sensor to the FrameTalk applications without installing any mobile app. Therefore, any person with arbitrary smartphone can enjoy the interaction with CalligraphyTalk.

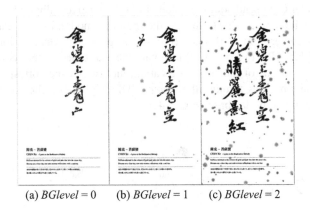

(a) *BGlevel* = 0   (b) *BGlevel* = 1   (c) *BGlevel* = 2

Fig. 6. The *BGlevel* for a sunny day.

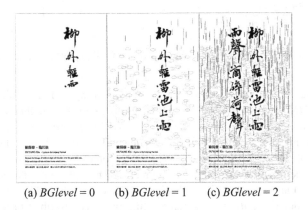

(a) *BGlevel* = 0   (b) *BGlevel* = 1   (c) *BGlevel* = 2

Fig. 7. The *BGlevel* for a rainy day.

4   PotraitGuess: Guess Historical Figures

This section describes PotraitGuess, a FrameTalk application that uses a smartphone (the input device) to interact with a digital painting frame (the output device) for portrait guess [9]. Like CalligraphyTalk, PotraitGuess includes two interaction modes with the painting frame: weather-painting and portrait-guess. The portrait-guess mode allows an audience to play a portrait guess game with the picture frame through a smartphone without installing any mobile app. This game has been used in the history classes to guess the historical figures, which significantly attract attention of the students. Figure 8(a) illustrates the frame hardware. We use Intel-Cam to detect if someone stands in front of the frame. If so, a QR code pops up to be scanned by a smartphone to enter the portrait guess game webpage called PG-web (Fig. 8(b)). In our design, only one smartphone can interact with the frame at a time. Therefore, after a smartphone has scanned the QR code, other people cannot access the QR code anymore to avoid interference of game playing.

When the person presses the "Play" button, the frame shows the first feature of a portrait (Fig. 9(a)). Then the smartphone displays the message "Who am I?" and lists five buttons labelled five names of historical figures to be chosen (Fig. 8(c)). If the person presses a button with a wrong name, the smartphone shows the message "Incorrect answer. Try again!" and the frame displays the next portrait feature (Fig. 9 (b)). When the person presses the button labelled with the correct name, the smartphone shows the message "You get it" and the "Play Again" button (Fig. 8(d)). The frame displays the remaining portrait features one by one (Fig. 9(b) → (c) → (d)), and stops at the final (complete) portrait (Fig. 9(d)). The person presses "Play Again" button to start a new game.

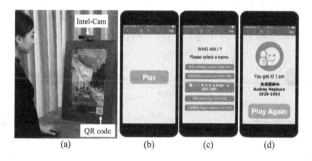

Fig. 8. (a) Entering the portrait-guess mode. (b)–(d) The smartphone portrait guess game webpage.

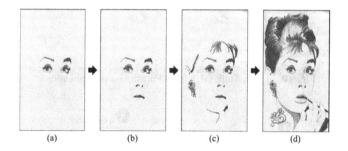

Fig. 9. The painting feature displaying.

Figure 10 illustrates the connections of the input and the output devices, where the corresponding functional block diagrams of the IoT devices are illustrated in Fig. 11. The Weather input device (Fig. 10(a)) and the Intel-Cam input device (Fig. 10(c)) are the same as those in Fig. 4. The portrait-guess mode is implemented as a FrameTalk application module (Fig. 11(2)). This module includes a painting database storing several portrait folders, where each folder has one or more picture files (e.g., (a)–(d) in Fig. 9). In Fig. 10(i), two default ODFs (Fig. 10(j) and (k)) are connected to Weather and Intel-Cam through Joins 1 and 2 just like CalligraphyTalk. PortraitGuess has two extra ODFs besides the default ODFs. The *Forward* ODF (Fig. 10(l)) triggers PortraitGuess to display the next picture in the portrait folder. The *End* ODF (Fig. 10(m)) triggers PortraitGuess to display all remaining pictures in the folder sequentially.

In the PotraitGuess application, the PotraitGuess webpage PG-web (Fig. 8(b)–(d)) is browsed by a smartphone. PG-web in the smartphone (Fig. 11(1)) implements three IDFs including *Name-I* (Fig. 10(f)), *Wrong* (Fig. 10(g)) and *Correct* (Fig. 10(h)). When the audience presses the "Play" button in Fig. 8(b), PG-web selects a portrait name, and sends the name through the *Name-I* IDF to the *Name-O* ODF of PortraitGuess via Join 3 in Fig. 10. When *Name-O* receives the portrait name, the frame shows the first feature of the corresponding portrait (Fig. 9(a)). Then PG-web displays five name buttons. If the audience presses a button with a wrong name, the *Wrong* IDF sends a signal to the *Forward* ODF via Join 4, and the frame displays the next portrait feature. When the audience presses the correct name button, the *Correct* IDF sends a signal to the *End* ODF via Join 5. Then the frame displays the remaining portrait features one by one, and finally stops at the complete portrait.

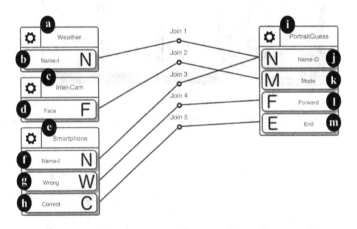

Fig. 10. The IoTtalk GUI for a PortraitGuess application.

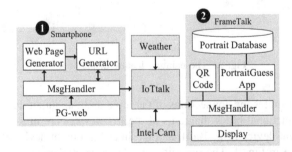

Fig. 11. The functional block diagrams of the PotraitGuess application.

5   Conclusion

This paper designed and developed a frame and human interaction platform based on an IoT service platform called IoTtalk. With digital frame hardware, we implemented an IoT device called FrameTalk, which can be connected to an arbitrary smartphone to perform frame art interaction without installing any mobile app. We used two examples to demonstrate how to create FrameTalk applications. The first example is CalligraphyTalk that allows the audience to interact with the frame to write the poems. Based on CalligraphyTalk, an artwork called "When the Cloud Rising" won a Finalists' Award in Asia Digital Art Award FUKUOKA in 2018 [11]. The second example is PortraitGuess that allows the audience to guess historical figures through the frame. PortraitGuess was one of the 2017 FarEastone IoT Contest award recipients [12]. A demo for CalligraphyTalk can be found in [13].

References

1. Nixplay. https://www.nixplay.com. Accessed 7 May 2018
2. Meural. https://meural.com. Accessed 7 May 2018
3. Yiiisu. http://yiiisu.com. Accessed 7 May 2018
4. FRAMED. https://frm.fm. Accessed 7 May 2018
5. Lin, Y.B., et al.: EasyConnect: a management system for IoT devices and its applications for interactive design and art. IEEE IoT J. **2**(6), 551–561 (2015)
6. Lin, Y.B., Lin, Y.W., Huang, C.M., Chih, C.Y., Lin, P.: IoTtalk: a management platform for reconfigurable sensor devices. IEEE IoT J. **4**(5), 1552–1562 (2017)
7. Lin, Y.W., Lin, Y.B., Hsiao, C.Y., Wang, Y.Y.: IoTtalk-RC: sensors as universal remote control for aftermarket home appliances. IEEE IoT J. **4**(4), 1104–1112 (2017)
8. Processing. http://www.processing.org. Accessed 7 May 2018
9. Hsiao, C.Y., Chen, L.K., Lin, Y.B.: Interacting with paintings using IoTtalk. In: IoT 2017, Proceedings of the Seventh International Conference on the Internet of Things. ACM, New York (2017). https://doi.org/10.1145/3131542.3140269
10. Hsiao, C.Y., Huang, C.C., Lin, Y.B., Lin, Y.W.: Flower sermon: an interactive visual design using IoTtalk. Mobile Netw. Appl. J. (2018). https://doi.org/10.1007/s11036-018-1086-z
11. Asia Digital Art Award FUKUOKA. https://adaa.jp/en/winners/winners2017.html. Accessed 7 May 2018
12. PortraitGuess demo video. https://www.youtube.com/watch?v=j7V7Xa0hf30&feature=youtu.be. Accessed 9 Oct 2018
13. CalligraphyTalk demo video. https://www.youtube.com/watch?time_continue=2&v=1KGgCiFVwPK. Accessed 9 Oct 2018

Physiognomy in New Era: A Survey of Automatic Personality Prediction Based on Facial Image

Xu Jia[1,2(⊠)], Weijian Tian[2], and Yangyu Fan[2]

[1] North China University of Science and Technology, Tangshan 063210, China
Xujiajia_2008@163.com
[2] Northwestern Polytechnical University, Xi'An 710129, China

Abstract. At present, personality computing technology facilitates the understanding, prediction, and management of human behavior. With the increasing importance of faces in personal daily assessments, establishing a relationship between facial morphological features and personality traits is a major breakthrough in personality computing technology. This paper is a survey of such technology of automatic personality prediction based on face and it aims at providing not only a solid knowledge base about the state-of-the-art in automatic personality prediction, but also to provide a conceptual model of automatic personality prediction, based on the literature. In addition, the analysis of the prediction results of the existing researches is emphasized, and there are still problems in the field, such as lack of information on research data, single age group of the sample population, incomplete design characteristics of the artificial design etc., and the potential applications and development directions are determined.

Keywords: Personality prediction · Face recognition · Machine learning

1 Introduction

1.1 Status Quo of Research on Personality Calculation

Since the Greek philosopher Theophrastus (371 BC - 287 BC), people have been keen on the study of personality [1]. As a construct, personality aims at capturing stable individual characteristics, typically measurable in quantitative terms, that explain and predict observable behavioral differences [2]. Current research on the personality model successfully predicts the relationship between 'thinking, emotions and behavioral patterns' [3] and the important aspects of life (including 'happiness, physical and mental health, quality of interpersonal relationships, career choices, professional satisfaction and professional performance, participation in social activities, criminal activities and political ideology,' etc. [4] have had some beneficial effects. Furthermore, attitude and social behavior towards a given individual depend to a significant extent, on the personality impression others develop about her [5].

Given all of this, the ability to reliably, effectively and efficiently evaluate personality is a valuable goal, and with the advent of the era of big data, modern computer

© ICST Institute for Computer Sciences, Social Informatics and Telecommunications Engineering 2019
Published by Springer Nature Switzerland AG 2019. All Rights Reserved
B. Li et al. (Eds.): IoTaaS 2018, LNICST 271, pp. 12–29, 2019.
https://doi.org/10.1007/978-3-030-14657-3_2

science has the practical potential to advance this effort. In recent years, research on personality has been moving in some interesting directions, such as the combination of personality analysis and human-computer interaction technology. Figure 1 shows the number of papers including the word "personality" in the title on IEEE Xplore and ACM Digital Library, probably the two most important repositories of computing oriented literature. While being only the tip of the iceberg, most articles revolving around personality do not mention it in the title, these papers clearly show that the interest for the topic is growing and that the trend promises to continue in the fore-seeable future.

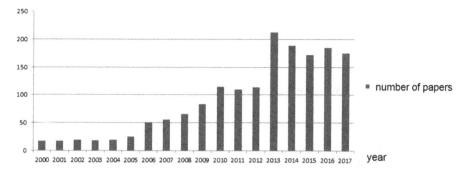

Fig. 1. The chart reports the number of papers per year with the word "personality" in their title (sum over IEEE Xplore and ACM Digital Library).

Some studies have shown that personality traits can predict user behavior by establishing an interaction between personality and automatic computation [6, 18, 32, 43, 52, 54]. Overall, personality is relevant to any computing area involving under-standing, prediction or synthesis of human behavior. Still, while being different and diverse in terms of data, technologies and methodologies, all computing domains concerned with personality consider the same three main problems: That is, (1) selec-tion of personality measurement methods, (2) selection and construction of data sets, and (3) establishment of calculation models. Through research, we found that there are many researches on automatic personality prediction, including: text-based [52], social media-based [6, 32, 54], mobile-based [16, 18], and computer games-based [43] and so on. However, there are few studies on automatic personality prediction based on human facial images. Figure 2 shows the number of papers containing the words 'face' and 'personality' in the headings of the IEEE Xplore and ACM digital libraries. Only a few articles have been searched since 2010. This paper mainly investigates the work of establishing the connection between personality and the use of computing technology based on face recognition. This is one of the main research issues in the development of personality calculation methods. Based on what we know, this is the first investigation of a solution to these problems.

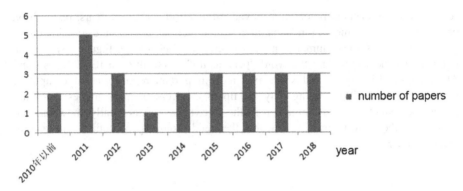

Fig. 2. The chart reports the number of papers per year with the word 'personality' and 'face' in their title (sum over IEEE Xplore and ACM Digital Library).

1.2   Status Quo of Face-Based Personality Prediction Research

Even in ancient China, Egypt and Greece, people had tried to establish the relationship between facial morphological features and personality traits of an individual. The modern psychological studies revealed that people tend to evaluate others on their appearance and then move on to interact with them based on these first impressions. Nowadays, it has been a well-established fact the face plays a central role in the everyday assessments of other people. For example, humans perform trait judgments from faces and the results of this unconscious behavior can sometimes decisively affect the results of important social events, such as an electoral process [7, 23], crime tracking [53], etc.

Currently, the following two points hold on automatic face evaluation: Firstly, some self-reported personality traits and the intelligence can be evaluated by the human based on the facial features to a certain extent. Secondly, the commonalities existing in the evaluation behavior of the human can be mined by the machine learning methods. The following are some of the relevant work in the literature.

The research on the first point mainly focuses on the following points: In [32], the authors studied the human tendency to evaluate others on their faces and identified some important facial features that generate first impressions. In [11], Humans can make valid inferences for at least four personality traits from facial features. In [52], the relationship between self-reported personality traits and first impressions was studied. Study on the second point: To investigate whether the trait evaluations performed by humans can be learned by computers, [41] used machine learning methods to construct an automatic trait predictor based on facial structural descriptors and appearance descriptors. They found that all the analyzed personality traits could be predicted accurately.

The rest of this paper is organized as follows: Sect. 2 introduces the concept of personality (with particular emphasis on trait based models and the Big Five) and the techniques for its measurement. Section 3 introduces the establishment of data sets in the automatic prediction model in different literatures. Section 4 introduces the extraction of face personality traits. Section 5 investigates the algorithm of face-based automatic personality prediction in different literatures. Section 6 aims to summarize the forecast results in the existing research and analyze the reasons; Sect. 7 analyze the problems that still exist in the field, and determine the potential applications and development directions.

2 Personality Model and Its Measurement

Personality psychology aims at predicting observable individual differences based on stable, possibly measurable, individual characteristics [2]. The theory of personality is representative of trait theory, type theory, and integration theory. Different theories use different' intrinsic attributes' as the basis of personality, including physiology (biological perspective), unconsciousness (psychological perspective), environment (behavioralist perspective), internal state (humanistic perspective), and mental (cognitive perspective), etc. (see extensive survey [8, 16]). However, the model that most effectively predicts the measurable aspects of people's lives is a model based on traits—the personality trait model, which is widely regarded as the 'structural one of the major achievements of psychology' [15]. On the other hand, several decades of research and experiments have shown that the same traits appear with surprising regularity across a wide spectrum of situations and cultures, suggesting that they actually correspond to psychologically salient phenomena [15]. This regularity and stability brings inspiration to the calculation of personality.

2.1 The Choice of Personality Trait Model

The trait theory holds that traits are the basic characteristics that determine individual behavior, are effective constituent elements of an individual, and are also the basic units commonly used for measuring personality. The main trait theories are: Alport's trait theory (common traits and personal traits), Cartel's personality trait theory (he believes that each person has these 16 traits), and Essen's three-factor model (outside (Pantastic, psychotic, and neurotic), the five-factor (large five personality model) model of Tarpez (external, pleasant, responsible, neurotic, and open), and the seven-factor model of Trigan (positive, negative, Titer, positive price, negative sentiment, reliability, appetite, and comparability). How to choose a trait model for our personality calculation is one of the issues discussed in many studies.

We found that the five factors of Tapes (the big five model can get relatively true and reliable personality traits, which laid the foundation for the accuracy of the experiment. Most of the works surveyed use the five-factor (Big Five) personality trait (More than 50% in the literature use the 'Big Five'). On the one hand, other theories can hardly reflect the dominance of feature-based models in personality psychology. On the other hand, the 'Five Factor Model (Big Five)' numerically expresses personality (see below), which is a form that is particularly suitable for computer processing. The Big-Five traits are as follows:

- Extraversion: Active, Assertive, Energetic, Outgoing, Talkative, etc.
- Agreeableness: Appreciative, Kind, Generous, Forgiving, Sympathetic, Trusting, etc.
- Conscientiousness: Efficient, Organized, Planful, Reliable, Responsible, Thorough, etc.
- Neuroticism: Anxious, Self-pitying, Tense, Touchy, Unstable, Worrying, etc.
- Openness: Artistic, Curious, Imaginative, Insightful, Original, Wide interests, etc.

Attempts were made to enrich the trait set with more dimensions, but 'Five- factor solutions were remarkably stable across studies, whereas more complex solutions were not' [16]. Other models considered less traits [13], but these still appeared to be linear combinations of theBig Five. In other words, the Big Five 'provide a set of highly replicable dimensions that parsimoniously and comprehensively describe most phenotypic individual differences' [43].

2.2 Existing Research Personality Model Selection

In the existing face character prediction works, their personality measurement methods are as follows: Qin et al. In the face-based personality and intelligence prediction research [37, 38], they uesd self-assessment, Cartel's Sixteen Personality Questionnaire (16PF) was used to measure participants' personality traits; Rojas et al. in the evaluation of personality based on facial feature points [40], they used others' assessments, each evaluator wrote an un-constrained description from a set of 66 standardized faces from the Karolinska [30] amateur actors face database. 1134 descriptions were collected, The researchers' classification of the unconstrained descriptions, resulted in 14 selected categories; In the study of Karin et al.'s interpretation of the impact of facial features on first impressions and personality [52], used the Cubiks Indepth Personality Questionnaire, CIPQ2.0, a normative self-report questionnaire scoring 17 personality traits covering the Big Five was used to measure the participants' personality traits; In his dissertation [54], Zeng also used the Big Five Personality Scale to test the Big Five personality traits using statistical analysis; In the study [6] of Al Moubayed et al., they used the method of others' assessments. The "Big Five Scale" was provided by 11 independent judges with white British ethnic background, normal vision, and no hearing problems; Oosterhof et al. [32] used Others' assessments, in the first phase of the project, they asked 55 participants to characterize the unrestricted faces, then classify these descriptions as feature dimensions, and extracted 14-dimensional feature dimensions, which are highly consistent with Rojas et al. [41]. The personality assessment methods used in all relevant studies are shown in Table 1.

Table 1. Personality assessment model in face-based personality prediction research.

Ref	Assessment type	Trait theory	Characteristic expression
[6]	Others' assessments	Big Five	Score(1–9)
[8]	Others' assessments	Manually select: Dominance, Warmth, Sociability and Credibility	Score
[9]	Others' assessments	Manually select: Intelligence, Maturity, Warmth, Sociality, Dominance and Credibility	Score(1–3)
[32]	Others' assessments	Manually select personality descriptors	Score(1–9)
[30, 41]	Others' assessments	Manually select personality descriptors	14-dimensional
[37, 38]	Self-assessments	16SP	Score(0–9)
[52]	Self-assessments	Big Five	Score(1–9)
[54]	Self-assessments	Big Five	Score(1–60)
[53]	Others' assessments	Manually select: Dominance, Attraction, Credibility and Extraversion	Score

3   Construction of Data Set

Focusing on face-based personality prediction, the database of experiments should include two parts: face image database and personality evaluation database. Their collection is the basis of the entire prediction process. The contents of this section will provide a brief summary of the current status of database construction in the field and the existing problems.

On the one hand, the current domestic and international research results on face-based personality prediction are few, and it is difficult to obtain open experimental data sets. On the other hand, since personality research involves participants' personal privacy, experimental data sets are not Suitable for public. Summing up the existing research results, we found that for the face database, the vast majority of them are self-built database with a small number of face databases [6, 8, 9, 23, 32, 37, 38, 41, 52–54]; For the personality evaluation results database, using the methods of self-assessment or others' assessment to measure and obtain the final personality data as described in the previous Sect. 2.2.

3.1   Construction of Face Database

The selection of face data is the key to verifying the face-based automatic personality prediction model and guaranteeing its generalization ability. Ideally, face databases should include face samples of different gender, race, and age under different personality traits. However, so far, in the field of personality calculation, there is still no such a recognized database. In the field, different jobs often build independent databases based on their actual situation. The number of faces, ages, genders, races, postures and expressions of these face samples are different. Table 2 briefly describes the basic situation of the face database built in the existing personality calculation study.

From Table 2, it can be seen that the existing face database based on automatic personality prediction study contains the number of samples ranging from tens of samples to several thousand samples. Rojas et al. [41], developed a personality prediction model based on a set of 66 standardized faces in the database of amateur actors' faces of Karolinska [30] as a preliminary attempt on the personality calculation of face-based. Most of the research is based on young people in colleges and universities. For example, Qin et al. [37, 38] collected facial images of Xiamen University of Technology students. Zeng et al. [54] collected facial images in a college in Jiangxi province. Wolffhechel et al. [52], also collected images of college students' faces in Danish universities. This is because the research scholars are basically in scientific research institutes such as colleges and universities. They are more convenient when collecting data in this area, which facilitates the development of experiments, but at the same time it also brings a problem of a single age structure. It is worth mentioning that in order to enhance the generalization ability of personality calculations in real applications, some researchers have abandoned facial samples from laboratory environments and downloaded diverse face images from social networking sites. For the first time, White [51] and others selected thousands of images from the www.hotornot. com website for building a face database. This can be considered as a new attempt in the area of personality calculation. [8, 9] uses software-generated face data, using the popular composite software program FACES [20] produced by Inter Quest and Micro. Through FACES, it is possible to randomly generate a large number of unique faces by manipulating individual facial features. Images can be limited by many external conditions.

3.2 Collection of Personality Trait Data

A major problem in the study of personality calculations is the lack of real personality traits data, and other relevant research, such as crime prediction [53], beautiful attraction calculation [28, 29], expression recognition [21, 50], gender [34], true and false face recognition [7], etc., the classification tag is easy to obtain, and the acquisition of personality traits is much more difficult. In the attractive attraction calculations, such as [28, 29], it is only necessary to find enough observers to score face samples. Each face sample corresponds to a score; In facial expression recognition [21, 50], the definition of expression is also fixed, such as 'happy', 'sad', etc. The observer also only needs to give a label to the face sample; Crime judgments, gender judgments, and true/false face judgments simplify calculations to a two-category problem, making it easier to classify. However, the collection of personality traits data, whether self-evaluation or other people's evaluation, must be evaluated by the personality rating scale. Many scales have a large number of questions, and it is possible to measure a single sample for several tens of minutes to obtain the results of personality traits. Each sample must correspond to the personality traits of multiple dimensions and need to be

Table 2. Attributes of Face data set in existing studies.

Ref	Num	Age	Gender	Race	Posture	Expression	Data sources
[1]	829	20–39	men& women	Mix	Positive	Neutral	Color-FERET Data set
[3]	220	Synthesis	men& women	Mix	Positive	Neutral	Synthesis by FACES software
[4]	480	Synthesis	men& women	Mix	Positive	Neutral	Synthesis by FACES software
[19]	650	30–50	men& women	Mix	Positive	Neutral	Image of politician
[29]	66	20–30	men& women	White	Positive	Neutral (Accessories and cosmetics are visible)	Karolinska's [26] amateur actor's face database
[35, 36]	186	18–22	men& women	Asia	Positive	Neutral	Students of Xiamen University of Technology
[39]	66	20–30	men& women	White	Positive	Neutral (Accessories and cosmetics are visible)	Karolinska's [26] amateur actor's face database
[51]	3998	18–25	men& women	Mix	Positive	different	Download in Social website
[52]	244	18–37	men& women	White	Positive	Neutral	Students of Danish Technical University
[53]	1856	18–55	men	Asia	Positive	Neutral	Contains two subsets of non-criminals and criminals
[54]	608	18–22	men& women	Asia	Positive	Neutral	Students of one University in Jiangxi

evaluated one by one. The process is complicated and slow, but of low precision. Therefore, the acquisition of personality traits data is very difficult, which is one of the important reasons for little research on automatic personality prediction.

In the existing studies, the assessors usually evaluate the personality of the face database with reference to different Likert scales, and finally calculate the scores of different personality traits through statistical calculations. Value-based, including 3-point system [9], 9-point system [6, 32, 41, 52], 10-point system [39, 40] and 60-point system [54]. Finally, the average score of each personality trait in each sample face is often used as a benchmark score for the machine prediction criteria. Table 1 summarizes the various attributes of the evaluators in the existing study including the scoring system.

4   Expression of Facial Character Traits

The expression of face personality traits is a crucial step in the face-based automatic personality prediction framework. On the one hand, because the pixel value feature of the face image is usually high in dimension and the amount of data is large, on the other hand, since the face similarity is high, the original pixel value data contains more redundant components. Before the experiment based on the face image is performed,

Table 3. Characterization of face character traits.

Ref.	Type of feature	Range of feature	Expression of feature
[6]	Geometric	Eyes, nostrils, chin tips, lip corners	34-dimensional area
	Holistic	Full face	Eigenface
[8]	Texture, geometry and overall three feature vectors	Full face	Pixels, PCA
[9]	Texture	Full face	Gabor, LBP, Pseudo-sliding window
[23]	Geometric	Head, eyebrows, eyes, nose and mouth	76 feature points and distances, ratios
	Texture	Full face	HOG
	Color	Full face	RGB
[30]	Texture	Full face	HOG
[32]	Holistic	Full face	Pixels, PCA
[37]	Texture	Full face	HOG, LBP, Gabor, GIST, SIFT
[38]	Geometric	Face profile, eyebrows, eyes, cheeks, nose, mouth	21 points, distance and ratio of 1134 dimensions
[41]	Geometric	Face profile, eyebrows, eyes, cheeks, nose, mouth	21 points, distance and ratio of 1134 dimensions
[52]	Texture	Full face	Pixels, PCA
	Holistic	Full face	PCA (32 PCs for male face, 35 PCs for female face)
[53]	Geometric	Face profile, eyebrows, eyes, cheeks, nose, mouth	FGM feature generation machine
	Texture	Full face	LBP, HOG
	Holistic	Full face	PCA
	Synthesis	Full face	Mosaic of
[54]	Geometric	Face profile, eyebrows, eyes, cheeks, nose, mouth	Improved ASM, distance and ratio of 32 feature points

features of the face image need to be extracted to reduce the dimension of the face feature, so as to extract information that can discriminate between faces.

In the existing research literature on face-based automatic personality prediction, the expression of face features can be divided into two categories: Feature-based Representation and Holistic Representation. Feature-based methods use geometric features, texture features, color features, and other local features to represent faces. The geometric features can be composed of coordinate values of the calibrated facial feature points, the distance between feature points, distance ratio, etc. [37, 38, 41]; the texture features can be Gabor filters [41] or local binary values. Local Binary Patterns (LBP) operator [39] to extract. Active Shape Model (ASM) and Active Appearance Model (AAM) parameters combine geometric shape and texture information and are also extracted as potential features [54]. The global expression method uses the original grayscale image to represent the human face, splicing it in rows as an input feature vector, and obtaining it through dimension reduction methods such as principal component analysis (PCA) or manifold learning. Required low-dimensional features such as feature faces [22, 51], manifolds [34], etc. Table 3 summarizes representative works of expression of face character traits.

4.1 Personality Prediction Based on Feature Expression

Face character traits are characterized and described by a diverse set of features, including geometry, textures, colors, shapes, etc., where geometric features are the most widely used features in the field. The basis for measuring the face geometry information lies in the accurate positioning of the critical areas of the face (i.e., face contours, eyebrows, eyes, nose, and mouth) and their feature points. Feature points can be automatically retrieved by the feature location method [27], or manually extracted from the graphical user interface [27]. The number of feature points selected in different research work varies from 21 points [30, 37, 38, 41], 39 points [54] to 76 points [27].

4.2 Personality Prediction Based on Overall Expression

The overall expression method is proposed to overcome the limitations of artificial design features based on feature expression methods. The feature-based method extracts features that are often local and discrete face features, while the holistic approach extracts global features from the entire face and studies the spatial relationships between features. Since the human brain's perception of personality is a holistic processing model, such methods are also widely used in face analysis applications.

In the overall expression method, a typical face expression is to splice the original grayscale image into a high-dimensional feature vector. As an effective subspace projection method, PCA reduces the dimensions of the original high-dimensional human face space to obtain a low-dimensional human face feature, such as a feature face [6, 41]. The literature [41] applied eigenfaces to personality prediction studies for the first time. The eigenface method was used to analyze pixel information [47]. Based on information theory, and intended to find the main components of face distribution, that is, the algorithm projects the image across the group. The most significant change in the feature space is in the image. [41] Rojas et al. used two kinds of feature extraction schemes to determine whether the information conveyed by the overall characterization is complementary to the information expressed by the structural information. The conclusion is positive. In [6], the eigenface employed is the principal components extracted from a set of training images, which form the basis in the space where each point corresponds to a different facial image. The eigenfaces are sorted by the amount of difference they occupy in the original data. In the experiment, the extracted contribution rate was 90% before the 103 feature faces. [8, 32] used principal component analysis (PCA) to extract components that account for most of the changes in facial shape.

Face character prediction based on the overall expression depends on the apparent features of the face, and does not require manual calibration, so that the prediction process is fully automated.

5 Face-Based Character Prediction Algorithm

Based on the previous research, almost all studies have obtained the necessary data, including the characteristics of face data and the scores of personality traits of different standards. Next we need to use these data to accurately infer a person's personality Traits.

5.1 Personality Inference Experiment Based on Classification Model

Classification of Personality Trait Scores. From Sect. 3.2, we summarize the acquisition of personality traits data in relevant literature. Whether it is the 'Big Five Personality' or '16PF' assessment method, the study will be divided into different dimensions of personality traits, including a 3-point system. [8], 9 points [6, 32, 41, 52], 10 points [37, 38], and 60 points [54]. Corresponding to the discrete scores within each interval, most of the studies are to N-value the score of personality traits (N generally takes 2 or 3) in order to construct a classification problem. For example, in [37, 38], for the 20 kinds of personality traits of '16PF', the score between 6–10 is defined as a category, indicating that the performance of such personality traits is more significant, and the score is between 1–5. The other category shows that this personality trait is not obvious. Similarly,in [41], first, the five with the highest personality traits are classified as having "traits", and the lowest 5 points as 'non-characteristic' categories, which then form two categories. [54] differs from the previous studies in that personality traits are divided into three categories (e.g., low-amenity, medium-amenity and high-amenity). In [6], in order to facilitate the classification, only the extreme values (highest and lowest) of the personality trait scores are retained, which are divided into two categories. The main reason is that a higher consensus was reached between the evaluation members of the extreme values. In [23], for the evaluation of the corresponding personality traits of politicians, a positive and negative two-classification method was performed. The prediction of criminality in [53] itself is a question of a two-category (criminal?).

Selection of Classifiers. Different studies have also used different classification methods to study the problem of inferring personality traits based on human facial features. Qin et al. [37, 38] used the Parzen window [35], the Decision Tree [17], K-Nearest Neighbor(KNN) [38], Naive Bayes [38] and Random Forest [10]. In Rojas et al. [30, 41], the classification in the experiment was performed using the most advanced classifier. Five algorithms are used: GentleBoost as an example of an iterative method, support vector machine(SVM) has a radial basis function kernel as an example of a non-linear classifier, and K-nearest neighbor as an example of a nonparametric classifier with Parzen window and binary decision tree of Random Subspace. The system was evaluated using a 20x cross validation strategy and the results were supported by confidence intervals calculated with 95% confidence. Zeng et al. [54] adopted a deep confidence network classification algorithm based on BP algorithm. A deep confidence network structure model including a 5-layer RBM is used, where the fifth RBM is the output layer and the output is three types (low, medium, and high of the personality trait). Al Moubayed et al. [1] also used a support vector machine(SVM) binary classification method. Jungseock et al. [19] used the (RankSVM) [18] method to predict the social dimension in order to train the model. Unlike the general SVM algorithm, RankSVM is more suitable for their task because it aims to retain the training examples. The pre-specified pairwisesort order. This advantage makes RankSVM very popular in the literature of information retrieval (web search) and the recent related properties in the field of computer vision [34]. Wu et al. [53] constructed four classifiers to identify the relationship between criminality and human face: logistic regression [14], K-nearest neighbor(KNN), SVM, and convolutional neural network [26]. In [9], Brahnam and Nanni et al. use SVM and neural networks as decision rules (Table 4).

Table 4. Summarizes the selection of classification methods in different literature.

Ref.	Classification type	Classifier selection
[6]	Binary classification	SVM
[8]	Binary classification	SVM
[9]	Binary classification	SVM & neural network
[23]	Binary classification	RankSVM
[30, 41]	Binary classification	GentleBoost, SVM, K-Nearest Neighbor, Parzen Window and Binary Decision Tree
[37, 38]	Binary classification	Parzen Window, Decision Tree, K Nearest Neighbor, Naive Bayes and Random Forest
[53]	Binary classification	Logistic regression, K-nearest neighbor, support vector machine and convolutional neural network
[54]	Three classification	Deep confidence network based on BP algorithm

In the classification experiments, the classification accuracy rate used as the evaluation criteria, we can also see that most of the literature classification results are good (see below).

5.2   Individual Inference Experiment Based on Regression Model

Section 5.1 mainly introduces the experiments based on the morphological character-istics of different faces in the existing literature, using a variety of classification methods to infer human personality traits. In this section, the regression method used in the literature is analyzed to estimate the actual values of personality traits based on the morphological characteristics of the face. Qin Rizhao and other scholars are the first to introduce regression models in personality prediction experiments. In [37, 38], the scores of each personality trait obtained by the 16PF test are integer discrete scores ranging from 1–10. They did not make any changes and they directly used these values as regression targets. In regression experiments, Mean Square Error (RMSE) was used to measure the performance of regression experiments. Qin Rixi and others used the facial features and texture features to perform regression experiments, respectively. Compared with the classification experiments, the regression algorithms obtained errors are relatively large, indicating that the use of human facial images to predict the specific score of personality traits is still difficult of.

6   Forecast Analysis

The question of how much personality trait information can be learned from human face images can be studied around from a computer science perspective. Table 5 summarizes the conclusions of different research results. In general, the existing studies

Table 5. Summary of accuracy results of personality prediction in different literatures

Ref.	Characteristics with better prediction results	The highest accuracy
[26, 27]	Responsibility(male)	81.56%
	Responsibility (female)	82.22%
	Skeptical (male)	72.64%
	Skeptical (female)	82.22%
[28, 29]	Dominant	91.23%
	Threatening	90%
	Extraversion	
[30]	Open, Striving, Dominant (female)	63%
	Reliability, friendliness, responsibility (male)	65%
[2]	Neuroticism	82.35%
	Extroverted	84.31%
	Rigor	84.31%
[31]	Openness, Extraversion, Neuroticism	65%
[42]	Criminality	89.51%
[53]	Warm	76%
	Reliability	81%
[54]	Intelligence, maturity, sociality, dominance, warmth, credibility	80%

based on the automatic personality prediction of human face mainly have the following conclusions:

1. The calculation and prediction of personality traits based on facial images is reliable to a certain extent.
2. Partial personality traits ('responsibility', 'dominant', 'threatening', 'attractiveness', 'openness', 'extroversion' and 'nervousness', etc.) are more relevant to face images. These personality traits can achieve better prediction results. The other part of the environmental impact of personality trait prediction results are not satisfactory.
3. The application of the latest deep learning method can automatically learn more advanced face expressions from face images. The prediction results are more breakthrough than the traditional whole and feature methods, but the interpretation of extracted features is not intuitive and easy to understand [27].

7   Limitations and Prospects

Automatic personality prediction based on face images is a new research topic in the field of computer vision. Since 2008, some scholars have started to engage in this research, and there are few research results. However, after a small amount of research work, some preliminary and valuable results have also been obtained. However, there are still many problems worthy of deep thinking and exploration. In particular, the most important question in face personality prediction, namely 'what factors do people face contribute to the judgment of personality, and how much contribution', is still far from being explained better.

7.1   Existing Problems in Existing Research

Summarizing the existing research, the prominent problems in the field of automatic personality prediction based on face images at the current stage are mainly reflected in the following aspects:

1. Lack of information on research data: In the current research on personality prediction, all studies reviewed in this paper are based on two-dimensional human faces, especially two-dimensional frontal face images. However, simply relying on positive images will lose a lot of personality-related information. From ancient times to the present, Chinese facial studies mentioned that personality-related descriptions of 'five features', 'three courts', and 'twelve houses' are related to certain prominent facial regions (e.g., Forehead (rich in the vestibule), nose (big and tall), cheekbones (broad and wide), chin (ground radius, etc.). The described key features can only be accurately located from the side face. Recent studies have found that lateral faces can describe face features well and can be used for identity recognition [20], gender and race recognition [43], and face recognition [34]. Therefore, future face character analysis based on 2.5D or 3D is an inevitable trend in future face character prediction research.

2. Sampling of the age structure of the sample population: In all the existing studies, the convenience of the experimental implementation and the authenticity of the data were fully taken into account. The majority of the sample population were college students. The age range was basically 18–35. Although many studies have fully considered the different disciplines [54], different genders [37, 38], and different occupations [30, 41], the diversity of samples has been reflected, but the well-known Chinese psychologist Song Chubo In the 'Heart to Heart' [45], it is said that 'there is no heart, no phase, no heart, no heart, no heart. The words are simple, and the truth of the human relations program', the face of the people above the age of 35 and their personality. Therefore, it is desirable to collect data for more than 35-year-olds and predict personality, in the hope that there will be more objective results in predicting accuracy.

3. Artificial design features are not comprehensive: In the existing research work, researchers usually manually design a set of features (geometry, texture, color, global appearance features, etc.) based on heuristic criteria to predict personality traits. A large number of diversified features have been constructed and analyzed for different databases, but so far there is no unified conclusion to prove which features constitute the main factors affecting personality traits. The artificial design features that have been proposed so far have generally been unsatisfactory in character prediction research, and it is therefore reasonable to conclude that these features are not comprehensive. Is there a specific area related to facial feature assessment on the person's face? How much is the contribution to character prediction? This important aspect of the issue is not well explained. This problem was only attempted in Rojas et al. [41], trying to identify the most important area of facial assessment of personality traits, but the results did not have deep guidance. significance. Therefore, whether or not we can find a more comprehensive description of face features and feature extraction methods, and unify them under an effective framework for overall research, is the key to final performance prediction.

7.2 Development Prospects of Automatic Personality Prediction

So far, the interest in computerized automatic personality prediction is still at a relatively early stage (see Fig. 2). Most of the work is devoted to establishing the field, collecting data, developing methods, and identifying related tasks. Research shows that the application of personality calculation in many fields brings convenience to human society, first in daily life [49]: the user's personality score improves the performance of the recommendation system [46, 47]; synthetic speech based on individual preferences is The proof can improve the acceptance of the GPS system [31, 44]; the correct positioning of advertising campaigns on potential users [25]; and retrieval techniques that match the user's personality [14]. Secondly, in terms of medical care: With advances in technology involving autism spectrum problems and other mental problems [42], personality calculations may be based on techniques designed to detect psycho-psychiatric disorders such as paranoia and schizophrenia that often interfere with personality. Plays an important role in [13]; Finally, in terms of social behavior, personality calculations may help establish the link between features and behaviors, and so far it has not been possible to achieve this connection for the prediction of major social events such as elections [23], criminal identification [53] and so on.

Future research is expected to establish a more extensive data set of age structure (35 years old or older) that satisfies the requirements of 2.5D-3D research, and collect scores of personality traits from reliable observers to fill existing limitations. At the same time, it should also try to introduce some new research models for face character prediction. For example, consider it as a marker distribution learning problem to cope with the problem of lack of training samples; explore the impact of group classification variables on personality scores, and customize the character prediction model of the corresponding group.

References

1. Edmonds, J.M.: The Characters of Theophrastus. W. Heinemann, London (1967)
2. Matthews, G., Deary, I., Whiteman, M.: Personality Traits, 3rd edn, pp. 1–568. Cambridge University Press, Cambridge (2009)
3. Funder, D.C.: Personality. Annu. Rev. Psychol. **52**(1), 197–221 (2001). https://doi.org/10.1146/annurev.psych.52.1.197. pMID: 11148304
4. Ozer, D.J., Benetmartinez, V.: Personality and the prediction of consequential outcomes. Annu. Rev. Psychol. **57**(1), 401–421 (2006)
5. Uleman, J.S., Saribay, S.A., Gonzalez, C.: Spontaneous inferences, implicit impressions, and implicit theories. Annu. Rev. Psychol. **59**(1), 329–360 (2008)
6. Al Moubayed, N., Vazquez-Alvarez, Y., McKay, A., Vinciarelli, A.: Face-based automatic personality perception. In: Proceedings of the 22nd ACM International Conference on Multimedia, pp. 1153–1156. ACM (2014)
7. Ballew, C.C., Todorov, A.: Predicting political elections from rapid and unreflective face judgments. Proc. Natl. Acad. Sci. **104**(46), 17948–17953 (2007)
8. Brahnam, S.: A study of artificial personality from the perspective of the observer, June 2018
9. Brahnam, S., Nanni, L.: Predicting trait impressions of faces using local face recognition techniques. Expert Syst. Appl. **37**(7), 5086–5093 (2010)
10. Breiman, L.: Random forests. Mach. Learn. **45**(1), 5–32 (2001)
11. Carney, D.R., Colvin, C.R., Hall, J.A.: A thin slice perspective on the accuracy of first impressions. J. Res. Pers. **41**(5), 1054–1072 (2007)
12. Cheng, X.H.: Research on human face detection and recognition. Masters thesis, Xidian University, September 2006
13. Corr, P.J., Matthews, G.: The Cambridge Handbook of Personality Psychology. Cambridge University Press, Cambridge (2009)
14. Cristani, M., Vinciarelli, A., Segalin, C., Perina, A.: Unveiling the multimedia unconscious: implicit cognitive processes and multimedia content analysis. In: Proceedings of the 21st ACM International Conference on Multimedia, pp. 213–222. ACM (2013)
15. Deary, I.J.: The trait approach to personality. In: The Cambridge Handbook of Personality Psychology, pp. 89–109 (2009)
16. Digman, J.M.: The curious history of the five-factor model (1996)
17. Duda, R.O., Hart, P.E., Stork, D.G.: Pattern Classification. Wiley, Hoboken (2012)
18. Eysenck, H.J.: Dimensions of personality: 16, 5 or 3? - Criteria for a taxonomic paradigm. Personality Individ. Diff. **12**(8), 773–790 (1991)
19. Freedman, D.A.: Statistical Models: Theory and Practice. Cambridge University Press, Cambridge (2009)
20. Freierman, S.: Constructing a Real-life Mr. Potato Head. Faces: The Ultimate Composite Picture. The New York Times 6 (2000)
21. Huang, J., Li, W.S., Gao, Y.J.: Research advance of facial expression recognition. Comput. Sci. (z2), 123–126 (2016)

22. Joachims, T.: Optimizing search engines using clickthrough data. In: Proceedings of the Eighth ACM SIGKDD International Conference on Knowledge Discovery and Data Mining, pp. 133–142. ACM (2002)

23. Joo, J., Steen, F.F., Zhu, S.C.: Automated facial trait judgment and election out-come prediction: social dimensions of face. In: Proceedings of the IEEE International Conference on Computer Vision, pp. 3712–3720 (2015)

24. Kakadiaris, I.A., Abdelmunim, H., Yang, W., Theoharis, T.: Profile-based face recognition. In: 8th IEEE International Conference on Automatic Face and Gesture Recognition, FG 2008, pp. 1–8. IEEE (2008)

25. Kosinski, M., Stillwell, D.: myPersonality Research Wiki (2012)

26. Krizhevsky, A., Sutskever, I., Hinton, G.E.: ImageNet classification with deep convolutional neural networks. In: Advances in Neural Information Processing Systems, pp. 1097–1105 (2012)

27. Liu, S.: Research on the prediction of facial beauty from 2D to 3D. Ph.D. thesis, Northwestern Polytechnical University (2017)

28. Liu, S., Fan, Y.Y., Samal, A., Guo, Z.: Advances in computational facial attractiveness methods. Multimedia Tools Appl. **75**(23), 16633–16663 (2016)

29. Liu, Shu, Fan, Yangyu, Guo, Zhe, Samal, Ashok: 2.5D facial attractiveness computation based on data-driven geometric ratios. In: He, Xiaofei, et al. (eds.) IScIDE 2015. LNCS, vol. 9242, pp. 564–573. Springer, Cham (2015). https://doi.org/10.1007/978-3-319-23989-7_57

30. Lundqvist, D., Flykt, A., Ohman, A.: Karolinska directed emotional faces (Department of Neurosciences, Karolinska hospital, Stockholm, Sweden). Karolinska Directed Emotional Faces 2(4) (1998)

31. Nass, C., Brave, S.: Wired for Speech: How Voice Activates and Advances the Human-Computer Relationship. MIT Press, Cambridge (2005)

32. Oosterhof, N.N., Todorov, A.: The functional basis of face evaluation. Proc. Natl. Acad. Sci. **105**(32), 11087–11092 (2008)

33. Pantic, M., Patras, I., Rothkruntz, L.: Facial action recognition in face profile image sequences. In: 2002 IEEE International Conference on Multimedia and Expo, 2002. ICME 2002. Proceedings, vol. 1, pp. 37–40. IEEE (2002)

34. Parikh, D., Grauman, K.: Relative attributes. In: 2011 IEEE International Conference on Computer Vision (ICCV), pp. 503–510. IEEE (2011)

35. Parzen, E.: On estimation of a probability density function and mode. Ann. Math. Stat. **33** (3), 1065–1076 (1962)

36. Pei, W.W.: Gender studies based on the feature extraction of face images analysis. Masters thesis, Taiyuan University of Technology, February 2010

37. Qin, R.: Personality analysis based on face image. Master's thesis, Chinese Academy of Sciences University (2016)

38. Qin, R., Gao, W., Xu, H., Hu, Z.: Modern physiognomy: an investigation on predicting personality traits and intelligence from the human face. Sci. China Ser. F Inf. Sci. **61**(5), 058105 (2018)

39. Ren, S., Cao, X., Wei, Y., Sun, J.: Face alignment at 3000 fps via regressing local binary features. In: Proceedings of the IEEE Conference on Computer Vision and Pattern Recognition, pp. 1685–1692 (2014)

40. Rish, I.: An empirical study of the naive Bayes classifier. In: IJCAI 2001 Workshop on Empirical Methods in Artificial Intelligence, vol. 3, pp. 41–46. IBM (2001)

41. Rojas, M., Masip, D., Todorov, A., Vitria, J.: Automatic prediction of facial trait judgments: appearance vs. structural models. PLoS ONE **6**(8), e23323 (2011)

42. Saint-Georges, C., et al.: Do parents recognize autistic deviant behavior long before diagnosis? Taking into account interaction using computational methods. PLoS ONE **6**(7), e22393 (2011)

43. Saucier, G., Goldberg, L.R.: The language of personality: lexical perspectives. In: The Five-Factor Model of Personality: Theoretical Perspectives, pp. 21–50 (1996)

44. Schwab, K., Marcus, A., Oyola, J., Hoffman, W., Luzi, M.: Personal data: the emergence of a new asset class. In: An Initiative of the World Economic Forum (2011)

45. Tariq, U., Hu, Y., Huang, T.S.: Gender and ethnicity identification from silhouetted face profiles. In: 2009 16th IEEE International Conference on Image Processing (ICIP), pp. 2441–2444. IEEE (2009)

46. Tkalcic, M., Tasic, J., Košir, A.: Emotive and personality parameters in multimedia recommender systems. In: Affective Computing and Intelligent Interaction, ACII 2009, vol. 33 (2009)

47. Tkalcic, M., Tasic, J., Košir, A.: The LDOS-PerAff-1 corpus of face video clips with affective and personality metadata. In: Proceedings of the Multimodal Corpora: Advances in Capturing, Coding and Analyzing Multimodality (Malta 2010), LREC, p. 111 (2010)

48. Turk, M., Pentland, A.: Eigenfaces for recognition. J. Cogn. Neuro- Sci. **3**(1), 71–86 (1991)

49. Vinciarelli, A., Pantic, M., Bourlard, H.: Social signal processing: survey of an emerging domain. Image Vis. Comput. **27**(12), 1743–1759 (2009)

50. Wang, Z.L., Chen, F.J., Xue, W.M.: A survey of facial expression recognition. Comput. Appl. Softw. (12), 63–66 (2003)

51. White, R., Eden, A., Maire, M.: Automatic prediction of human attractiveness. UC Berkeley CS280A Project 1, 2 (2004)

52. Wolffhechel, K., et al.: Interpretation of appearance: the effect of facial features on first impressions and personality. PLoS ONE **9**(9), e107721 (2014)

53. Wu, X., Zhang, X.: Automated inference on criminality using face images. arXiv preprint arXiv:1611.04135 (2016)

54. Zeng, Z.: An analysis of students' personality traits based on face features and deep learning. Masters thesis, Jiangxi Normal University, March 2017

The Design and Implementation of Dive Maneuver

Weihao Liang[1]($\boxtimes$), Jianhua He[1], Lei Yang[2], Siqi Tao[1], and Libin Chen[3]

[1] School of Electronics and Information, Northwestern Polytechnical University, Xi'an, China
lwh@mail.nwpu.edu.cn
[2] School of Mechanical Engineering, Northwestern Polytechnical University, Xi'an, China
[3] China State Shipbuilding Corporation, Beijing 100094, China

Abstract. When it comes into air-to-air battlefield, using the dive maneuver, for an aircraft, has great realistically profound meaning in dominance of air domain and higher flexibility in attacking enemy planes below. Taking the task requirement and overload constraint into account, a mathematical model of dive maneuver is established and, further, the dynamic changing law of the fighter's parameters with smooth damp is designed. Simulation result shows that the model we used could shorten the time cost of dive maneuver compared to those of classic ones, leaving fighter more advantages in air combat.

Keywords: Dive maneuver · Overload constrain · Flight control

1 Introduction

1.1 A Subsection Sample

In aerial combat, the pilot steers the current position and posture of the aircraft through maneuver, so as to occupy the advantageous position in attack or avoid the attack from its opponent [1]. The victory in an air battlefield also depends on the maneuvering strategy of an airplane [2]. For example, an aircraft which has smaller mass and stronger maneuver could take advantage from its smaller turning radius to avoid attack or obtain favorable position; Through dive maneuver or strong thrust, an aircraft with bigger mass and faster speed could further improve its flight speed to get rid of the pursuit.

Dive maneuver is indispensable for modern fighters [1], since it plays a key role in the air battle and has a strong practical significance to compete for the airpower in the battlefield. On the one hand, missiles or bombs often use dive maneuver to attack at the end of flight, whose trajectory show as a sharp turn in the vertical plane at that time, and then they will accelerate in a straight line [3]. During that, the trajectory of dive maneuver directly determines the operational effectiveness of weapon system [4]; on the other hand, pilots change the current position and posture of the aircraft through dive maneuver, making it occupy a favorable position to attack or escape attack from its opponent.

B. Li et al. (Eds.): IoTaaS 2018, LNICST 271, pp. 30–39, 2019.
https://doi.org/10.1007/978-3-030-14657-3_3

The realization of maneuver capable of altering height and direction of the airplane, such as dive maneuver and hover maneuver, mainly considers the trajectory traits of aircraft. Meanwhile, actually for better performance, it needs to consider the maneuver task requirements (such as desired height and expected speed) and the limitation of aircraft hardware and mechanical conditions (such as overload). Therefore, assuming the aircraft could meet instantaneous torque equilibrium condition, we use dynamics course equations presenting the motion of aircraft, at the mercy of overloads in 3-DOF. Then, in consideration of overload limitation and digital signal processing technology, the changing law of course elements with smooth damp (course roll, course yaw and course pitch etc.) are designed. In the end, those laws above are as the expected input of the flight control system to implement the dive maneuver.

1.2 Maneuver Model Establishing with Smooth Damp

The generation of maneuver focuses on the discussion of course characteristics in which overload constraints act as a primary factor. To this end, the navigation model of vehicle in three degrees is considered in geography coordinate system (x , y and z lies northward, upward and eastward, respectively).

$$\begin{cases} \dot{v} = g(n_x - \sin\theta) \\ \dot{\theta} = g(n_y \cos\gamma_s - \cos\theta)/v \\ \dot{\psi}_s = -gn_y \sin\gamma_s/(v\cos\theta) \\ \dot{x} = v\cos\theta\sin\psi_s \\ \dot{y} = v\sin\theta \\ \dot{z} = -v\cos\theta\sin\psi_s \end{cases} \tag{1}$$

where v is the velocity vector; g is gravity acceleration; n_x is the tangential overload; θ is the course pitch; n_y is the normal overload; γ_s is the course roll; ψ_s is the course yaw; x is the displacement northward; y is the displacement upward; z is the displacement eastward. It should be pointed out that the sideslip angle and the lateral force are assumed to be zero in (1). Generally, this assumption is reasonable, for lateral force do not need to be considered in the maneuver such as BTT (Bank to Turn), hovering and diving.

In order to generate the desired maneuver, the variation rules of the course elements $(\theta, \psi_s, \gamma_s)$ are given under the limitation of overload. The overload is calculated with the following equation

$$\begin{cases} n_x = \dot{v}/g + \sin\theta \\ n_y = (v\dot{\theta}/g + \cos\theta)/\cos\gamma_s \end{cases} \tag{2}$$

with

$$\begin{cases} n_x \leq n_{xm} \\ n_y \leq n_{ym} \end{cases} \tag{3}$$

In general, the overload limitation includes four aspects: the available thrust of engine, the available moment of rudder, the endurance of the aircraft structure and the physiological endurance of pilots.

Considering the principle that jerk (jerk is a physical term used to describe the changing rate of acceleration or force. The bigger the jerk, the more likely it renders material fatigue.) should not change in a rocket way, according to mechanics of materials, parameters with smooth damp are given directly by the model in order to reduce the difficulty of the design of the flight control system. In the process of modeling, thus, the design principle of flight control system is just to improve the tracking accuracy as much as possible, without considering additional damp characteristics. For this purpose, the following model constraints are applied

$$\frac{dn}{dt} < \infty (n = n_x, n_y) \tag{4}$$

Which means $|\dot{n}| < +\infty$; It will be no longer re-state in the latter section.

Thus, the problem of maneuver generation is converted into solving nonlinear functional problems satisfying certain requirements in a particular space. Specifically, below gives the mathematical model:

$$\theta^* = \theta(t); \psi_s^* = \psi_s(t); \gamma_s^* = \gamma_s(t) \tag{5}$$

$$s.t. \begin{cases} \theta^* \in C^2[D(\theta)] \\ \psi_s^* \in C^2[D(\psi_s)] \\ \gamma_s^* \in C^2[D(\gamma_s)] \\ (1) \sim (4) \end{cases} \tag{6}$$

Where $D(\theta)$ is the definition domain of θ, which is usually obtained by experience. For example, in dive maneuver the value of θ ranges within $[0, \pi/2)$. $C^2[\cdot]$ here is an operator representing all the functions with its second order derivative not only existential but consecutive on $D(\cdot)$.

1.3 Design of Maneuver

The dive maneuver could be divided into three stages: entry, straight flight and recovery [5–7], which are shown in Fig. 1.

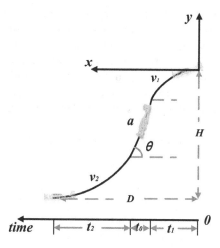

Fig. 1. Schematic diagram of dive maneuver.

where v_1, v_2 represents the velocity vectors at beginning of entry sand recovery respectively; a is the acceleration improving the velocity in straight state; t_1, t_0, t_2 is the time cost in the entry, the straight and the recovery, respectively; x is the displacement northward; y is the displacement upward.

Obviously, the expected values of the course yaw and the course roll should be zero in the whole process.

To analyze the varying traits of course element, we consider

$$\begin{cases} n_y = v\dot{\theta}/g + \cos\theta \\ \dot{n}_y = (v\ddot{\theta} + \dot{v}\dot{\theta})/g - \dot{\theta}\sin\theta \end{cases} \tag{7}$$

Clearly, in order to ensure boundedness of $\dot{n}_y$, neither $\dot{\theta}$ nor v can change suddenly.

To this end, the upcoming problem is that whether we could simplify the modeling process getting the anticipant course parameters in assumption of that the overload changes linearly.

Theorem 1. It is impossible to get the apt changing laws of course elements by randomly assigning a bounded known function to $\dot{n}_y$.

Proof: as Fig. 1 illustrates, the changing rule of θ is supposed to alter gradually from 0 to a negative number, and then back to zero, thus the statement above holds.   ∎

According to Theorem 1, in order to meet the overload limitations, the variation rule of the course pitch must be designed indirectly. To simplify the analysis, the following assumptions are established:

Hypothesis 1: during the dive maneuver, the velocity's magnitude during the entry phase and the recovery phase keep unchanged; in the straight flight stage, the velocity increases with an acceleration that would not change abruptly. Under the limitations above, define that the speed is maintained as at the beginning of the entry stage, say, v_1; likewise, it should be v_2 during the recovery phase; the maximum acceleration is a_m in the straight flight stage. At last, in order to simplify the analysis process, the course pitch is supposed to be zero before entry stage and after the recovery stage.

The speed changing process is dominated by n_x, according to (1). At this point, the problem (5) is simplified as

$$\theta^* = \theta(t) \tag{8}$$

As is stated above, $\ddot{\theta} < \infty$ is requested to satisfy $\dot{n}_y < \infty$, thus we can assume that alters in a most simple way, say a constant number or zero accordingly. More specifically,

$$\ddot{\theta} = \begin{cases} 2h_1/t_1 & 0 < t < t_1/2 \\ -2h_1/t_1 & t_1/2 < t < t_1 \\ 0 & t_1 < t < t_1 + t_0 \\ -2h_2/t_2 & t_1 + t_0 < t < t_1 + t_0 + t_2/2 \\ 2h_2/t_2 & t_1 + t_0 + t_2/2 < t < t_1 + t_0 + t_2 \end{cases} \tag{9}$$

where $h_i (i = 1, 2)$ is a constant. Then the change rate of course pitch is given by the following equation.

$$\dot{\theta} = \begin{cases} 2h_1 t/t_1 & 0 < t < t_1/2 \\ -2h_1 t/t_1 + 2h & t_1/2 < t < t_1 \\ 0 & t_1 < t < t_1 + t_0 \\ -2h_2 t/t_2 + 2h_2 \frac{t_0 + t_1}{t_2} & t_1 + t_0 < t < t_1 + t_0 + t_2/2 \\ 2h_2 t/t_2 - 2h_2 \frac{t_0 + t_1 + t_2}{t_2} & t_1 + t_0 + t_2/2 < t < t_1 + t_0 + t_2 \end{cases} \tag{10}$$

then we can get the course pitch:

$$\theta = \int_0^{t_0 + t_1 + t_2} \dot{\theta} dt \tag{11}$$

with the limitation of

$$0 < \theta < \max(\theta) =: \theta_E \tag{12}$$

where θ_E is dive angle, as defined in Fig. 1. The general variation rule of θ is shown as Fig. 2.

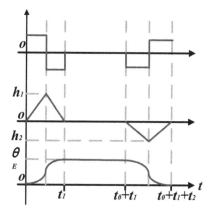

Fig. 2. Change law of course pitch and its first order and second-order differential.

Obviously

$$\begin{cases} \dfrac{h_1 t_1}{2} = \theta_E \\ \dfrac{h_2 t_2}{2} = \theta_E \end{cases} \qquad (13)$$

From the limitations of overload that

$$\begin{cases} \dfrac{v_1 h_1}{g} + \cos \theta_1 \leq n_{ym} \\ \dfrac{v_2 h_2}{g} + \cos \theta_2 \leq n_{ym} \end{cases} \qquad (14)$$

where $0 \leq \theta_1, \theta_2 \leq \theta_E < \pi$, and the condition that

$$\sup_{0 \leq \theta_i \leq \theta_E} \cos \theta_i = 1, (i = 1, 2) \qquad (15)$$

we can get

$$v_i h_i \leq g(n_{ym} - 1) = const \quad i = 1, 2 \qquad (16)$$

In order to reduce the time cost of maneuver, (16) can hold as equation, namely left equals to right. Thus, (16) is changed into

$$h_i = g(n_{ym} - 1)/v_i \qquad (17)$$

thus

$$t_i = 2\theta_E/h_i \quad i = 1, 2 \qquad (18)$$

In the stage of straight flight, there exists

$$\begin{cases} n_x = a/g + \sin\theta_E \leq n_{xm} \\ a \leq g(n_{xm} - \sin\theta_E) \end{cases} \tag{19}$$

It can take its equality sign to reduce the motorized time. Therefore,

$$a_m = g(n_{xm} - \sin\theta_E) \tag{20}$$

Assuming $\dot{a}$ and $\dot{\theta}$ have the similar change law, we can obtain

$$v_2 = v_1 + S(a \sim t) = v_1 + a_m t_0/2 \tag{21}$$

where $S(a \sim t)$ represents the area between the curve and the t axis in the $acc - time$ figure. See Fig. 1 (21) can then be changed into

$$t_0 = \frac{2(v_2 - v_1)}{a_m} \tag{22}$$

Obviously, the time cost of maneuver depends on its specific purpose of how much the alteration of height and velocity should be. And there comes the following conclusions:

Theorem 2. Under the condition of smooth damp and the limitation of maximum available overload, the dive maneuver designed by the scheme (9)–(21) is optimal which consumes the least time in a sense, and has the best stationarity (smoothly maneuvering process).

Proof: it is obvious from the deduction.   ∎
    With the expected variation of height and velocity,θ_E can then be designed.
    At this point, it is necessary to consider whether any two of the expected velocity variation, the expected height variation and the expected northward variation can be quantitatively designed simultaneously during the dive maneuver.

Theorem 3. It is impossible to design the expected velocity variation, the expected height variation and the expected northward variation through scheme (9)–(21) at the same time. However, any two of the three expected can be optionally designed, and the one left would be uniquely determined by the formal two. That is to say, $\Delta v, \Delta X, \Delta H$ are not independent to each other.

Proof: the conclusion of Theorem 3 is obvious from the derivation process.   ∎
    It means,

$$\Delta X = \Delta X(\Delta v, \Delta H) \tag{23}$$

Or

$$\Delta v = \Delta v(\Delta X, \Delta H) \tag{24}$$

$$\Delta H = \Delta H(\Delta X, \Delta v) \tag{25}$$

1.4   Simulation of Dive Maneuver

The parameter settings in the dive maneuver process are shown in Table 1.

Table 1. The parameter settings in the dive maneuver process.

Variable	Value	Variable	Value
v_1	200 m/s	v_2	300 m/s
θ_E	−60 deg	n_{xm}	6.0
n_{ym}	8.0	H_0	7000 m

where H_0 is the initial height of the aircraft in the beginning of the dive maneuver. Other parameters recorded during the dive maneuver process are shown in Table 2.

Table 2. Other parameters during the dive maneuver process.

Variable	Value	Variable	Value
t_0	3.9751 s	t_1	7.1238 s
t_2	10.6857 s	a_m	50.3130 m/s^2
ΔH	−3009.5 m	ΔX	4217.5 m

where ΔH is the decrement in height, ΔX is the progress made in the direction of x axis.

The variation curves of each course characteristic variables are shown in Fig. 3.

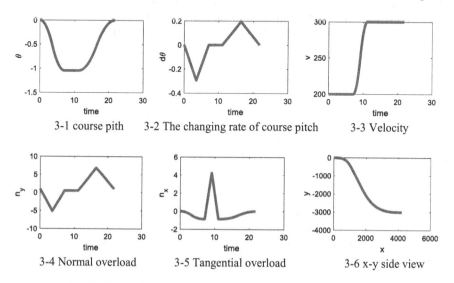

3-1 course pith   3-2 The changing rate of course pitch   3-3 Velocity

3-4 Normal overload   3-5 Tangential overload   3-6 x-y side view

Fig. 3. The variation curves of each course characteristic variables

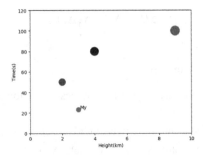

Fig. 4. Comparision results with other dive maneuver algorithms. The red circle is the algorithm result of this paper. The blue circle is the algorithm result of [8]. The black and green circles are the algorithm results of [9] (Color figure online).

Results shows aircraft consumes about 23 s when the height changes 3000 m. However, in [8], the scheme designed for dive maneuver used about 50 s when the height changes 2000 m. The results validate the effectiveness of Theorem 2 from another picture; in [9], the height change with 9000 m uses about 100 s, with 4000 m takes about 80 s (Fig. 4).

1.5   Conclusion

Dive is the process of converting the gravitational potential energy into kinetic energy. Fighter attacks enemy planes below through changing the height difference between itself and the enemy. In the article, the mathematical model of dive maneuver is established and, further, the dynamic changing law of the fighter's parameters with smooth damp is designed. After that, the simulation is carried out according to the UAV data in the air combat process. Comparing the simulation results with other related paper, we find that the time cost is shorter than the classic ones. Besides, the comparison also proves the rationality and effectiveness of this model.

References

1. Yang, Y.: Static instability flying wing UAV maneuver flight control technology research. Nanjing University of Aeronautics & Astronautics (2015)
2. AIAA: Flying on air: UAV flight testing of inflatable wing technology. AIAA J. (2000)
3. Liu, C.: Research on strong maneuvering target tracking algorithm in complex environment. Shanghai Jiao Tong University (2015)
4. Zhu, J., Liu, L., Tang, F., et al.: Optimal guidance method for hypersonic vehicle dive maneuver. J. National Univ. Defense Technol. (6), 25–30 (2013)
5. Zhang, Y., Gu, X., Zhang, W., et al.: Research on weapon delivery planning of combat aircraft based on tactical maneuvering action. J. Syst. Simul. **22**(01), 158–162 (2010)
6. Ricardo, H.M.: Overload control model design for conventional maneuver simulation of aircraft. Flight Mech. **31**(3), 213–216 (2013)

7. Xue, H., Shen, L.: Performance analysis and calculation of tactical maneuvering flight for weapon delivery of combat aircraft. Command Control Simul. **32**(5), 49–53 (2010)
8. Wang, J.: Automatic flight simulation of several typical maneuvers. Northwestern Polytechnical University (2007)
9. Yuan, J., Zhang, X., Cunkun, Lin, et al.: Simulation study of stealth aircraft penetration and penetration to radar target detection. Comput. Simul. **34**(4), 21–24 (2017)

Security Enhancement for IoT Video Streaming via Joint Network Coding and Retransmission Design

Pengxiang Qin[1,2], Pinyi Ren[1,2(✉)], Qinghe Du[1,2], and Li Sun[1,2]

[1] School of Electronic and Information Engineering, Xi'an Jiaotong University,
28 West Xianning Road, Xian 710049, Shaanxi, China
qinpengxiang@stu.xjtu.edu.cn, pyren@mail.xjtu.edu.cn
[2] Shaanxi Smart Networks and Ubiquitous Access Research Center,
28 West Xianning Road, Xian 710049, Shaanxi, China

Abstract. Relying on the development of technology, the communication Internet has included not only the traditional Internet but also the Internet of Things (IoT). However, a large number of IoT applications especially video streaming confront kinds of security challenges. In this paper, we consider the requirements of video streaming such as sufficient reliability, security, real-time and investigate the trade-off among them. Based on the above consideration, a security scheme for IoT video streaming via joint network coding and retransmission is proposed. The scheme relates the independent packets and ensures a part of them to be reliably transmitted by ARQ protocol simultaneously. Moreover, the secrecy performance is evaluated by probability analysis. And simulation results which make comparison with the noise aggregation scheme further corroborate the performance in our scheme.

Keywords: IoT video streaming · Physical layer security · Network coding · ARQ protocol

1 Introduction

With development and popularization of Internet of Things technology, various applications of IoT have been implemented gradually, which impels people's life more convenient and efficient [1]. In the meantime, wireless communication has been the main key technology of IoT. However, openness, an inherent nature of wireless transmission environment, makes information in IoT applications exposed to security threats [2]. For few years, traditional encryption schemes [3] based on computational complexity are confronting a huge challenge due to

The research reported in this paper was supported in part by: Key Research and Development Program of Shannxi Province under Grant 2017ZDXM-GY-012, National Natural Science Foundation of China under Grant No. 61431011, Fundamental Research Funds for the Central Universities.

B. Li et al. (Eds.): IoTaaS 2018, LNICST 271, pp. 40–47, 2019.
https://doi.org/10.1007/978-3-030-14657-3_4

higher performance chips. Depend on Shannon information theory, physical layer security utilizes the physical nature of wireless channel to degrade the wiretap channel quality to enhance security with no relying on computational complexity. On contrast of traditional encryption schemes, physical layer technologies are more applicable to enhance security for IoT.

Automatic repeat request scheme has been widely implemented in secure communications to provide high reliability. That's suitable and feasible for secure transmission of IoT streaming data because the round-trip time (RTT) is relatively small compared to the allowed delay most of time. The issue of quality of service (QoS) for real-time traffic over a wireless channel deploying ARQ error control was studied in [4]. A novel and simple loss impact estimation based ARQ algorithm was proposed in [5]. Furthermore, a number of schemes based on ARQ protocol has been studied. For instance, an information-theoretic perspective of retransmission protocols for reliable packet communication under a secrecy constraint was considered in [6]. A packet coding scheme in [7] relates all the packets and used ARQ to achieve secure file delivery. Noise aggregation scheme [8] used ARQ to degrade the wiretap channel quality in immersive system. In addition, the average secrecy rate in noise aggregation scheme has been analyzed in [9]. Though the noise aggregation scheme meets real-time transmission but can't well ensure the security due to its encoding method. Our basic idea is to relate the original independent packets and use ARQ protocol to achieve sufficiently reliable, secrecy and real-time transmission. Considering other problems ARQ protocol brings, the balance between reliability and security would be made up in IoT video streaming. On the basis of the above thoughts, a security scheme via joint network coding and retransmission is proposed in this paper.

The rest of this paper is organized as follows: Sect. 2 presents the system model in wireless physical layer security transmission. Section 3 describes the proposed scheme and analyzes the performance. Section 4 evaluates the average performance by probabilistic analysis. Finally, the paper concludes with Sect. 5.

2   System Model

The system model is illustrated in Fig. 1. A legitimate transmitter (Alice), a legitimate receiver (Bob) and an eavesdropper (Eve) constitute this model. On contrast of wire-tap model [10], there is an extra noiseless feedback link between Alice and Bob to ensure Bob's reliable transmission. In this case, Bob expects to receive the confidential information without being overheard by Eve. Nevertheless, Eve always exists in the network and is hard to be eliminated. When Alice is transmitting a packet to Bob over the legitimate channel, Eve also is passively receiving the data over the wiretap channel. Different from Eve, Bob can request the retransmission of lost or wrong packets via feedback link. In other words, if Bob has correctly received a packet, Alice starts to transmit the next one whether Eve successfully received the previous packet or not.

In this paper, we assume that both the legitimate channel and the wiretap channel undergo independent quasi-static fading, where the channel gains remain

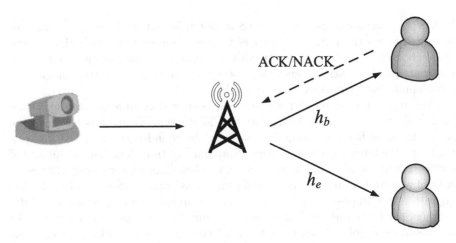

Fig. 1. System model for secrecy transmission.

constant in a packet slot and change independently at random from one slot to another. A transmitting and receiving antenna is equipped at Alice and Bob. However, Eve just has one receiving antenna as a passive node. When Alice transmits a symbol x, the received signals at Bob and Eve are

$$y_{b,i} = h_{b,i}x + n_{b,i} \tag{1}$$

and

$$y_{e,i} = h_{e,i}x + n_{e,i} \tag{2}$$

respectively. The legitimate channel gain is denoted as $h_{b,i}$ in slot i and the wiretap channel gain is $h_{e,i}$. The symbols $n_{b,i}$ and $n_{e,i}$ represent additive white noises with variances σ_b^2 and σ_e^2, respectively.

When receiving the signals, both Bob and Eve try to recover the confidential information. Assuming Eve has known the principle of the encoding method deployed at Alice, Eve and Bob recover information by the corresponding decoding method. Besides, both Eve and Bob use the optimal decision rules. The proposed scheme is discussed in detail in Sect. 3.

3   Joint Network Coding and Retransmission Design

3.1   Principles of the Scheme

Although ARQ protocol ensures the reliability of data, it also brings some problems. In common or bad transmission environment, Bob would request retransmission in several times, which means the same packet would be transmitted by Alice until Bob correctly receives this packet. As a passive node, Eve would also receive these packets which contain the same confidential information. It would bring Eve more available information to correctly receive this packet. In other words, Eve would have more probability to recover information in this case in contrast to not deploying ARQ protocol. In this way, it obviously violates the principle of secrecy transmission if no other measures are implemented. Additionally, the delay which retransmission in overmany times brings may not meet the real-time requirement. Based on the above consideration, we propose a security scheme for IoT video streaming, where Eve can't decode the current packet which is related to other previous packets ensured by ARQ protocol.

We assume that Alice needs to transmit plenty of packets $S_1, S_2, ..., S_N$ and each packet contains the same length binary data. The original packets are encoded to $X_1, X_2, ..., X_N$ one by one. In our scheme, Alice performs bitwise exclusive-or (XOR) operation on packets. If Alice starts to transmit S_i where i is odd and greater than 1, the corresponding packet X_i is given by

$$X_i = S_{i-2} \oplus S_i. \tag{3}$$

Specifically, X_1 is equal to S_1 when $i = 1$. Furthermore, X_{i+1} is given by

$$X_{i+1} = S_i \oplus S_{i+1}. \tag{4}$$

Assuming $Y_1, Y_2, ..., Y_N$ denote the received packets at receiver after demodulation, Bob or Eve try to recover the packets by the decoding method inverse to the encoding method. It's easy to describe the corresponding decoding method. We assume $\tilde{S}_1, \tilde{S}_2, ..., \tilde{S}_N$ denotes decoded packets and i is odd. Then, $\tilde{S}_i$ is given by

$$\tilde{S}_i = Y_1 \oplus Y_3 \oplus \cdots \oplus Y_{i-2} \oplus Y_i. \tag{5}$$

Furthermore, the even packet $\tilde{S}_{i+1}$ is given by

$$\tilde{S}_{i+1} = Y_1 \oplus Y_3 \oplus \cdots \oplus Y_{i-2} \oplus Y_i \oplus Y_{i+1}. \tag{6}$$

It's noted that the odd packets are transmitted by ARQ protocol to ensure the reliability but the even not. In fact, each received packet is influenced by inherent noise and random fading for Bob or Eve. Nevertheless, Bob can request retransmission for the odd packets via feedback link in contrast to Eve. According to the decoding method and ARQ protocol deployed on the odd packets for Bob, every odd original packet can be correctly recovered. And whether Bob can recover an even original packet only depends on the quality of transmission environment including noise and fading. On the contrary, Eve can correctly recover an odd original packet relying on all the previous odd packets. Moreover, an even packet can be correctly recovered also depends on the previous odd packets. Considering the different and random position in each packet in error and the huge bits a packet contains in reality, any odd packet in error greatly influences information recovering for Eve. The probabilistic model for the security capacity of the scheme is presented in Sect. 3.2.

3.2   Performance Analysis

According to the principle of the scheme in Sect. 3.1, Eve can correctly recover the information under many difficult conditions so that there exists the probability that the original packet or the original bit error occurs. Assuming α and

β are the packet error probabilities of the legitimate channel and the wiretap channel, respectively. Then, the probability that Eve can get the correct packet before Bob gets the correct packet is given in [8] by

$$P_{\text{p,odd}}(C_{\text{eve}}) = \frac{1 - \beta}{1 - \alpha\beta}. \tag{7}$$

Because just one bit error results in a packet error and the feature of XOR, it's more essential to analyze the bit error probability (BER) in our scheme. For Bob, the BER in the odd original packets is zero due to ARQ protocol. Assuming both the legitimate channel and the wiretap channel are independent identically distributed (idd) Rayleigh fading, the receiving BER with binary phase shift keying (BPSK), which is equal to the BER in the even original packets at Bob, is given in [11] by

$$P_{\text{b,odd}}(E_{\text{bob}}) = \frac{1}{2}\left(1 - \sqrt{\frac{\bar{\gamma}_b}{1 + \bar{\gamma}_b}}\right), \tag{8}$$

where $\bar{\gamma}_b$ is the average signal-to-noise ratio (SNR) at Bob. Then, we consider maximal ratio combining (MRC) reception diversity method is equipped at Eve when Eve can't get the correct packet before Bob gets the correct packet. In this case, the BER in the odd packets is given in [11] by

$$P_{\text{b,odd,MRC}}(E_{\text{eve}}) = \left(\frac{1 - \Gamma_e}{2}\right)^M \sum_{m=0}^{M-1}\binom{M-1+m}{m}\left(\frac{1+\Gamma_e}{2}\right)^m. \tag{9}$$

In Eq. (9), M is the number of the odd packet's transmission and $\Gamma_e = \sqrt{\bar{\gamma}_e/(1 + \bar{\gamma}_e)}$, where $\bar{\gamma}_e$ is the average SNR at Eve. In the even slots, the BER at Eve can also be expressed in Eq. (8) except the difference of $\bar{\gamma}$. Assuming receiving the odd packet in slot $2n - 1$ and one packet just contains one bit, Eve can correctly recover the $(2n - 1)$th original packet if an even number of error occurs in the previous n packets due to the feature of XOR. On the contrary, Eve can't correctly recover it if an odd number of error occurs. According to the specific feature of XOR, it's easy to prove

(a) When $0 < P_{\text{b,odd,MRC}}(E_{\text{eve}}) < 0.5$, the BER at Eve is always less than 0.5. when $P_{\text{b,odd,MRC}}(E_{\text{eve}})$ is constant, the BER approaches 0.5 from 0 gradually as n increases. When n is constant, the BER also approaches 0.5 gradually as $P_{\text{b,odd,MRC}}(E_{\text{eve}})$ increases.
(b) When $P_{\text{b,odd,MRC}}(E_{\text{eve}}) = 0.5$, the BER at Eve is always 0.5.
(c) When $0.5 < P_{\text{b,odd,MRC}}(E_{\text{eve}}) < 1$, it should be divided to two parts to discuss. When n is odd, the BER at Eve is always greater than 0.5. When $P_{\text{b,odd,MRC}}(E_{\text{eve}})$ is constant, the BER approaches 0.5 from 1 gradually as n increases. When n is constant, the BER approaches 1 from 0 as $P_{\text{b,odd,MRC}}(E_{\text{eve}})$ increases. On the contrary, When n is even, the BER at Eve is always less than 0.5. When $P_{\text{b,odd,MRC}}(E_{\text{eve}})$ is constant, the BER approaches 0.5 from 0 as n increases. When n is constant, the BER approaches 0 from 1 as $P_{\text{b,odd,MRC}}(E_{\text{eve}})$ increases.

According to the above conclusion, the BER would approach 0.5 if the sufficient packets are related. Although the BER is less than 0.5 in some cases, we can't promise these harsh conditions in reality. Hence, an efficient and effective way is increasing the related packets resulting in the probability of correctly decoding one bit for Eve is almost only 0.5. In the meantime, one packet contains so many bits in reality, which leads that error occurs in different positions in one packet. Considering the above analysis, our scheme can ensure secrecy transmission obviously.

Besides error probability analysis, the delay as another considerable part should be considered. Firstly according to the encoding method, each encoded packets to be transmitted relates to the previous original or current packet, which means Alice can transmit signals in real time. Secondly, Eve or Bob can decode the current packet with the previous and current received packets, which meets the real-time condition at receivers. The feasibility of real-time proves our scheme is suitable for IoT video streaming. The only odd packets are ensured by ARQ protocol but the even ones not, which reduces us nearly half of delay in whole transmit. There are many other measures can be implemented to achieve various objectives during the extra time which the scheme brings.

4   Simulation Results

To evaluate the performance of the proposed scheme, simulation results are present in this section. Note that the secrecy performance is related to the size of bits which one packet contains. In this paper, we consider one packet contains 512 bits. In addition, maximum likelihood hard decision decoding is implemented at Bob and Eve. Furthermore, maximal rate combination is deployed at Eve when not correctly received the packet before Bob gets the correct packet in one packet slot. Then, Rayleigh fading and additive white Gaussian noise (AWGN) also been considered.

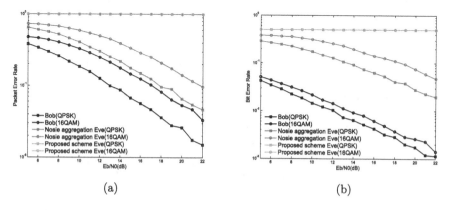

(a)   (b)

Fig. 2. (a) PER of Bob and Eve for same channel conditions in two schemes. (b) BER of Bob and Eve for same channel conditions in two schemes.

Fig. 3. (a) Screenshots of original video. (b) Video screenshot decoded by Bob. (c) Video screenshot decoded by Eve in the noise aggregation scheme. (d) Video screenshot decoded by Eve in the proposed scheme.

Figure 2(a) shows the packet error rate (PER) of two schemes for Bob and Eve in QPSK and 16QAM modulation. It should be noted that the PER in the noise aggregation scheme in [8] is equal to the proposed scheme because of the same proportion of reliable transmission packets due to ARQ protocol. For Eve in the proposed scheme, we can see PER is still zero that means Eve can't correctly decode any confidential packet. Comparing to Noise Aggregation, it greatly improves the security of confidential information. In order to prove the effectiveness more convincingly, the BER is also analyzed. Figure 2(b) shows the BER of two schemes for Bob and Eve. There is an error floor where the BER is 0.5 for Eve in the proposed scheme but the other not. Besides, the BER is always higher in the proposed scheme. That also proves the scheme has higher secrecy performance.

To further verify the effectiveness of the proposed scheme, we compare the received video screenshot in different ways. From Fig. 3(b) at Bob, we can clearly recognize the content although there is some noise on the screenshot. Comparing to the screenshot via noise aggregation in Fig. 3(c) at Eve, the screenshot in Fig. 3(d) at Eve is illegible and inundated with noise in the proposed scheme. Intuitively, our scheme also can preferably ensure the security.

5   Conclusions

The paper proposes a security scheme for IoT video streaming, which employs joint network coding and retransmission. Based on wire-tap model with a feedback link, the scheme degrades the wiretap channel quality. Furthermore, it also provides sufficient reliability, security, real-time and low delay simultaneously. Then, we theoretically study the performance. Besides, the comparison between the noise aggregation scheme and the proposed scheme has been given. And simulation results show that the proposed scheme has more high secrecy performance than the other and still supports real-time secure transmission and prove the effectiveness of our proposed scheme.

References

1. Khan, R., et al.: Future internet: the internet of things architecture, possible applications and key challenges. In: International Conference on Frontiers of Information Technology, pp. 257–260. IEEE (2013)
2. Xu, Q., et al.: Security enhancement for IoT communications exposed to eavesdroppers with uncertain locations. IEEE Access **4**, 2840–2853 (2016)
3. Keoh, S.L., Kumar, S.S., Tschofenig, H.: Securing the internet of things: a standardization perspective. IEEE IoT J. **1**(3), 265–275 (2014)
4. Quan, Z., Chung, J.M.: Analysis of packet loss for real-time traffic in wireless mobile networks with ARQ feedback. In: 2004 IEEE Wireless Communications and Networking Conference, WCNC 2004, vol. 1. IEEE (2004)
5. Ge, X., Liu, H., Wang, G.: A novel loss-impact-estimating based ARQ for wireless real-time H.264/SVC video stream. In: International Conference on Electronics Information and Emergency Communication, pp. 131–135. IEEE (2015)
6. Tang, X., et al.: On the throughput of secure hybrid-ARQ protocols for Gaussian block-fading channels. IEEE Trans. Inf. Theory **55**(4), 1575–1591 (2009)
7. He, H., Ren, P.: Secure ARQ protocol for wireless communications: performance analysis and packet coding design. IEEE Trans. Veh. Technol. **PP**(99), 1 (2018)
8. Hussain, M., et al.: Security enhancement for video transmission via noise aggregation in immersive systems. Multimed. Tools Appl. **75**(9), 5345–5357 (2016)
9. Xu, Q., et al.: On achievable secrecy rate by noise aggregation over wireless fading channels. In: IEEE International Conference on Communications, pp. 1–6. IEEE (2016)
10. Wyner, A.D.: The wire-tap channel. Bell Labs Tech. J. **54**(8), 1355–1387 (2014)
11. Stuber, G.L.: Principles of Mobile Communication. Kluwer Academic Publishers, Dordrecht (2001). 98C106

Actor-Critic for Multi-agent System with Variable Quantity of Agents

Guihong Wang [ID] and Jinglun Shi[✉] [ID]

South China University of Technology, Guangzhou 510641, China
eew.guihong@mail.scut.edu.cn, shijl@scut.edu.cn

Abstract. Reinforcement learning (RL) has been applied to many cooperative multi-agent systems recently. However, most of research have been carried on the systems with fixed quantity of agents. In reality, the quantity of agents in the system is often changed over time, and the majority of multi-agent reinforcement learning (MARL) models can't work robustly on these systems. In this paper, we propose a model extended from actor-critic framework to process the systems with variable quantity of agents. To deal with the variable quantity issue, we design a feature extractor to embed variable length states. By employing bidirectional long short term memory (BLSTM) in actor network, which is capable of process variable length sequences, any number of agents can communicate and coordinate with each other. However, it is noted that the BLSTM is generally used to process sequences, so we use the critic network as an importance estimator for all agents and organize them into a sequence. Experiments show that our model works well in the variable quantity situation and outperform other models. Although our model may perform poorly when the quantity is too large, without changing hyper-parameters, it can be fine-tuned and achieve acceptable performance in a short time.

Keywords: Multi-agent · Reinforcement learning ·
Variable quantity of agents · Communication · Fine-tune

1 Introduction

In recent years, owning to the great progress in deep learning, reinforcement learning (RL) has attracted a lot of attention from researchers [1]. By combining with deep neural network, it has been applied to a variety of fields and solved many problems, such as game playing including Atari video games and Go game high-dimensional robot control and etc.

Previous works have extended reinforcement learning to multi-agent domain. In cooperative systems, where all the agents share the goal of maximizing the discounted sum of global rewards, most researchers fix the quantity of agents in the systems and pay more attention to the communication between them. The CommNet [2] uses a single network to control agents. Each agent sends its hidden state as communication message to the embedded communication channels. The averaged message from other agents then is sent to the next layer of a specific agent. Unlike CommNet, developed from DQN, ACCNet [3] and MADDPG [4] are both extended from actor-critic policy

© ICST Institute for Computer Sciences, Social Informatics and Telecommunications Engineering 2019
Published by Springer Nature Switzerland AG 2019. All Rights Reserved
B. Li et al. (Eds.): IoTaaS 2018, LNICST 271, pp. 48–56, 2019.
https://doi.org/10.1007/978-3-030-14657-3_5

gradient method. They collect actions from all agents, and put the concatenation of them into the critic network, using the critic as a communication medium. However, these methods should fix the quantity of agents before training, and when the quantity is changed, both of them should be retrained. In lots of practical applications, we cannot know the quantity of agents in the environment in advance. Additionally, the number may change over time. Take the urban traffic control as an example, the cars in the road are always moving, so it's impractical to fix the quantity of agents in the learning model.

Tampuu [5] etc. simply use independent Deep Q-learning Network (IDQN) to control agents. This approach avoids the scalability problem, but because of the experience replay, a thorny problem appears that the environment may become non-stationary from the view of each agent. To solve this problem, Leibo et al. [6] have limited the size of experience replay buffer to keep track of the most recent data, while Foerster [7] uses a multi-agent variant of importance sampling and fingerprint to naturally decay obsolete data in the experience replay memory. Similarly, DIAL [8] also uses a single network for each individual agent, but in their model, each network has an extra output stream for communication actions. When communication is to be performed, the source agent outputs a communication signal and puts it into the target network for the next timestep. However, the environment may still become non-stationary since the message needs to be delayed for one timestep.

Recurrent neural network (RNN) has also been an effective method for coordinating variable quantity agents in some research [9]. The actor network in our proposed method is similar to the BiCNet [10] which uses BLSTM [11, 12] unit as a communication medium between agents. With BLSTM, it shares all parameters so that the number of parameters is independent of the number of agents and allows it to train using only a smaller number of agents, while freely scaling up to any larger number of agents during the test. However, RNN should be used in the sequence situation while in most natural systems, it can't directly regard the agents as a sequence.

In this work, we propose a model extended from actor-critic framework to process the multi-agent systems with variable quantity agents. In our model, we add a feature extractor to embed variable length states. The actor networks play the role in making decision for agents, and similar to BiCNet [10], by employing BLSTM, the agents can communicate and coordinate with each other. However, it is noted that the BLSTM is generally used to process sequences, so we use the critic network as an importance estimator for all agents and organize them into a sequence, sorting by their importance. Besides, because of partial observability, we embed a long short term memory (LSTM) layer in the critic network for single agent to maintain its historical states. Our experiments show that our model can still work when the quantity of agents is changed, and if the model cannot perform well in the systems with too many agents, it can be fine-tuned in the new system and get acceptable performance in a short time.

2 Proposed Method

2.1 Preliminaries: Multi-agent Markov Games

In this paper, we consider a fully cooperative multi-agent setting in which the system is composed of a set of states $\mathcal{S}$, a set of actions $\mathcal{A}_1, \mathcal{A}_2, \ldots, \mathcal{A}_N$ and a set of rewards $\mathcal{R}_1, \mathcal{R}_2, \ldots \mathcal{R}_L$. Each agent i uses a stochastic policy $\pi_{\theta_i} : \mathcal{S}_i \times \mathcal{A}_i \mapsto [0, 1]$ to choose actions, and later the next state will be produced according to the state transition function $\mathcal{T} : \mathcal{S} \times \mathcal{A}_1 \times \mathcal{A}_2 \times \ldots \times \mathcal{A}_N \mapsto \mathcal{S}$. Simultaneously, each agent i will obtain rewards as a function of the state and agent's action $r_i : \mathcal{S} \times \mathcal{A}_i \mapsto R$. In this setting, all agents should cooperate with each other to maximize the global expected return:

$$R = \sum_{i=1}^{N} \sum_{t=0}^{T} \gamma^t r_{it} \tag{1}$$

where γ is a discount factor, T is the time horizon and N is the total quantity of agents. In addition, we use local observation setting in which each agent has its own observations. The states of an selected agent $\mathcal{S}_i$ can be divided into $\mathcal{S}_s$, $\mathcal{S}_e$ and $\mathcal{S}_o$. $\mathcal{S}_s$ is the property of itself, while $\mathcal{S}_o$ is a set of states of its observed agents and $\mathcal{S}_e$ stands for a set of states of observed objects that can't be controlled in the system.

In reinforcement learning, there are several terminologies. State value function, denoted $V_\pi(s)$, is defined as the expected return when starting in s and following the policy π thereafter. It can be formulated as:

$$V_\pi(s) = E_\pi[R_t | S_t = s] = E_\pi \left[\sum_{k=0}^{\infty} \gamma^k r_{t+k+1} | S_t = s \right] \tag{2}$$

Similarly action value function or Q-value, denoted $Q_\pi(s, a)$ is defined as the expected return starting from s taking the action a, and thereafter following policy π:

$$Q_\pi(s, a) = E_\pi[R_t | S_t = s, A_t = a] = E_\pi \left[\sum_{k=0}^{\infty} \gamma^k r_{t+k+1} | S_t = s, A_t = a \right] \tag{3}$$

2.2 Feature Exactor

In our setting, the states contain three parts: the selected agent's states Ss, the other observed agents' states $So = \{So_1, \ldots, So_{N-1}\}$ and observed objects' states in the system $Se = \{Se_1, \ldots, Se_M\}$. Because of partial observability, both the quantity of other observed agents N and observed objects M may change when the selected agent move, which means that the length of So and Se are variable.

To achieve the goal that the network can deal with the variable length features, we add a module to preprocess the states. Let $So_i \in R^d$ and $So = \{So_1, \ldots So_n\} \in R^{n \times d}$. As the Fig. 2 shows, similar to textCNN [13], we apply a filter $w \in R^{h \times d \times c}$ with c channels to a window of h agents and produce a new feature map $f_m \in R^{(n-h+1) \times c}$ after a convolution operation. Besides, we apply a mean pooling operation over all windows and then get a feature vector $f_{vo1} \in R^c$. This pooling scheme naturally deals with variable lengths and reduce the influence of the operation that we organize the other agents' states into *So* randomly (Fig. 1). As the pooling operation may compress and lose useful information, we use a number of filters with different window length and then make a concatenation of them as $f_{vo} = [f_{vo1}, f_{vo2}, \ldots]$ to get more rich information.

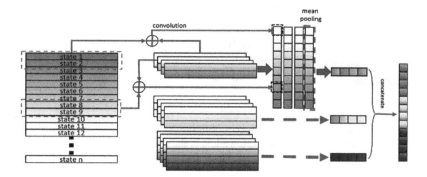

Fig. 1. The structure of feature exactor 1-dimension filters is applied to states and after the convolution operation, new feature maps are processed by mean pooling operation, followed by a concatenation among different filters.

The operation for the $Se = \{Se_1, \ldots, Se_M\}$ is the same to that for the $So_i \in R^d$. Let the output of the feature extractor for Se as $f_{ve} = [f_{ve1}, f_{ve2}, \ldots]$. At last, we concatenate f_{vo}, f_{ve} and the feature of the selected agent as the input for the subsequent deep neural network.

2.3   Critic Network

The critic network is used to estimate the current state value of a single agent. In order to make the estimations as accurate as possible, we need to collect all the useful states that the agent can obtain.

As the Fig. 2a shows, the inputs to the critic network is composed of three parts: the state of the environment, the state of the current agent, and the state of the other agents processed by the feature exactor mentioned in the previous section. The three parts are joined together and put into deep neural network. The structure of the network is shown in the Fig. 2b. Considering the local observation setting, we add a LSTM layer over time before the output layer to remember the historical states which is beneficial to estimate current state value more accurately. The output of the network is a continuous value that represents the state value of the selected agent in the current state.

The training of the network uses a supervised learning approach, where the time differential loss is shown as follows:

$$loss = E_{r,s}[(y - V(s_t))^2] \qquad (4)$$

$$y = r_{it} + V(s_{t+1}) \qquad (5)$$

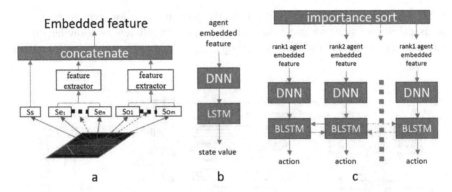

Fig. 2. Our proposed model. a is the process of feature extractor dealing with variable length states. b is the structure of a critic network while c is the processes of the actor networks.

Wherein r_{it} represents the reward obtained by the agent i at time t; s_t represents the states of the agent at time t.

2.4 Actor Network

Actor networks are used to map agents' states into actions. In our framework, the actor networks are still decentralized, indicating that each agent makes decisions based on its local information.

The structure of the actor network is shown in the Fig. 2c. The input module is the same to the critic network, where feature extractors are applied. Furthermore, we use a BLSTM layer to help the agents communicate with each other. However, the structure of the RNN is used to process sequence, and if the rankings in the sequence is changed, the obtained result may be different. In multi-agent systems, multi-agents cannot be naturally considered as a sequence in most situations, and different ranking of the agents may output different result. So we need a criterion to depend which order is proper.

In the multi-agent decision-making process, important agents should have high priorities for decision-making. Each agent can be sorted according to its importance, and then makes decision successively by order. Maximizing global score is the ambition for every agent, so the agents that have stronger scoring ability should be given greater importance. With this idea, we use the critic network to evaluate the importance of the agents in the system. We apply the critic network to all agents and get their state value. Then we sort them according to their state value and successively input them into the actor networks. It is noted that the critic network is only used in training steps. The state value describes the potential scoring ability in certain states and thus it can be regarded as the importance of the agents.

In single-agent actor-critic algorithms [14], agent update its parameters with local rewards. The gradient can be formulated as follows:

$$\nabla_\theta J(\theta) = E_{s \sim \rho^\mu, a \sim \pi}[\nabla_\theta \log \pi(a|s)(r + V_{\theta_v}(s') - V_{\theta_v}(s))] \tag{6}$$

Wherein ρ^π is the distribution of state, θ_v is the parameter of value network, and θ_v is the parameter of actor network.

However if we directly use this update in the multi-agent setting, it may encourage the agents to maximize their local return and ignore the global return leading to a local optimum. To eliminate this contradiction, we update the network with a global temporal difference loss, aiming to stimulate all agents coordinate to maximize the global return. The parameters of the actor network for agent i update as follow:

$$\nabla_{\theta_i} J(\theta_i) = E_{s \sim \rho^\mu, a \sim \pi_i}[\nabla_{\theta_i} \log \pi_i(a_i|s_i)\delta] \tag{7}$$

$$\delta = \delta_1 + \ldots + \delta_N = \sum_1^N r_i + V_{\theta_{vi}}(s_i) - V_{\theta_{vi}}(s_i) \tag{8}$$

3 Experiments

3.1 Experiment Setup

Environment. To perform our experiments, we modify the environment proposed in pysc2 [15], a challenging environment for reinforcement learning. In our task, there is a large map with some agents and 50 mineral shards. Rewards will be earned when an agent touch a mineral shard. To achieve the optimal score, the agents should split up and move independently. Whenever all mineral shards in the map have been collected, a new set of Mineral Shards are spawned at random locations. The collection time is limited to 3 min. Besides, the agents only have local vision, and can just perceive the presence of other agent and mineral shards within its scope (Fig. 3).

a   b

Fig. 3. The figure a is global view of our environment, where red objects are mineral shards and green objects are agents. The figure b is a local view. (Color figure online)

Baselines. We implement two baseline networks: independent DQN (IDQN) and BiCNet. This two networks are able to process the variable quantity issue, but without local information processing units, they work poorly in the system with variable quantity of agents. We fix the length of the local state, and train and test this two models in the system with fixed quantity agents. Differently, we train the proposed model with certain quantity and test with variable quantity. For example, the IDQN and BiCNet will train on the system with 3 agents, and test on that system to evaluate its ability. While our model will train on the system with 3 agents, and test on the system with 2, 3, 5 or more agents. Our model shares all parameters so that the number of parameters is independent of the number of agents and allows it to train using only a smaller number of agents, while freely scaling up to any number of agents during the test.

3.2   Results

We train our model with four systems, and test them in 6 systems with different agents. As it is shown in the Fig. 4, with the increase of the quantity, the reward become larger.

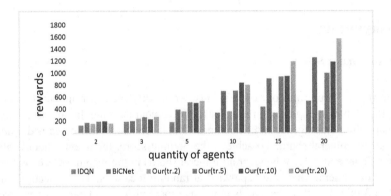

Fig. 4.   Results of the models. tr.2 means that the model is trained on the system with 2 agents.

It is because that more agents can get more mineral shards. But the growth rate is different. The tr.20 model can outperforms the other models in most scenarios, In most of the scenarios that the train quantity is larger than the test quantity, indicating that the agents can co-ordinate with each other and split well very well to increase the global reward. It can work well in different systems with different quantities of agents. The agent collect local information in the task, where the local scope can be regarded as a small system with variable quantities of agents, so it can have strong generalization ability.

It should be noted that as the quantity increase, model tr.2 and tr.3 work worse, and tr.5 and tr.10 also not work as well as tr.20. It is because that when the agents in the system increase, conflict between agents also increase. Model tr.20 is trained with 20 agents, so it can adapt for more complicated situations. While tr.2 and tr.3 only is trained with little agents, so the model haven't master the knowledge for complicated situations. That is to say, when an agent meet with ten or more agents, it may not know how to perform efficiently.

Although our model have limitation on generalization in some respects, with highly scalable structure, it needn't change any hyper parameters and the networks can be fine-tuned in the systems with more quantity of agents. As the Fig. 5 shows, we choose model tr.2, tr.3 and tr.3 to be fine-tuned in the system with 20 agents, and compare the learning curve to other models.

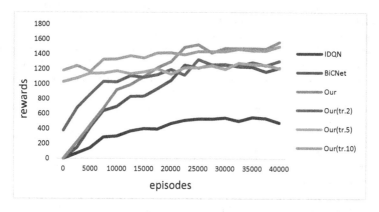

Fig. 5. Learning curve of different models trained with 20 agents.

Compared to other models that are retrained from scratch, the fine-tuned models can spend less time to reach the acceptable performance. Because the ability of dealing with variable quantities of agents, the three fine-tuned have a stronger ability to score in the beginning. Besides, the learning curve of them are steep before about 15000 episodes and become gentle after that. Without adjusting hyper parameters, previous knowledge in the networks can be transferred to the new one, so they can win at the starting line. In general, the fine-tuned models outperform IDQN and BiCNet, but they are not as good as our model that is retrained from scratch. It seems that fine-tuning may cause the network get into local optimum while retraining can help the model break away from local optimum and go farther. However, in some time-critical systems, our model can save a lot of time especially when the quantity of agents is very large and achieve acceptable performance.

4   Conclusion

In this paper, we proposed a model extended from actor-critic framework for the systems with variable quantities of agents. We not only design a feature exactor for networks to deal with variable quantity issue and embed a BLSTM layer in the actor networks, enabling agents to co-ordinate with each other. Furthermore, we use the critic network to compute the importance of all agents and sort them into a sequence. Experiments show that our model work well in the variable quantity situation and outperform other models. Although our model may perform poorly when the quantity is too large, without changing hyper-parameters, it can be fine-tuned and achieve acceptable performance in a short time.

References

1. Li, Y.: Deep reinforcement learning: An overview. arXiv preprint arXiv:1701.07274 (2017)
2. Sukhbaatar, S., Fergus, R.: Learning multiagent communication with backpropagation. In: Advances in Neural Information Processing Systems (2016)
3. Mao, H., et al.: ACCNet: Actor-Coordinator-Critic Net for Learning-to-Communicate with Deep Multi-agent Reinforcement Learning. arXiv preprint arXiv:1706.03235 (2017)
4. Lowe, R., et al.: Multi-agent actor-critic for mixed cooperative-competitive environments. In: Advances in Neural Information Processing Systems (2017)
5. Tampuu, A., et al.: Multiagent cooperation and competition with deep reinforcement learning. PLoS ONE 12(4), e0172395 (2017)
6. Leibo, J.Z., Zambaldi, V., Lanctot, M., Marecki, J., Graepel, T.: Multi-agent reinforcement learning in sequential social dilemmas. arXiv preprint arXiv:1702.03037 (2017)
7. Foerster, J., et al.: Learning to communicate with deep multi-agent reinforcement learning. In: Advances in Neural Information Processing Systems (2016)
8. Foerster, J., Assael, Y.M., de Freitas, N., Whiteson, S.: Learning to communicate with deep multi-agent reinforcement learning. In: Advances in Neural Information Processing Systems, pp. 2137–2145 (2016)
9. Foerster, J.N., Assael, Y.M., de Freitas, N., et al.: Learning to communicate to solve riddles with deep distributed recurrent q-networks. arXiv preprint arXiv:1602.02672 (2016)
10. Peng, P., Wen, Y., Yang, Y., et al.: Multiagent Bidirectionally-Coordinated Nets: Emergence of Human-level Coordination in Learning to Play StarCraft Combat Games. arXiv preprint arXiv:1703.10069 (2017)
11. Schuster, M., Paliwal, K.K.: Bidirectional recurrent neural networks. IEEE Trans. Sig. Process. 45(11), 2673–2681 (1997)
12. Hochreiter, S., Schmidhuber, J.: Long short-term memory. Neural Comput. 9(8), 1735–1780 (1997)
13. Kim, Y.: Convolutional neural networks for sentence classification. Eprint Arxiv (2014)
14. Konda, V.R., Tsitsiklis, J.N.: Actor-critic algorithms. In: Advances in Neural Information Processing Systems (2000)
15. Vinyals, O., Ewalds, T., Bartunov, S., et al.: Starcraft ii: A new challenge for reinforcement learning. arXiv preprint arXiv:1708.04782 (2017)

Video Captioning Using Hierarchical LSTM and Text-Based Sliding Window

Huanhou Xiao⬩ and Jinglun Shi$^{(\boxtimes)}$ ⬩

South China University of Technology, Guangzhou 510641, China
x.huanhou@mail.scut.edu.cn, shijl@scut.edu.cn

Abstract. Automatically describing video content with natural language has been attracting a lot of attention in multimedia community. However, most existing methods only use the word-level cross entropy loss to train the model, while ignoring the relationship between visual content and sentence semantics. In addition, during the decoding stage, the resulting models are used to predict one word at a time, and by feeding the generated word back as input at the next time step. Nevertheless, the other generated words are not fully exploited. As a result, the model is easy to "run off" if the last generated word is ambiguous. To tackle these issues, we propose a novel framework consisting of hierarchical long short term memory and text-based sliding window (HLSTM-TSW), which not only optimizes the model at word level, but also enhances the semantic relationship between the visual content and the entire sentence during training. Moreover, a sliding window is used to focus on k previously generated words when predicting the next word, so that our model can make use of more useful information to further improve the accuracy of forecast. Experiments on the benchmark dataset YouTube2Text demonstrate that our method which only uses single feature achieves superior or even better results than the state-of-the-art baselines for video captioning.

Keywords: Multimedia · Sentence semantics · Long short term memory · Sliding window · Video captioning

1 Introduction

With the rapid development of Internet technology, huge amounts of videos are uploaded online every day, which need to be quickly retrieved and understood. Driven by this challenge, automatically generating video caption has recently received increased interest and become an important task in computer vision. Moreover, video captioning provides the potential to bridge the semantics connection between video and language. A wide range of applications can benefit from it such as multimedia recommendation [1], assist the visually impaired [2], and human-robot interaction [3].

Before exploring the video captioning, previous work predominantly focused on describing images with natural language. Owing to the rapid development of deep learning, significant improvements have been made in image captioning. Then, researchers have extended these approaches to video. However, compared to describing images, video captioning is more challenging as the diverse information of objects, actions, and scenes.

© ICST Institute for Computer Sciences, Social Informatics and Telecommunications Engineering 2019
Published by Springer Nature Switzerland AG 2019. All Rights Reserved
B. Li et al. (Eds.): IoTaaS 2018, LNICST 271, pp. 57–66, 2019.
https://doi.org/10.1007/978-3-030-14657-3_6

Recently, the Long Short Term Memory (LSTM) [4] based encoder-decoder framework has been explored to generate descriptions for videos. LSTM is able to learn when to forget previous hidden states and when to update hidden states. Therefore, it can naturally deal with sequences of frames and learn long-range temporal patterns. In order to make a soft-selection over visual signals during sentence generation, attention mechanism is proposed to compute a categorical distribution of visual features, which further improve the quality of the descriptions.

Although previous encoder-decoder approaches have shown promising improvements, most of them ignore the semantic relationship between the video content and the complete sentence during training, which may cause the resulting model to generate incorrect semantics such as objects or verbs. In addition, they are trained to predict the next word given the previous ground truth word as input, while the other generated words are not holistically exploited. Therefore, the model is easy to "run off" if the last generated word is ambiguous during testing.

To tackle the above issues, we propose a Hierarchical Long Short Term Memory Model with Text-based Sliding Window (HLSTM-TSW), which utilizes an extra loss to bridge the video content and the entire sentence, as shown in Fig. 1. As a result, the relationship between visual content and sentence semantics can be explored during training. Simultaneously, a sliding window is proposed to make use of k previously generated words when predicting the next word, so that our model is able to exploit more useful information in the decoding stage. The popular video captioning dataset, Microsoft Research Video Description Corpus (YouTube2Text) [5] is used in our experiments, which demonstrates the effectiveness of the proposed method.

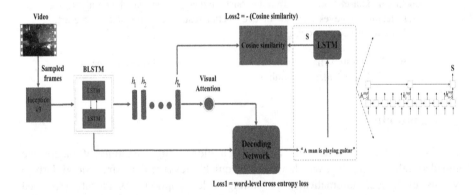

Fig. 1. The overall framework of our proposed HLSTM-TSW. Loss1 that represents the word-level cross entropy loss and Loss2 that represents the semantic relationship between video content and entire sentence are utilized together to optimize the captioning model.

2 Related Work

Early works for captioning task mainly focus on rule based systems, which detect the visual attributes (subjects, verbs, and objects) firstly, and then generate description using the template-based approach. For example, early work in [6] predicts phrases with a bilinear model and generates sentence using simple syntax statistics. However, the expansibility and richness of the natural language generated by these methods are limited by the language template.

With the rapid development of deep learning, the encoder-decoder framework has been widely applied to image captioning and video captioning. Recent works make a combination of convolutional neural network (CNN) [7] and recurrent neural network (RNN) [8] to translate the visual input to the textual output. In the case of image captioning, Vinyals et al. [9] utilize the LSTM to generate sentences with CNN features extracted from the image. Xu et al. [10] use an attention mechanism to obtain correspondences between the feature vectors and image regions. The authors of [11] propose a deep multimodal similarity model to project image features and sentences into a joint embedding space.

In video captioning, Venugopalan et al. [12] transfer knowledge from image caption models via adopting the image CNN as the encoder and LSTM as the decoder. Pan et al. [13] use the mean-pooling caption model with joint visual and sentence embedding. However, they ignore the temporal structures of video. To address this issue, Yao et al. [14] incorporate the local C3D features and a global temporal attention mechanism to select the most relevant temporal segments. Venugopalan et al. [15] present a sequence to sequence video captioning model which incorporates a stacked LSTM to read the CNN outputs firstly and then generates a sequence of words. Pan et al. [16] propose a hierarchical recurrent video encoder to exploit multiple time-scale abstraction of the temporal information.

In order to generate high-quality description for a target video, Chen et al. [17] combine the multi-modalities such as visual and audio contents to predict video topics as guidance to further improve the video captioning performance. A hierarchical structure that contains a sentence generator and a paragraph generator for language processing is introduced in h-RNN [18]. In addition, Gan et al. [19] use the Semantic Compositional Network (SCN) which extends each weight matrix of the LSTM to an ensemble of tag-dependent weight matrices to generate captions. More recently, the authors in [20] propose a multi-model stochastic RNNs network (MS-RNN) which models the uncertainty observed in the data using latent stochastic variables to improve the performance of video captioning. Song et al. [21] design an adjusted temporal attention mechanism to avoid focusing on non-visual words during caption generation. In [22], a novel encoder-decoder-reconstruction network is proposed to utilize both the forward and backward flows for video captioning.

Though the video captioning approaches mentioned above have achieved excellent results, the semantic relationship between the video content and the complete sentence is not fully exploited. Inspired by [13], in this paper, we design an extra loss to bridge the video content and sentence. Moreover, our proposed HLSTM-TSW contains a sliding window with window length of k, which enables it to focus on k previously generated words during the decoding stage.

3   Proposed Method

In this section, we introduce our approach for video captioning, as shown in Fig. 1. Firstly, the encoding stage with visual attention mechanism is presented. Then, we propose a textual attention in decoding network to calculate the contribution of words contained in the sliding window. Finally, we introduce our mixed-loss model, which simultaneously considers the context relationship between previous words and future words and the semantic relationship between visual content and entire sentence.

3.1   Encoding Network

Given a video $\mathbf{v}$ with N sampled frames, the visual features and the textual features can be represented as $v = \{v_1, v_2, \ldots, v_i, \ldots v_N\}$ and $w = \{w_1, w_2, \ldots, w_i, \ldots w_T\}$, where $v_i \in R^{D_v \times 1}$, $w_i \in R^{D_w \times 1}$, and T is the length of the sentence. Specifically, D_v and D_w are the dimension of frame-level features and the dimension of vocabulary respectively. We use a bi-directional LSTM (Bi-LSTM) which can capture both forward and backward temporal relationships to encode the visual features. The activation vectors are obtained as:

$$h_t = h_t^{(f)} + h_t^{(b)} \tag{1}$$

where $h_t^{(f)}$ and $h_t^{(b)}$ are the forward and backward hidden activation vectors.

The attention mechanism is realized by using attention weights to the hidden activation vectors throughout the input sequence, so the output context vector at time step t can be represented as:

$$a_t = \sum_{i=1}^{N} \alpha_{t,i} h_i \tag{2}$$

and

$$\alpha_{t,i} = \frac{\exp(e_{t,i})}{\sum_{k=1}^{N} \exp(e_{t,k})} \tag{3}$$

$$e_{t,i} = w^T \tanh(W_a h_i + V_a h_{t-1} + b_a) \tag{4}$$

where w, W_a, V_a, b_a are learned parameters, and h_{t-1} is the hidden state of the decoder LSTM at $(t-1)$-th time step.

3.2   Decoding Network

In our decoding network, we use hierarchical LSTM to generate the description, as described in Fig. 2. During the sentence generation process, we use a sliding window to focus on k nearest generated words when predicting the next word. Following it, a

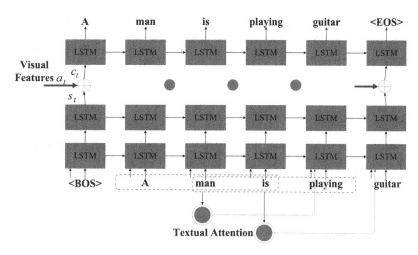

Fig. 2. The schematic diagram of our decoding network. When predicting the next word, a sliding window is utilized to focus on k nearest generated words, and the textual attention calculates their corresponding contributions.

textual attention is used to calculate the corresponding contributions of these k words. The output of it is:

$$q_t = \sum_{i=t-k}^{t-1} \beta_{t,i} w_i \tag{5}$$

and

$$\beta_{t,i} = \frac{\exp(u_{t,i})}{\sum_{m=1}^{k} \exp(u_{t,m})} \tag{6}$$

$$u_{t,i} = w_c^T \tanh(W_c w_i + V_c a_t + b_c) \tag{7}$$

where w_c, W_c, V_c, b_c are learned parameters.

Once the above operations are completed, the concatenation of w_{t-1} and q_t will be utilized as input to the bottom LSTM. Therefore, our model can focus on k previously generated words instead of only the last generated word. In addition, a visual adjusted gate is designed to avoid the problem that imposing visual attention on non-visual words, which is introduced in [21]. It can be computed as:

$$g_t = sigmoid(W_g r_t) \tag{8}$$

where W_g is learned parameter, r_t is the output of the bottom LSTM. Suppose the output of the middle LSTM is s_t, Then the input of the top LSTM is:

$$c_t = g_t a_t + (1 - g_t)s_t \tag{9}$$

3.3   Mixed-Loss Model

According to the above analysis, at time step t, our model utilizes **v** and the previous words $w_{<t}$ to predict a word w_t with the maximal probability $P(w_t|w_{<t}, \mathbf{v})$, until we reach the end of the sentence. So the word-level cross entropy loss can be defined as:

$$loss1 = -\sum_{t=1}^{T} \log P(w_t|w_{<t}, \mathbf{v}; \theta) \tag{10}$$

where θ is the model parameter set.

To explore the semantic relationship between the visual content and the entire sentence, the last hidden activation vector h_n that represents the visual information of the video content and the sentence vector S that represents the semantic information of the entire sentence are utilized to calculate the cosine similarity, as shown in Fig. 1. In particular, S is the final output of another LSTM whose inputs are the corresponding hidden activation vectors of the top LSTM of decoding network. It is worth noting that the input at $t = 1$ will only flow through $T/5$ steps to the final output rather than T steps, which prevents the loss of information during long-distance transmissions, especially for short sentences. The cosine similarity between h_n and S can be computed as:

$$\cos(h_n, S) = \frac{h_n \bullet S}{\|h_n\| \|S\|} \tag{11}$$

Aiming to pull the corresponding video-sentence pairs closer in the mapping space, we define our loss2 as follow:

$$loss2 = -\cos(h_n, S) \tag{12}$$

and the final loss of our model is:

$$loss = loss1 + \delta loss2 \tag{13}$$

where δ is the tradeoff parameter.

4   Experiments

4.1   Dataset

The YouTube2Text dataset consists of 1,970 short video clips collected from You-Tube, which is well suited for training and evaluating an automatic video captioning model. This dataset contains about 80,000 clip-description pairs and each clip has multiple sentence descriptions. Following [14] and [15], we split 1200 videos for training, 100 videos for validation, and 670 videos for testing.

4.2 Data Preprocessing

We convert all descriptions to lower cases, and then utilize the WordPunct function from NLTK[1] toolbox to tokenize sentences and remove punctuations. Therefore, it yields a vocabulary of 13374 in size for the dataset. In our experiments, we use the one-hot vector (1-of-N decoding, where N is the vocabulary size) to represent each word, and use the inceptionv3 [23] to extract frame-level features. In addition, we uniformly sample 60 frames from each clip.

4.3 Training Details

In our experiments, with an initial learning rate 10^{-5} to avoid the gradient explosion, we set all the LSTM unit size and the word embedding size as 512, empirically. In addition, we train our model with mini-batch 64 using ADAM optimizer [24], and the length of sentence T is set as 20. For sentence with fewer than 20 words, we pad the remaining inputs with zeros. Moreover, beam search with beam width of 5 is used to generate descriptions during testing process. To regularize the training and avoid overfitting, we apply dropout with rate of 0.5 on the outputs of LSTMs.

4.4 Metrics

We evaluate our model on the following widely-used metrics: BLEU [25], METEOR [26] and CIDEr [27], and use the Microsoft COCO evaluation server [28] to obtain our experimental results reported. BLEU is defined as the geometric mean of n-gram precision scores multiplied by a brevity penalty for short sentences. CIDEr measures the consensus between the candidate descriptions and the reference sentences. METEOR is defined as the harmonic mean of precision and recall of unigram matches between sentences.

4.5 Results and Analysis

In this subsection, we firstly explore the effect of the tradeoff parameter δ. We adjust it from 0.1 to 0.9 at intervals of 0.2. The performance curves with a different tradeoff parameter are shown in Fig. 3. We normalized METEOR and BLEU scores using the following function:

$$Q_{norm} = \frac{Q - \min(Q)}{\min(Q)} \tag{14}$$

where Q and Q_{norm} are the original and normalized performance values, respectively.

---

[1] [Online]. Available: https://www.nltk.org/index.html.

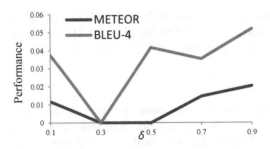

Fig. 3. The effect of δ on YouTube2Text dataset.

Table 1. Caption performance of HLSTM-TSW and other state-of-the-art methods on YouTube2Text dataset in terms of BLEU-4, METEOR, and CIDEr scores (%). HLSTM (single) represents that it was trained by cross entropy loss only, and HLSTM (mixed) represents that it was trained using mixed loss. The symbol "–" indicates such metric is unreported.

Model	BLEU-4	METEOR	CIDEr
S2VT [15]	–	29.8	–
SA [14]	41.9	29.6	51.7
h-RNN [18]	49.9	32.6	–
HRNE-SA [16]	46.7	33.9	–
hLSTMat [21]	53.0	33.6	73.8
MS-RNN [20]	**53.3**	33.8	74.8
RecNet [22]	52.3	34.1	80.3
HLSTM-TSW (single)	50.2	34.5	80.0
HLSTM-TSW (mixed)	50.5	**35.0**	**82.8**

From Fig. 3 we can see that our captioning model achieves the best performance when $\delta = 0.9$, which proves that enhancing the semantic relationship between the visual content and the entire sentence is conducive to boost the captioning model.

Then, we compare our HLSTM-TSW approach with other state-of-the-art methods, including the baseline sequence to sequence model (S2VT, MS-RNN), and the attention-based LSTM Model (SA, h-RNN, HRNE-SA, hLSTMat, RecNet).

Table 1 shows the quantitative results of the comparison. We can observe that our HLSTM-TSW performs best on METEOR and CIDEr metrics, verifying the effectiveness of our proposed method. In addition, HLSTM (mixed) performs better than HLSTM (single) on all metrics, which demonstrates that exploring the semantic relationship between video content and entire description benefits the captioning model.

Besides, some representative captions are presented in Fig. 4. Six videos are used for demonstration and two frames are extracted from each video. We notice that the sentences generated from our model are able to describe the salient contents of videos, such as woman-applying-makeup, man-shooting-gun, and monkey-pulling-dog's tail, which proves the superiority and reliability of our approach. In some of the cases, our model correctly identifies parts of the sentences, but fails to find the correct object. For example, for the top video in the right column, the generated caption is "a man is playing a piano keyboard" while the reference is "a boy is playing a keyboard". This is due to the reason that our training data does not provide training samples to distinguish "man" and "boy". Therefore, existing datasets for video captioning still require further refinement.

Fig. 4. Example results of YouTube2Text dataset.

5   Conclusion

In this paper, we propose a novel framework HLSTM-TSW to make use of the semantic relationship between video content and the entire description. In our hierarchical structure, an extra loss is utilized to map the video-sentence pairs closer in the embedding space. Moreover, the combination of the text-based sliding window and the textual attention mechanism enables the model to exploit k previously generated words instead of only the last generated word in next-word generation. Experimental results on YouTube2Text dataset show that our HLSTM-TSW achieves superior performance compared with the current start-of-the-art models. In the future work, we will combine the reinforcement learning algorithms to further improve our caption model.

References

1. Sun, L., Wang, X., Wang, Z., Zhao, H., Zhu, W.: Social-aware video recommendation for online social groups. IEEE Trans. Multimedia **19**(3), 609–618 (2017)
2. Wu, S., Wieland, J., Farivar, O., Schiller, J.: Automatic Alt-text: computer-generated image descriptions for blind users on a social network service. In: CSCW, pp. 1180–1192 (2017)
3. Das, A., et al.: Visual dialog. In: CVPR (2017)
4. Hochreiter, S., Schmidhuber, J.: Long short-term memory. Neural Comput. **9**(8), 1735–1780 (1997)
5. Chen, D.L., Dolan, W.B.: Collecting highly parallel data for paraphrase evaluation. In: ACL (2011)

6. Lebret, R., Pinheiro, P.O., Collobert, R.: Phrase-based image captioning. arXiv preprint arXiv:1502.03671 (2015)
7. Simonyan, K., Zisserman, A.: Very deep convolutional networks for large-scale image recognition. arXiv preprint arXiv:1409.1556 (2014)
8. Sutskever, I., Vinyals, O., Le, Q.V.: Sequence to sequence learning with neural networks. In: NIPS (2014)
9. Vinyals, O., Toshev, A., Bengio, S., Erhan, D.: Show and tell: a neural image caption generator. In: CVPR (2015)
10. Xu, K., et al.: Show, attend and tell: neural image caption generation with visual attention. In: ICML (2015)
11. Fang, H., et al.: From captions to visual concepts and back. In: CVPR (2015)
12. Venugopalan, S., Xu, H., Donahue, J., Rohrbach, M., Mooney, R., Saenko, K.: Translating videos to natural language using deep recurrent neural networks. In: NAACL HLT (2015)
13. Pan, Y., Mei, T., Yao, T., Li, H., Rui, Y.: Jointly modeling embedding and translation to bridge video and language. arXiv preprint arXiv:1505.01861 (2015)
14. Yao, L., et al.: Describing videos by exploiting temporal structure. In: ICCV (2015)
15. Venugopalan, S., Rohrbach, M., Donahue, J., Mooney, R., Darrell, T., Saenko, K.: Sequence to sequence - video to text. In: ICCV (2015)
16. Pan, P., Xu, Z., Yang, Y., Wu, F., Zhuang, Y.: Hierarchical recurrent neural encoder for video representation with application to captioning. arXiv preprint arXiv:1511.03476 (2015)
17. Chen, S., Chen, J., Jin, Q.: Generating video descriptions with topic guidance. In: ICMR (2017)
18. Yu, H., Wang, J., Huang, Z., Yang, Y., Xu, W.: Video paragraph captioning using hierarchical recurrent neural networks. In: CVPR (2016)
19. Gan, Z., et al.: Semantic compositional networks for visual captioning. In: CVPR (2017)
20. Song, J., Guo, Y., Gao, L., Li, X., Hanjalic, A., Shen, H.T.: From deterministic to generative: multi-modal stochastic RNNs for video captioning. arXiv preprint arXiv:1708.02478 (2017)
21. Song, J., Guo, Z., Gao, L., Liu, W., Zhang, D., Shen, H.T.: Hierarchical LSTM with adjusted temporal attention for video captioning. arXiv preprint arXiv:1706.01231 (2017)
22. Wang, B., Ma, L., Zhang, W., Liu, W.: Reconstruction network for video captioning. arXiv preprint arXiv:1803.11438 (2018)
23. Szegedy, C., Vanhoucke, V., Ioffe, S., Shlens, J., Wojna, Z.: Rethinking the inception architecture for computer vision. In: CVPR (2016)
24. Kingma, D.P., Ba, J.: Adam: a method for stochastic optimization. arXiv preprint arXiv:1412.6980 (2014)
25. Papineni, K., Roukos, S., Ward, T., Zhu, W.J.: BLEU: a method for automatic evaluation of machine translation. In: ACL (2002)
26. Lavie, A., Agarwal, A.: METEOR: an automatic metric for MT evaluation with improved correlation with human judgments. In: Proceedings of the ACL Workshop on Intrinsic and Extrinsic Evaluation Measures for Machine Translation and/or Summarization, pp. 65–72 (2005)
27. Vedantam, R., Lawrence Zitnick, C., Parikh, D.: CIDEr: consensus-based image description evaluation. In: CVPR (2015)
28. Chen, X., et al.: Microsoft COCO captions: data collection and evaluation server. arXiv preprint arXiv:1504.00325 (2015)

Missing Data Imputation for Machine Learning

Shaoqian Wang, Bo Li, Mao Yang$^{(\boxtimes)}$, and Zhongjiang Yan

School of Electronics and Information, Northwestern Polytechnical University,
Xi'an, China
`wangshaoqian@mail.nwpu.edu.cn`, {`libo.npu,yangmao,zhjyan`}`@nwpu.edu.cn`

Abstract. The imputation of missing values in datasets always plays an important role in the data preprocessing. In the process of data collection, because of the various reasons, the datasets often contain some missing values, and the excellent missing data imputation algorithms can increase the reliability of the dataset and reduce the impact of missing values on the whole dataset. In this paper, based on the Artificial Neural Network (ANN), we propose a missing data imputation method for the classification-type datasets. For each record which contains missing values, we make a list of the values that can be used to replace the missing data from the complete dataset. Our ANN model uses the complete records as the train dataset, and selects the most appropriate value in the list as the final result based on the label categories of the missing data. In our experiments, we compare our algorithm with the traditional single value imputation method and mean value imputation method with the Pima dataset. The result shows that our proposed algorithm can achieve better classification results when there are more missing values in the dataset.

Keywords: Data imputation · Machine learning ·
Artificial Neural Network

1 Introduction

With the development of data mining technology and the rise of machine learning technology, various datasets are becoming more and more important. Although everyone wants to get perfect datasets, in the process of data collection, any problem, such as human error and machine failure, will affect the datasets, resulting in abnormal or even missing values. In fact, many large datasets contain missing values. A large number of missing values will reduce the reliability of the whole dataset. For some datasets which are difficult to collect, the missing values problem will make enormous loss. At the same time, the missing data will also bring huge challenges to the data processing and analysis process and have an impact on the results of the experiment.

© ICST Institute for Computer Sciences, Social Informatics and Telecommunications Engineering 2019
Published by Springer Nature Switzerland AG 2019. All Rights Reserved
B. Li et al. (Eds.): IoTaaS 2018, LNICST 271, pp. 67–72, 2019.
https://doi.org/10.1007/978-3-030-14657-3_7

Because it is difficult to ensure that there is no error in the whole process of data collection, so the imputation algorithms of missing values are very necessary. Aiming at the problem of the missing values in the datasets, many researchers have proposed various data-filling algorithms. The simpler way to solve the problem is to delete records which contain missing values, single value imputation algorithm and mean value imputation algorithm, but all these algorithms have great limitations. The rest of the filling algorithms can be divided into statistical imputation methods and data mining based imputation methods. Commonly used statistical imputation methods include EM imputation, regression model imputation, multiple erasing difference imputation, et al. Classical data mining methods include neural network, decision tree, Bayesian network, KNN algorithm, etc. These algorithms have been widely applied to various kinds of missing values problems according to their advantages and characteristics, and have achieved remarkable results.

In this paper, we propose imputation algorithm based on neural network for classification-type datasets. We use the complete data in the dataset as training dataset to train our ANN model. Our ANN model can show probabilities of a record being classified as different categories. For each record which contains missing values, we use the different appropriate values from other complete records to replace missing values. In this way, we get a list of new records and we will select the best one from these records as the imputation result. Based on our ANN model, We can achieve the probability of the records which are classified as the correct categories. The record which achieve the highest probability will be select. Compared with other data imputation algorithms, this method takes advantage of the characteristics of the classified dataset label, thus further improving the imputation effect in the classified dataset. In our experiment, compared with the single value filling and mean filling method with Pima datasets, our proposed method can achieve better classification results when there are many missing values in the dataset.

In the rest of this article, the second part introduces the background knowledge and some related work of this algorithm. The third part introduces the concrete implementation process of our algorithm. The fourth part explains the experimental process and result analysis based on Pima dataset. The fifth part summarizes the full text.

2   Related Work

In some datasets that contain very few missing data, ignoring method, deleting method and zero value method are often used for data filling [1, 2]. These methods have an effective effect when the missing values have little effect on the dataset, but these methods are no longer applicable when there are more missing data in the dataset. In a variety of complex missing data problems, various data imputation algorithms based on machine learning have achieved good results.

The KNN method is often used in the data imputation algorithm, and the Batista [3] proposed the KNNI algorithm. For a record Ri containing the missing value, the KNN algorithm is used to find the most similar k records to Ri in the whole dataset, and then mend the missing values in Ri based on the value in k records. The processing of KNNI algorithm is simple, but when the dataset is large, it will take a long time to calculate.

Rahman [4] proposed an imputation method named DMI based on decision tree. For the record Ri which contains missing values, DMI algorithm will select all the records belonging to the same leaf node with the record Ri in the dataset as a new dataset Di. Based on the dataset Di the EMI algorithm will be used to finish the imputation work. The quality of decision tree model has great influence on the effectiveness of DMI algorithm. If the classification effect of leaf nodes is not good enough, the DMI algorithm will be affected.

Neural network algorithm is also widely used in data mining based incomplete data imputation problem. The main idea is to minimize the error between the simulated value and the actual value of the network output, and use the error to adjust the weight. The neural network algorithm has a strong generalization as its advantage.

Silvaramírez et al. [5] designs a 3 layer perceptual network for data filling. The number of neurons in the network input and output layer are equal to the number of attributes of the dataset. By artificially deleting some data, the disturbance dataset is generated as input, the original complete dataset is trained as the output for the network, and then the data is filled with the obtained network. Compared with most machine learning based filling algorithms, the method can get higher filling effect. But this method will take a long time to train the model, and it don't take advantage of the labels in the datasets.

In fact, there are few cases of labels missing in classification-type datasets. The label is an important attribute to judge the similarity between different records. In this paper, our proposed algorithm builds a classification model based on ANN, and takes advantage of different labels in dataset to complete data imputation. Although it can only be used with classification-type datasets, experiments show that this method can significantly improve the classification accuracy of incomplete classification datasets.

3   Data Imputation Method

The algorithm in this paper constructs an artificial neural network with three hidden layers and a Softmax classifier. Softmax classifier can show the probability of the records which are classified as different categories. The process is mainly divided into the training of the model and the selection of the appropriate filling value by the trained model.

3.1   Training Step

From the dataset which contain missing values, all the complete records are selected to form a new dataset as the training dataset. In this paper, in order to test the imputation effect of datasets on different data missing percentage, we artificially deleting some data randomly. All the records which have been deleted form the missing-value dataset Dm, the remaining complete records constitute the complete dataset D, then the dataset D is used as the training dataset to the train our ANN model, thus a model for classification is obtained. The process is shown in Fig. 1.

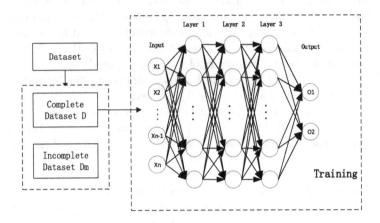

Fig. 1. ANN training with complete datasets

3.2 Data Imputation

After the network training is completed, our algorithm can solve the data imputation based on our ANN model. For a record Xi which have missing values, the first step in our method is find out the attributes which have missing values in this record, the values of these attributes from other complete records form the dataset Vm. The values in Vm will be used as the different imputation alternatives for the missing values of record Xi. The second step is to combine the various values in Vm with the complete part of the record. The missing values of the record Xi will be filled in respectively with all the values in Vm to form a dataset Dxi, which contain the final imputation result for the record Xi. In the third step, ANN model will calculate a list of outputs by the Softmax classifier using the records in Dxi. These outputs show the probabilities that the records are classified as different categories. In fact, the labels of the records in Dxi are the same with Xi's label Yi. The record Zi which achieve the highest probability to be classified as the category of Xi will be the best imputation result for the record Xi. By using the 3 steps above for all records in the incomplete dataset Dm, the final imputation dataset can be obtained. The whole process is shown in Fig. 2.

4 Evaluation

In the paper, we use the Pima Indians Diabetes to test the proposed method. This dataset contains the incidence of diabetes in hundreds of people within 5

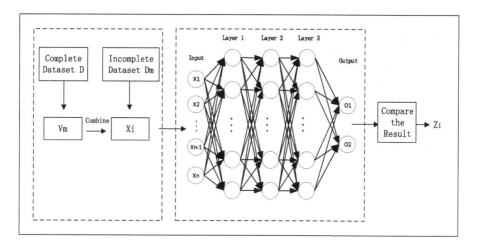

Fig. 2. Imputation process

years. It contains 8 types of physical condition variables for each person and 1 label which indicate whether a person is suffering from diabetes.

We first divide the whole dataset into training dataset and test dataset. We take 80% of the dataset as the training dataset, and the remaining 20% of the dataset as the test dataset. The Pima dataset is not a complete dataset, there have been some missing values in some attributes. We build an ANN model with 3 hidden layers, the number of neurons in each layer is 12, 24 and 16. Base on this ANN model, we achieved a result of 82.285% classification accuracy with our training dataset and test dataset. In order to test the classification results of datasets after data imputation with different missing rate, we delete the values in the datasets artificially, and get 3 new datasets which have 30%, 50%, 70% missing rate respectively. Each record in the new datasets contains a maximum of one missing value. Then, according to the imputation process described in the last section, we achieve 3 complete datasets after the imputation work. We achieve the classification accuracy of these 3 new complete datasets with our ANN model, and compare the results with the mean value imputation method and zero value imputation method. The mean imputation method replaces the missing values with the mean of the whole values of the missing values' attribute and the zero value method replaces all the missing locations with zero values.

Table 1. Classification accuracy with different imputation methods

Missing rate	Zero value method (%)	Mean value method (%)	Our method (%)
30%	80.12	80.56	84.96
50%	78.57	79.67	84.28
70%	77.14	78.42	84.20

From Table 1, we can find that with the increase of missing proportion, the classification accuracy of the 3 imputation methods has decreased. At the same time, through the lateral contrast, we can find that the classification accuracy of the zero value imputation is slightly higher than the mean imputation for the dataset. Meanwhile, the classification accuracy of our proposed method is greatly improved compared with the other two kinds of methods. For all the kinds of missing proportion, the accuracy of classification increased by more than 4%. Moreover, with the increase of data missing proportion, the accuracy of classification achieved by our method has not decreased a lot.

5   Conclusion

This paper proposes an artificial neural network-based incomplete data imputation method for categorical datasets. A classification neural network model is obtained by training with the complete records, and the missing values are fixed with possible values. The optimal imputation value is selected as the final imputation result by our ANN model. We test our method on the Pima dataset and compare the classification accuracy with zero value imputation method and mean imputation. The results show that our method has a greater improvement in the classification accuracy compared with zero imputation method and mean imputation method at different data missing degrees.

Acknowledgement. This work was supported in part by the National Natural Science Foundations of CHINA (Grant No. 61771390, No. 61771392, No. 61501373, and No. 61271279), the National Science and Technology Major Project (Grant No. 2016ZX03001018-004, and No. 2015ZX03002006-004), and the Fundamental Research Funds for the Central Universities (Grant No. 3102017ZY018).

References

1. Cheng, Y., Miao, D., Feng, Q.: Positive approximation and converse approximation in interval-valued fuzzy rough sets. Inf. Sci. **181**, 2086–2110 (2011)
2. Meng, Z., Shi, Z.: Extended rough set-based attribute reduction in inconsistent incomplete decision systems. Inf. Sci. **204**(20), 44–69 (2012)
3. Batista, G.E.A.P.A., Monard, M.C.: An analysis of four missing data treatment methods for supervised learning. Appl. Artif. Intell. **17**(5–6), 519–533 (2003)
4. Rahman, G., Islam, Z.: A decision tree-based missing value imputation technique for data pre-processing. In: The Australasian Data Mining Conference, pp. 41–50 (2010)
5. Silvaramírez, E.L., et al.: Missing value imputation on missing completely at random data using multilayer perceptrons. Neural Netw. Official J. Int. Neural Netw. Soc. **24**(1), 121–129 (2011)

Development of 3D Exhibition System for IoT-Oriented Simulation Platform

Wenjun Yang[1,2], Qinghe Du[1,2(✉)], Xinlei Xiong[1,2], Li Sun[1,2], and Pinyi Ren[1,2]

[1] Department of Information and Communications Engineering,
Xian Jiaotong University, Xian 710049, China
yangwenjun2014@stu.xjtu.edu.cn, {duqinghe,lisun,pyren}@mail.xjtu.edu.cn,
xiongxinleixxl@163.com
[2] National Simulation Education Center for Communications and Information
Systems, Xian 710049, China

Abstract. This paper develops an exhibition system based on the system-level simulation platform for massive wireless access of IoT. The system appends a complete data output mechanism based on the original simulation platform, which provides a convenient environment for development and debugging, observation of system process and performance evaluation for the NB-IoT simulation platform. The exhibition system includes multiple modules, such as a data processing module, a driving engine module, a drawing module and a 3D scene module. Meanwhile, the system designs multiple interaction modules to restore the simulation details. Finally, this paper completes the debugging of the system, analyzes and discusses the results.

Keywords: NB-IoT · Exhibition system · Qt

1 Introduction

In order to meet the needs of future users for delay and speed, the fifth-generation mobile communication system (5G) [1–3] proposes the objective of high-speed, low-latency and massive access to realize the Internet of Everything. In this context, the Narrow Band-Internet of Things (NB-IoT) has attracted more and more attention around the world. In order to carry out comprehensive evaluation and verification of NB-IoT technology, and provide reference and basis for NB-IoT technology selection, the research team developed a system-level simulation platform for NB-IoT, but had not yet developed a simulation demonstration system and failed to provide a complete data output mechanism. In order to build a convenient environment for development and debugging, observation of

Supported by the National Natural Science Foundation of China under the Grants No. 61671371, the Key Research and Development Program of Shaanxi Province under Grant No. 2017ZDXM-GY-012, and the Fundamental Research Funds for the Central Universities.

B. Li et al. (Eds.): IoTaaS 2018, LNICST 271, pp. 73–80, 2019.
https://doi.org/10.1007/978-3-030-14657-3_8

system working process and performance evaluation for NB-IoT simulation platform, this paper sorts out the function structure of NB-IoT simulation platform and adds a data output module. Based on the simulation record file, this paper designs and implements the 3D demonstration system to match with the simulation platform. And the cross-platform C++ application development framework Qt [4] is used to construct the 3D model. Based on the signal and slot mechanism [5] provided by Qt, the event-driven engine is designed. When the interactive mode of the system is demonstrated, the system perspective, the base station perspective as well as the user perspective are designed for the users to observe. In the analysis and demonstration of the simulation data, the exhibition system designed the real-time rendering curve function, and the observer can analyze the key performance indicators of the simulation platform through observing the curve.

The rest of the paper is structured as follows: Sect. 2 briefly introduces the general structure and function of the NB-IoT simulation platform, and Sect. 3 summarizes the system structure of the exhibition system. Section 4 explains the concrete implementation of the 3D exhibition system, including the structural design and the specific implementation of each function. Section 5 analyzes and discusses the demonstration results, and finally Sect. 6 summarizes the full text and gives conclusions.

2   Simulation Platform for NB-IOT

In order to carry out comprehensive evaluation and verification of NB-IoT technology and provide reference and basis for NB-IoT technology selection, the research team designed and developed NB-IoT system-level simulation platform based on 3GPP TR45.820 [6]. The platform is written in C++, and the modules of service generation module, drive engine module, resource allocation module, channel module, random access module, power control module and coverage type division module are designed and implemented to complete the simulation of capacity of NB-IoT network and the probability of access failure. In order to reduce the probability of access failure and avoid congestion, the backoff mechanism must be optimized. Therefore, in the development process of the simulation platform, a different type of backoff mechanism based on coverage type is proposed. Thanks to this mechanism, the simulation platform can avoid large-scale access failures.

3   System Architecture

The exhibition system is roughly composed of four major functional modules: a data reading module, a data processing module, a 3D scene module and an engine module. System perspective, base station perspective and user perspective are designed in the 3D scene demo module. By switching of the angle of view, the different content in the simulation process can be showed. Three functions of real-time drawing function, 3D scene model and event list function are designed in various perspectives. The real-time drawing function can draw curves in real time for some key indicators during the simulation process, and the user can further deepen the understanding of the simulation through the drawn curves. The 3D scene model is based on the system perspective, the base station perspective and the user perspective respectively establish their own 3D models. As the simulation progresses, some events will trigger the animation effects in the 3D model. The function of event list is a demonstration of some of the key events that occur during the simulation. The engine module connects the above modules in series, and schedules each module by reading the data in the simulation log file, and finally achieves the driving effect on the entire demonstration system. The structure of the specific 3D exhibition system is shown in Fig. 1.

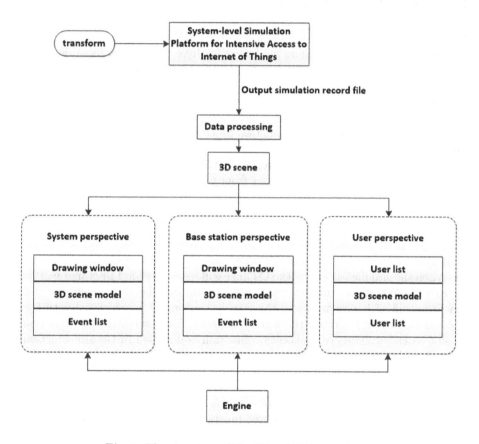

Fig. 1. The structure of the 3D exhibition system.

The operation process of exhibition system: firstly, the simulation record file of the original simulation platform transformation output is input into the exhibition system, and the file is read by the data reading module. Then, the data processing module processes the data of the file, and reads the file according to the format definition of the file, and it specifically reads and stores according to the format of the event structure. Eventually, all data will be stored in the set of events in a structured format. Then the drive engine module detects that the process of data reading is complete and waits for the instruction to start the simulation. Once the instruction of starting the simulation is received, the timer will be triggered. And each time the timer is full, an event will be taken from the set of events. Then the event processing function in which the multiple signals are designed will handle the event. After that, different signals will be emitted according to different types of events, which triggers different slot functions. By designing these slot functions, the 3D scene presentation module is driven.

4   System Implementation

4.1   Engine

The function of the drive engine module is to drive the various function modules to coordinate operation after reading the simulation record file. The engine module used in the exhibition system is the event-driven engine, and the core part of the event-driven engine is the event. After reading the record file and processing the data, the exhibition system gets an set of events, which arranges various events occurring in the simulation process in chronological order. During the simulation process, various functions and operations of the device are stored in the set of events in the form of events. The function of processing event takes the events and executes them in order from the set of events. Firstly, the type of the event is judged, and the corresponding signal is selected according to the different types of the event, and the corresponding slot function is triggered. Then, the engine module continuously processes the event, mobilizes other functional modules, and coordinates the cooperation to complete the demonstration. A detailed schematic diagram of the drive engine is shown in Fig. 2.

When the simulation starts, the QTimer will be started and the timer will be set to 1 ms. When the timer expires 1 ms, the timeout signal is issued, the slot function "run()" is designed, and the signal "timeout()" is associated with the slot function "run()". When the signal "timeout()" is issued, the slot function "run()" is called. The event execution function is designed in the slot function "run()". Take the event from the set of events and execute it, determine the type of the event firstly, and then send the corresponding signal according to the type of event. The slot function for drawing, the slot function for showing 3D models and the slot function for showing event list will be executed after receiving the signal, and the respective functions are implemented accordingly. The event-driven engine designed by the exhibition system relies on the powerful signal and slot mechanism provided by Qt to complete the scheduling of each functional module.

4.2   Event List

The function of the event list is to reflect the event that is happening in the current exhibition system by displaying the event name and the time of occurrence,

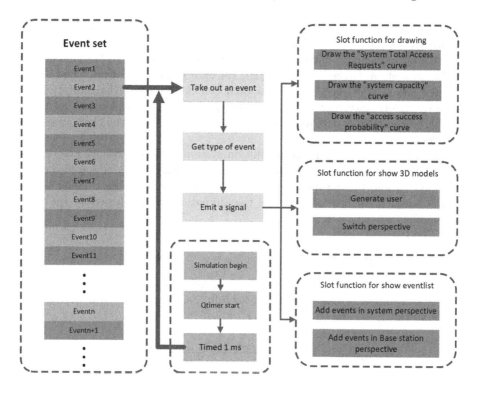

Fig. 2. Engine

visualizing the entire simulation process, and the event list is an indispensable part of the exhibition system. From the perspective of the system, the event list needs to display all events that occur during the entire simulation. From the perspective of the base station, the event list needs to display events related to the base station. Every time an event occurs, the time when the event occurred and the name of the event are displayed in the event list. When the event list is empty, the button for the first empty text is searched from top to bottom, and the name of the event and the time of occurrence are edited into the text of the button. When the event list is full, the event name and the time of occurrence are added to the bottommost button, and all the original text of all the buttons are moved up one space. The event list under the system view and the event list under the base station perspective share the same event list class, and the design principles are basically the same.

4.3   3D Scene Modeling

The 3D scene module is mainly to establish a corresponding 3D model for the simulation application scenario. The model needs to conform to the overall background of the simulation, and the animation effect is added on the basis of the model, so that the simulation process is more vivid. The 3D scene module is mainly divided into two parts: scene modeling and animation effects.

Scenario modeling: a large-scale access cell scenario needs to be built in the exhibition system. Based on the 3GPP TR45.820 recommendation, 52,547 NB-IOT devices need to be deployed in each cell. Considering the limitation of machine performance, the building is not modeled in the cell scene, and the NB-IOT equipment is deployed in a large amount to reflect the characteristics of large-scale access.

Animation effect: When the exhibition system runs, the occurrence of the event needs to reflect the user access action. In actual operation, you also need to be able to select a specific user. Therefore, the animation effect of the user's connection with the base station and the animation effect selected by the user are added.

4.4   Drawing Window

Draw the curve of "system capacity": System capacity is defined as the number of data transmissions that can be successfully completed per hour over a 200 kHz bandwidth, as calculated by Eq. (1):

$$N_{syscapacity} = \frac{N_{total} - N_{failure}}{N_{site} \times N_{200\,KHz}} \times \frac{60 \times 60}{TimeStamp \times 10^{-3}} \qquad (1)$$

In the formula:

N_{total}——The total number of transmissions per cell, the data comes from the total number of access requests(double) stored in the "nb-iot data transmission" event.

$N_{failure}$——The number of failed transmissions per cell, the data comes from the number of access failure requests(double) stored in the "nb-iot data transfer" event.

N_{site}——The number of cells set during the simulation, set to 1 during debugging.

$N_{200\,KHz}$——Number of 200 KHz transmission bandwidths per cell.

Draw the curve of "access success probability": The probability of system random access success can be calculated by Eq. (2).

$$P_{success} = \frac{N_{total} - N_{failure}}{N_{total}} \qquad (2)$$

In the formula:

N_{total}——The total number of transmissions per cell, the data comes from the total number of access requests(double) stored in the "nb-iot data transmission" event.

$N_{failure}$——The number of failed transmissions per cell, the data comes from the number of access failure requests (double) stored in the "nb-iot data transfer" event.

5 Results

Based on the recommendations of 3GPP TR45.820, the exhibition system configuration parameters are 1 cell, 3 base stations, 52547 NB-IOT devices, the devices are evenly distributed, and the device is set to a stationary state. Run the exhibition system in the current configuration. The results are shown in Fig. 3.

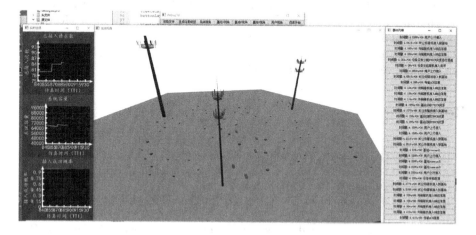

Fig. 3. Operation result

After the simulation starts, the event is displayed in the event list every time an event is executed. At the same time, according to the type of the event, the state of each function module of the exhibition system is updated, some events control the drawing window, and some events control the 3D scene window. Under the scheduling of the drive engine, each functional module cooperates to complete the demonstration.

6 Conclusions

The exhibition system visualizes the simulation platform oriented to the IoT dense access by constructing a 3D model, plotting curves in real time and displaying simulation events, and vividly displaying the simulation process in the 3D model. By plotting the curve, the user can observe the state of the system in real time and control the key performance indicators of the system. The demonstration results show that when the system bandwidth is 20 MHz, the system can support more than 1 million connections, which can meet the needs of millions of connections in the future 5G network. The running result of the demonstration system can also play a feedback role on the development of the simulation platform. When a node has a problem, the time and approximate location of the problem can be found through the exhibition system, which greatly facilitates the debugging program.

Add a few points, the current exhibition system only supports a single cell configuration, and on this basis, the configuration of the cell can be expanded, and the maximum can be expanded to 19 cells. And the number of deployments of multiple NB-IOT devices such as 12857, 25714, 52547, 64285, and 77142 is supported by a single cell. The above test results are obtained under the configuration of 52547 NB-IOT devices deployed. The probability of access failure measured under this configuration is extremely small, indicating that the coverage type-based backoff algorithm proposed by the simulation platform effectively reduces the access failure probability of the system. The currently deployed NB-IOT devices are all static. In the future, when upgrading the simulation exhibition system, we can consider adding a dynamic device model.

References

1. IMT-2020(5G), pp. 1–18 (2014). http://www.imt-2020.cn/zh
2. IMT-2020(5G), pp. 1–17 (2014). http://www.imt-2020.cn/zh
3. Andrews, J.G., Buzzi, S., Wan, C., et al.: What will 5G be? IEEE J. Sel. Areas Commun. **32**(6), 1065–1082 (2014)
4. Blanchette, J., Summerfield, M.: C++GUI Programming with Qt 4, Second Edn (2013)
5. Gong, L.: Research on signal and slot mechanism in Qt framework. Comput. CD Softw. Appl. **16**(11), 281–283 (2013)
6. 3GPP: Cellular system support for ultra low complexity and low throughput internet of things. TR 45.820, p. 1–495, December 2016

An Extreme Learning Approach for Electronic Music Classification

Jing Wang[✉]

Qilu Normal University, Ji'nan, Shandong Province, China
wangjing2986@163.com

Abstract. In order to recognize different kinds of electronic music, an extreme learning based method is proposed. Firstly, the feature of different electronic music data are extracted from cepstrum coefficient. Secondly, the kernel principal component analysis is adopted to reduce the dimension of features. Thirdly, in order to select appropriate parameters for an extreme learning machine, the genetic algorithm is used. Finally, experiments are carried out to verify the performance of the proposed electronic music classification method. In the experiments, we have established a database including four kinds of electronic music, i.e., "Guzheng", "Lute", "Flute", and "Harp". The experimental results show that the classification accuracy of the proposed method can reach up to 96%, while the wrong classification rate of the proposed method is only 14% which is much lower than existing electronic music classification models.

Keywords: Electronic Music Classification ·
Kernel Principal Component Analysis · Extreme Learning Machine ·
Feature Extraction

1   Introduction

With the continuous development of information technology, and the combination of music more tight Dense, there have been many kinds of electronic music which are able to relieve people's life and work pressure. However, each user likes different types of electronic music, so it is important to find the music from the mass of the electronic music library [1]. Electronic music classification is the key to improve the efficiency of electronic music query and is becoming the focus of attention [2].

Electronic music classification studies can be divided into two phases: traditional stage and modern stage. For the traditional stage, it is a manual classification that they are divided into the corresponding category through the analysis of electronic music by some experts and professionals [3]. When the electronic music data is very small, the classification of traditional classification of high accuracy, can be a good explanation of the classification results. However, the defects of traditional method including high error rate, low classification efficiency, are gradually reflected with the continuous increase of music data [4]. For the modern stage, the classification of electronic is accomplished automatically by computer [5]. Electronic music automatic classification belongs to a pattern recognition problem. It is necessary to extracts the characteristic information that reflects the electronic music. It is prolonged and inefficient to classify

B. Li et al. (Eds.): IoTaaS 2018, LNICST 271, pp. 81–89, 2019.
https://doi.org/10.1007/978-3-030-14657-3_9

the electronic music automatically according to the original feature information because of the large number of original features. Therefore, the technology of principal component analysis (PCA) is used to screen out the most important features to reduce the dimension of feature vector in order to speed up the automatic classification of electronic music [6]. Principal component analysis is a linear approach and can't be used to extract nonlinear information that describes an electronic music label [7]. The kernel principal component analysis is an improved principal component analysis method. By introducing the kernel function, the nonlinear information is extracted, and the feature is better than the principal component analysis [8]. Electronic music automatic identification also need to design electronic music classifier including Hidden Markov Model [9], Neural Network [10] and Support Vector Machine [11]. However, these methods exist some deficiencies. For example, the hidden Markov model can only be linearly classified and the results of electronic music classification are unreliable. Although the artificial neural network can classify the electronic music non-linearly, it requires sufficient electronic sample data. Once the sample can't meet the sufficient conditions, the electronic music classification effect dropped sharply. Although the support vector machine does not have the requirements of the neural network for the sample data, the learning process is complicated, the time complexity is high, and the speed requirement of the mass electronic music classification can't be satisfied.

For the reason of the shortcomings of the traditional model in the process of electronic music classification, an electronic music classification model for improving the limit learning machine is proposed. Firstly, the cepstrum coefficient of electronic music is extracted, the characteristics of electronic music are selected by kernel principal component analysis, and then the classifier of electronic music is improved by the limit learning mechanism. Finally, the simulation results show that the improved extreme learning machine improves the average classification rate of electronic music, and electronic music classification performance is better than the other.

2 An Electronic Music Classification Model for Improving Extreme Learning Machine

2.1 The Extraction of the Characteristics of Electronic Music

The current electronic music has many characteristics to describe its type, and electronic music is actually a kind of sound, Mel cepstrum coefficient can describe the sound frequency of energy changes and extract features quickly. The Mel cepstrum coefficient is selected as electronic music Classification of the characteristics for this article, the specific steps are as follows:

(1) The collected electronic music data is framed to remove the invalid frame.
(2) The frame signals of electronic music are processed by Fourier transform to obtain their amplitude spectrum.
(3) Through the Mel scale transformation of the amplitude spectrum, and the filter group is used to filter the spectrum, the energy value of the jth filter is

$$e[j] = \log\left(\sum_{k=0}^{N-1} w_j[k] \times |s[k]|\right), \ j = 1, 2, \cdots, p \tag{1}$$

Where $w_j[k]$ is the weight of the filter, $|s[k]|$ is the Fourier transform spectrum amplitude and p represents the number of filters. The crosstalk MFCC coefficient of electronic music is obtained by performing cosine transform on Eq. (1), the specific formula is:

$$x_i = \sqrt{\frac{2}{p}} \sum_{j=1}^{p} (e[j] \times \cos(\frac{i\pi}{p}(j - 0.5))), \ j = 1, 2, \cdots, L \tag{2}$$

where L describes the dimension of the coefficient.

2.2 The Selection of Characteristics of Electronic Music with PCA

A collection of electronic music samples is described by $X = \{x_1, x_2, \ldots, x_m\}$, $x_k \in R^N$, and then $\varphi(x)$ is used for Non-linear mapping and $\sum_{k=1}^{m} \varphi(x_k) = 0$, the electronic music feature selection problem can be described as

$$\lambda w_\varphi = C_\varphi w_\varphi \tag{3}$$

C_φ represents the covariance matrix of all samples, computed by

$$C_\varphi = \frac{1}{m} \sum_{k=1}^{m} \varphi(x_k) \varphi^T(x_k) \tag{4}$$

and

$$w_\varphi = \sum_{i=1}^{m} a_i \varphi(x_i) \tag{5}$$

By introducing a kernel function $K_{i,j} = k(x_i, x_j) = \langle \varphi(x_i) \cdot \varphi(x_j) \rangle$, the original electronic music feature selection problem is transformed into:

$$m\lambda K\alpha = K^2\alpha \tag{6}$$

So

$$m\lambda\alpha = K\alpha \tag{7}$$

The feature vector of the characteristic value of electronic music $\lambda_1, \lambda_2, \ldots, \lambda_m$ is described with $a = (a_1, a_2, \ldots, a_m)$.

In order to speed up the selection of the feature, the eigenvector is normalized, and for the test sample, the w_φ projection can be calculated by:

$$h_k(x) = \langle w_\varphi \cdot \varphi(x_j) \rangle = \sum_{i=1}^{m} a_i^k (\varphi(x_i) \cdot \varphi(x)) = \sum_{i=1}^{m} a_i^k K(x_i, x) \tag{8}$$

2.3   Improved Extreme Learning Machine

In order to solve the traditional neural network convergence speed is slow and easy to result in over-fitting, complex network structure defects, Huang et al. propose the method of extreme learning machine. As long as the simple random set of weights and thresholds, and the hidden layer of nodes, the training process can be successfully completed and get the optimal solution to solve the problem [12]. Let the number of nodes with hidden layers be L, then the output function of the limit learning machine is:

$$f_L(x) = \sum_{i=1}^{L} \beta_i G(a_i, b_i, x), \ x \in R^d, \beta_i \in R^m \tag{9}$$

Where g_i is the Implicit layer node output function $G(a_i, b_i, x)$. a_i, b_i are the Learning parameters and β_i is the Weight vector.

For the data $D = \{(x_i, t_i) | x_i \in R^d, t_i \in R^m, i = 1, 2, \ldots, N\}$, we can get

$$H\beta = T \tag{10}$$

Where H is the hidden layer output matrix of the network and is defined as:

$$H = \begin{bmatrix} G(a_1, b_1, x_1) & L & G(a_L, b_L, x_L) \\ M & 0 & M \\ G(a_1, b_1, x_N) & L & G(a_L, b_L, x_N) \end{bmatrix} \tag{11}$$

Where β is the weight matrix between the hidden layer and the output layer and T is the output matrix, defined as follows:

$$\beta = \begin{bmatrix} \beta_1^T \\ M \\ \beta_L^T \end{bmatrix} \tag{12}$$

$$T = \begin{bmatrix} t_1^T \\ M \\ t_L^T \end{bmatrix} \tag{13}$$

The least squares algorithm is used to solve the matrix of β:

$$\beta = \|H\beta - T'\| \tag{14}$$

And then

$$\widehat{\beta} = H^+T' \tag{15}$$

In the process of the extreme learning machine, the parameters a_i, b_i directly affect its learning performance. In order to solve this problem, this paper presents the genetic algorithm to determine the parameters of the limit learning machine as Fig. 1:

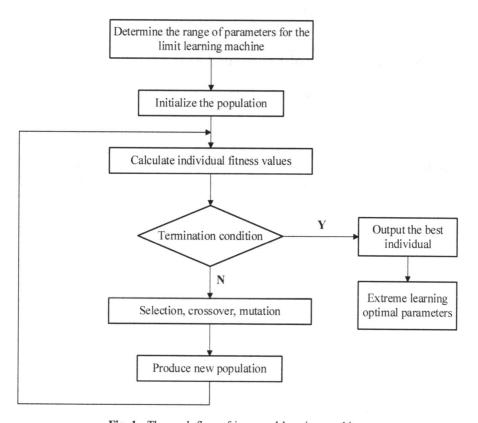

Fig. 1. The work flow of improved learning machine.

2.4   The Electronic Music Classification Steps of Improved Extreme Learning Machine

(1) Collecting electronic music sample data to form an electronic music database.
(2) Extracting the characteristics of the electronic music database which Composition feature vector library.
(3) The characteristics of the electronic music database are normalized.

(4) The kernel principal component analysis is used to select the characteristics of the electronic music database to form the optimal feature subset.

(5) According to the optimal feature subset, the training samples and test samples are dimensioned to reduce the size of the data.

(6) The training samples are input to the extreme learning machine for learning, and genetic algorithm is used to determine the optimal extreme learning machine parameters.

(7) According to the optimal parameters, extreme learning is able to establish electronic music classification model.

(8) Testing and analyzing the performance of electronic music classification models using electronic music test data.

3 Performance Testing of Electronic Music Classification

3.1 The Source of Electronic Music Data

In order to analyze the effect of improving the electronic music classification of the extreme learning machine, we choose a lot of data for simulation test. The data can be divided into four types of electronic music, such as GuZheng, Lute, Flute, and Harp. The number of samples is shown in Table 1. In order to make IELM electronic music classification effect is comparable, we designed two kinds of contrast model, described as follows:

Table 1. Number of training and testing samples in the experiments

Electronic music type	Number of training samples	Number of test samples
GuZheng	1000	250
Lute	800	200
Flute	500	125
Harp	200	50

(1) Principal component analysis and ELM electronic Music Classification Model (PCA-ELM).

(2) Kernel principal component analysis and Support Vector Machine's Electronic Music Classification Model (KPCA-SVM).

3.2 Results and Analysis

Each model runs 10 times to calculate their average. As are shown in Figs. 2, 3, and 4, the results of accuracy rate, error classification rate and average training time of electronic music classification between the KPCA-ELM and the contrast model is presented.

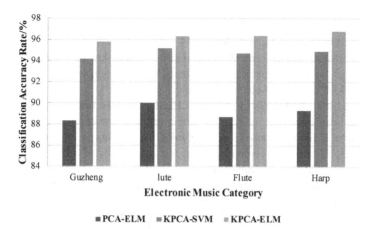

Fig. 2. The correct rate of electronic music classification.

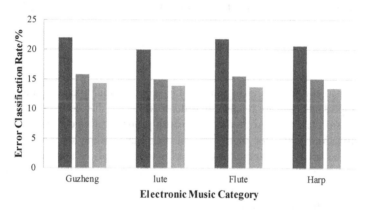

Fig. 3. The error rate of electronic music classification.

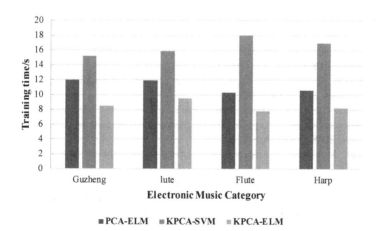

Fig. 4. The training time of electronic music classification.

Then we can make the conclusions that

(1) Compared to PCA-ELM, KPCAELELM's electronic music classification rate has been improved, and the error classification rate is smaller. This is because KPCA can extract better non-linear features than PCA, making the feature more accurately reflect the type of electronic music.

(2) Compared with KPCA-SVM, KPCAELELM electronic music classification rate has also been improved, effectively reducing the electronic music error classification rate. This is because ELM integrates the advantages of traditional neural networks and support vector machines, and establishes of a better performance electronic music classification model.

(3) In all electronic music classification models, KPCAELELM has the least number of electronic music classified training sessions. This is because KPCA can effectively reduce the feature dimension, and ELM can obtain faster learning speed than support vector machine. In addition, it speeds up the training speed of electronic music classification and improve the classification efficiency of electronic music, and more suitable for mass electronic music classification.

4   Conclusion

Electronic music classification can broaden the scope of multimedia applications and has a very important application value. In order to solve the shortcomings of the current electronic music classification model, this paper proposes an electronic music automatic classification model for improving the extreme learning machine. The model integrates the advantages of kernel principal component analysis and extreme learning machine. At the same time, the genetic algorithm is used to select the parameters of the limit learning machine, which improves the accuracy of electronic music classification and has wide application prospect.

References

1. Tzanetakis, G., Cook, E.: Musical genre classification of audio signals. IEEE Trans. Speech Audio Process. **10**(5), 293–302 (2001)
2. Costa, Y.M.G., Oliveira, L.S., Koefich, A.L., et al.: Music genre classification using LBP textural features. Sig. Process. **92**(11), 2723–2737 (2012)
3. Naghsh-Nilchi, A.R., Kadkhoda mohammadi, A.R.: Cardiac arrhythmias classification method based on MUSIC, morphological descriptors, and neural network. EURASIP J. Adv. Sig. Process. **2008** (2008). Article No. 202
4. Conklin, D.: Multiple viewpoint systems for music classification. J. New Music Res. **42**(1), 19–26 (2013)
5. Lippens, S., Martens, J., Mulder, T.D., Tzanetakis, G.: A comparison of human and automatic musical genre classification. In: Proceedings of International Conference on Acoustics, Speech and Signal Processing, ICASSP 2004, May 2004, vol. 4, pp. 233–236 (2004)

6. Turnbull, D., Elkan, C.: Fast recognition of musical genres using RBF networks. IEEE Trans. Knowl. Data Eng. **17**(4), 580–584 (2005)
7. Lee, I., Sokolsky, O., Chen, S., et al.: Challenges and research directions in medical cyber-physical systems. Proc. IEEE **100**(1), 75–90 (2012)
8. Cartwright, R., Cheng, A., Hudak, P., et al.: Cyber-physical challenges in transportation system design. In: Proceedings of 2008 National Workshop for Research on High-Confidence Transportation Cyber-Physical Systems: Automotive, Aviation and Rail, pp. 220–224. National Science Foundation, Washington, D.C. (2008)
9. Ikhsan, I., Novamizanti, L., Ramatryana, I.N.A.: Automatic musical genre classification of audio using Hidden Markov Model. In: ICoICT, pp. 397–402, May 2014
10. Scott, P.: Music classification using neural networks. Technical report, Stanford University, Stanford, CA 94305 (2001)
11. Xu, C., Maddage, N.C., Shao, X., Cao, F., Tian, Q.: Musical genre classification using support vector machines. Technical report, Laboratories for Information Technology, 21 Heng Mui Keng Terrace, Singapore 119613 (2003)
12. Huang, G.B., Zhu, Q.Y., Siew, C.K.: Extreme learning machine: theory and applications. Neurocomputing **70**(1), 489–501 (2006)

Quality Assessment for Networked Video Streaming Based on Deep Learning

Jinkun Guo and Shuai Wan[✉]

Northwestern Polytechnical University, Xi'an, China
guojinkun@mail.nwpu.edu.cn, swan@nwpu.edu.cn

Abstract. There arises the need for quality assessment in networked video streaming since video services have great significance for both users and providers. In this paper, a neural network is proposed to realize networked video streaming quality assessment. Firstly the key parameters of video streaming are extracted, including the bit-rate, the coded bits of each frames, the number of lost packet and so on. Then the neural network is built to study the mapping of these parameters and video quality. The influence on the video quality assessment by different network depth and different layer settings in the neural network is also taken into comparison. The performance of the proposed neural network has been compared with other methods and evaluated by the quality assessment experiment of videos in different resolutions. The results demonstrate the effectiveness and efficiency of video quality assessment based on the neural network.

Keywords: Video quality assessment · Deep learning · Video streaming

1 Introduction

Nowadays, more and more people use online video services. Therefore, quality assessment for networked video streaming is important for both video users and service providers. The traditional video subjective and objective quality assessment methods have been developing for a long time. Each has advantages, but the disadvantages are also prominent. The traditional subjective video quality assessment can achieve the most accurate result but it requires rigorous experiment environments. The traditional objective video quality assessment can avoid the great manpower involved in the experiment but it is generally not very accurate in reflecting the feelings of the viewers.

According to the researches related to the networked video streaming quality assessment, most of the methods and algorithms are based on the no-reference objective video quality assessment. Authors in [1] proposed a reconstruction-based no-reference video objective quality assessment algorithm. It can give out pleasant performances but the features of the testing frames have to be deeply mined first. Wei and Zhang proposed a no-reference video quality assessment method by utilizing the hybrid parameters extracted from compressed video frames [2]. This method could be effective, but the computation burden was high. An objective no-reference video quality assessment method was presented in [3], where the pictures have to be split into small blocks which will introduce much complexity in video decoding. A packet-layer

B. Li et al. (Eds.): IoTaaS 2018, LNICST 271, pp. 90–97, 2019.
https://doi.org/10.1007/978-3-030-14657-3_10

assessment model for video quality was introduced in [4]. The information including bit-rate, frame rate, etc., were provided by packet headers so the payload information was not needed for video quality assessment. Authors in [5] used the packet-layer video quality assessment model with the characteristics of the High Efficiency Video Coding (HEVC) standard which achieved a present correlation between subjective scores and objective scores. Quality of Experience (QoE) reflects the impact of factors that affect the satisfaction of users. Ref. [6] provided a structured way to build an objective QoE model by Principal Component Analysis (PCA) and Analytic Hierarchy Process (AHP) analysis. Ref. [7] was devoted to build a perceptual video quality metric based on features which were used for semantic task and human material perception. In this method, several estimators which could reflect the neuroscientific and psychophysical evidence were gathered together to achieve video quality assessment.

Combined with the deep learning method, which is popular and most suitable for dealing with large data problems at present, the traditional problems encountered in video quality assessment can be effectively avoided and even better results can be achieved. The literature [8–10] put forward different video quality assessment based on back propagation (BP) neural network. In [8], human visual regions of interest were selected and temporal and spatial features were extracted as the input of BP network. Literature [9] presented one kind of BP-based estimate on the network video QoE to construct the mapping model between QoE and quality of service (QoS). Ref. [10] provided an automated and computational video quality assessment method which employed offline deep unsupervised learning processes and inexpensive no-reference measurements at server side and client side, respectively.

The rest of this paper is organized as follows. Section 2 proposes a novel quality assessment method for networked video based on deep learning. Performance evaluation and the comparison of this method with the existed method are provided in Sect. 3 in detail. Conclusions are drawn in Sect. 4.

2 Quality Assessment for Networked Video Streaming Using Deep Learning

The general framework of the proposed method is shown as Fig. 1. Firstly, the useful parameters in a video stream are extracted. Then they are sent to a neural network to get the video quality. The neural network has to be carefully designed and trained for quality assessment. In the experiment, videos in three different resolutions from YUV video sequence are used. For training purposes, video sequences Slide Editing, Johnny in 720P; Crowdrun, Harbour in 4cif; Bus, Carphone in cif are used. For testing purposes, video sequences Shield in 720P, Crew in 4cif and Claire in cif are used.

The subjective video quality can be calculated in terms of the 5-point absolute category rating (ACR) mean opinion score (MOS) scale according to the recommendation P.1201.1. The ITU-T P.1201-series of Recommendations specifies models for monitoring the audio, video and audiovisual quality of IP-based video services based on packet-header information. MOS used in the experiment as the label of videos represents the mean opinion score of video quality by viewers with the consideration of influence brought by video encoding, packet-loss in the transmission, rebuffering and the screen size.

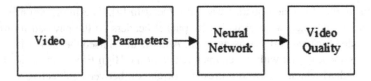

Fig. 1. Framework of proposed video quality assessment method.

2.1 Extract Key Parameters

Firstly, the original reference video is encoded and formatted by ffmpeg. The quantization parameters range from 0 to 51, producing 52 compressed videos which quality are successively reduced. Secondly, the Wireshark software [11] is used to simulate the network transmission of compressed video and capture each video packet during the transmission process.

After the completion of the Wireshark sending and receiving video data packets, each of the original video will generate a pcap file. A pcap file contains all the data packets in the process of video transmission. The detailed information of each video packet can be got, which contains comprehensive information for video streaming transmission on the network.

Since there are so many parameters in every data packet and most of them are not relative to the video quality assessment, the parameters below are chosen as the key values to assess the video quality: Br (the coded bits of each frames), bit-rate, Psize, Isize, Vplef (video packet-loss event frequency), Vir (impairment rate of video stream), Vairf (average impairment rate of video frequency), Vdsize (the screen size), Vdiag (number of pixels on the diagonal of screen) since they represent video impairment from different aspects. For example, Vir reflects the video impairment among frames and Vairf reflects video impairment within each frame.

Because of the video size, there is a difference in the number of packets that can be transmitted. Matlab software is used to obtain the mean value of the key parameters extracted from each video. The data sets of key parameters and MOS of compressed video are divided into train set and test set. To be precise, the key parameters extracted from videos of two different scenes in each resolution and the corresponding quality value MOS are selected as the train set. The key parameters and MOS of one video scene in every resolution are divided as test set.

2.2 Construct the Neural Network

Fully connected neural network minimizes the mean square error between network output and label by back-propagating error, which can adjust the weights and biases of network effectively. It has great performance of non-linear modelling. In order to achieve the optimal network model, the experiments try to build fully connected networks in different depth with the comparison among different activation function types based on the parameters and MOS extracted and calculated in advance.

The fully-connected back-propagation neural network is built based on TensorFlow. The loss function is the mean square error between the output of the network and the ground truth of MOS calculated by recommendation P.1201.1. In the back forward propagation process, Adam optimization algorithm is chosen to optimize the loss function. The Adam algorithm dynamically adjusts the learning rate of each parameter according to the first and second order moment estimate of the gradient of the loss function. It is also based on the gradient descent method, but in each iteration parameters have a definite range of learning step so that it will not lead to a large learning step size because of a large gradient and the parameter values are relatively stable. In this experiment, the training iterations are 200000 epochs and after every 500 epochs the current error value will be displayed, which is the difference between the network output and the video real quality. After the training process, the network structure and parameters will be saved which can be used directly in the testing process. For all the experiment, we have used the computer with two GeForce GTX 1080Ti Graphics Cards and Intel i7 7700k CPU and it will take around 330 s for the whole training and testing process.

3   Experimental Results and Analysis

The input of the fully connected neural network are key parameters Br, Psize, Isize, Vplef, Vir, Vairf, Vdsize, Vdiag, which are extracted from the videos in YUV video sequence. This dataset is commonly used in the video coding and decoding. Videos in the YUV video sequence are in the 4:2:0 YUV format. And also, MOS is computed as the label of each video. The Pearson correlation coefficient (PCC) is used to test the linear correlation between experimental video quality and real video quality. The range of PCC is $[-1,1]$. The greater the PCC is, the higher the linear correlation degree will be.

3.1   Comparison Among Different Neural Network Depth

The results from three-layer fully connected network 1 and four-layer fully connected network 2 are compared in this experiment. The number of neurons in each layer of network 1 are 10, 20, 30, respectively. The number of neurons in each layer of network 2 are 10, 20, 25, 30, respectively. The training set used in each training session is the same, and videos in each resolution are tested. Figures 2 and 3 show the results and errors comparison between two networks testing video Shield in 720P. The x-axis represents 52 distorted video sequences. The y-axis indicates the experimental video quality ranging from 0 to 5. It can be seen that increasing depth can enhance the ability of studying the mapping of key parameters and video quality by fully-connected back-propagation neural network.

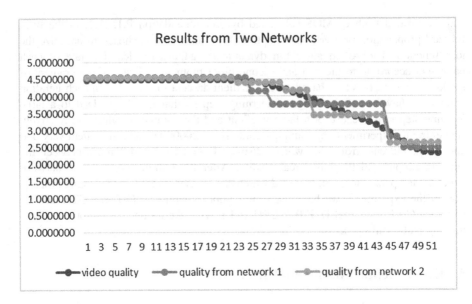

Fig. 2. The results of video Shield in 720P.

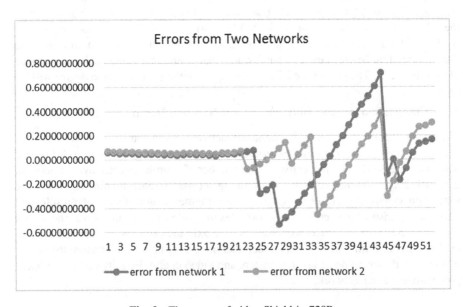

Fig. 3. The errors of video Shield in 720P.

3.2   Comparison Among Different Activation Function

In this experiment a four-layer fully connected neural network is chosen. The number of neurons in each layer are 10, 20, 25, 30, respectively. The second and third layers use ReLU activation function. The fourth layer uses softsign activation function. The experiment compares the results from two different structures: the first layer in network 1 uses softsign and network 2 first layer does not use any activation function. The training sets are the same in each training session, and videos in each resolution is tested. The x-axis represents 52 distorted video sequences. The y-axis indicates the experimental video quality ranging from 0 to 5. Figures 4 and 5 show the results and errors from two networks testing video Shield in 720P. The results and errors indicate that softsign activation function can improve the fitting capacity of the neural network and ReLU can ensure the convergence of computation results.

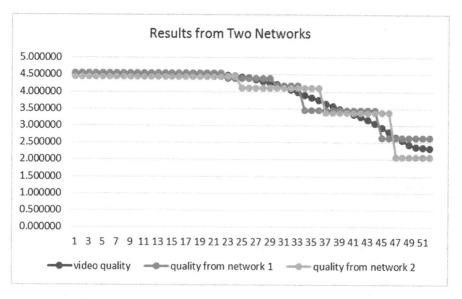

Fig. 4. The results of video Shield in 720P.

3.3   Analysis

After a large number of experiment, the optimal fully-connected back-propagation neural network is in four layers, of which the second and third network layer use softsign activation function, and the first, fourth network layer use ReLU. The PCC of video Crew in 4cif is 98.02% and the average error is 0.1502; the PCC of video Coastguard in cif is 89.47% and the average error is 0.2302; the PCC of video Shield in 720P is 97.57% and the average error is 0.1217. Compared with the performance of the method FRAME-FEBP proposed by Wei and Zhang in [2], which relevance is 75.20% in such video sequence, the results show that the four-layer fully-connected back-propagation neural network can perform the subject video quality assessment quite well.

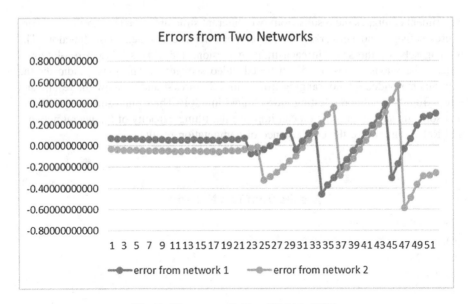

Fig. 5. The errors of video Shield in 720P.

4   Conclusions

This paper proposed a method of quality assessment for networked video streaming based on deep learning, where a fully connected neural network is designed. The network can conclude results which are very close to the traditional ways by learning the mapping between key parameters extracted from the networked video streaming and MOS calculated according to the recommendation P.1201.1. Experimental results have shown that the proposed model provides a high-performance and feasible solution to evaluate the quality of networked video streaming.

References

1. Wu, Z., Hu, H.: Reconstruction-based no-reference video quality assessment. In: Region 10 Conference (TENCON), pp. 3075–3078. IEEE (2016)
2. Wei, B., Zhang, Y.: No-reference video quality assessment with frame-level hybrid parameters for mobile video services. In: 2016 2nd IEEE International Conference on Computer and Communications (ICCC), pp. 490–494. IEEE (2016)
3. Jing, W., Zedong, W., Fei, W., et al.: A no-reference video quality assessment method for VoIP applications. In: 2016 IEEE 13th International Conference on Signal Processing (ICSP), pp. 644–648. IEEE (2016)
4. Yang, F.Z., Wan, S.: Overview of state-of-the-art and future of networked video quality assessment. J. Commun. (2012)
5. Guo, J., Zheng, K., Hu, G., et al.: Packet layer model of HEVC wireless video quality assessment. In: 2016 11th International Conference on Computer Science and Education (ICCSE), pp. 712–717. IEEE (2016)

6. Youssef, Y.B., Mellouk, A., Afif, M., et al.: Video quality assessment based on statistical selection approach for QoE factors dependency. In: 2016 IEEE Global Communications Conference (GLOBECOM), pp. 1–6. IEEE (2016)
7. Deng, B.W., et al.: Video quality assessment based on features for semantic task and human material perception. In: IEEE International Conference on Consumer Electronics-China. IEEE (2017)
8. Jiangbo, X., Xiuhua, J.: No-reference high definition video quality assessment based on BP neural network. In: Proceedings of the 2011 International Conference on Future Computer Science and Application (FCSA 2011 V1), p. 4. Intelligent Information Technology Application Association (2011)
9. Yao, H., Huang, Y.: BP-based estimate on network video QoE. Comput. Eng. Des. **38**(1), 1–6 (2017)
10. Vega, M.T., et al.: Deep learning for quality assessment in live video streaming. IEEE Signal Process. Lett. **PP**(99), 1 (2017)
11. https://www.wireshark.org/

Networking Technology for IoT

A Novel Next-Hop Selection Scheme Based on GPSR in VANETs

Ruiyan Han[1], Jinglun Shi[1](✉), Farhad Banoori[1],
and Weiqiang Shen[2]

[1] School of Electronic and Information Engineering,
South China University of Technology, Guangzhou 510641, China
eeeehry@mail.scut.edu.cn, shijl@scut.edu.cn
[2] College of Information Science and Technology,
Jinan University, Guangzhou 510632, China

Abstract. Vehicle ad-hoc networks (VANETs) are becoming the mainstream of network research recently. However, due to the high mobility of vehicles, dynamically changing topology and highly partitioned network, it is a challenge to overcome these shortcomings to guarantee a reliable link and efficient data delivery. In this paper, we present an improved scheme for the selection of next-hop based on greedy perimeter stateless routing (GPSR). The routing decision zone based on the concept which combines node mobility with message forwarding direction is introduced to avoid the node selected as next hop driving away from the neighborhood of the forwarder during communication. Our proposed protocol adaptively sets the size of the routing decision zone in each hop based on the dynamic forward neighbor information, including speed and one-hop transmission delay, and then consequently selects the next hop based on the position information and the routing decision zone. Simulation results show that our algorithm with low-delay performance is effective in some cases.

Keywords: VANETs · Next-hop selection · GPSR

1 Introduction

Vehicle ad-hoc networks (VANETs) have been proved for its great potential in various application especially enhancement of road safety, the optimization of traffic efficiency and the infotainment services. VANETs are a special type of Mobile ad-hoc networks (MANETs), which use vehicles as mobile nodes, transceivers and routers, including Vehicle-to-Vehicle (V2V) and Vehicle-to-Roadside (V2R) communication modes. However, VANETs have its own characteristics including the extremely dynamic topology, short-lived communication links, variable node density and highly partitioned network [1]. Therefore, it is very challenging to establish an optimal routing protocol in VANETs.

Among the MANETs routing protocols, the geographic routing protocol is now considered to be the most appropriate choice for VANETs [2]. The GPSR [3] protocol is a well-known position-based routing protocol in MANETs, which can minimize the end-to-end hop number and achieve faster data forwarding. GPSR adopts two position-

© ICST Institute for Computer Sciences, Social Informatics and Telecommunications Engineering 2019
Published by Springer Nature Switzerland AG 2019. All Rights Reserved
B. Li et al. (Eds.): IoTaaS 2018, LNICST 271, pp. 101–110, 2019.
https://doi.org/10.1007/978-3-030-14657-3_11

based packet forwarding strategies, one is Greedy Forwarding (GF) and the other is Perimeter Forwarding (PF). In GF, a forwarding node selects its neighbor closest to the destination node as the next hop. If the forwarder encounters a void region, the PF is used to choose the next forward hop by flooding. GPSR protocol requires each node to know its own position through GPS and share it with its one-hop neighbors, besides this each node maintains the knowledge of its one-hop neighbors by periodically exchanging Hello packets.

When GPSR is applied to highway scene in VANETs, it does not perform well in vehicular environment. Since the GF only uses the stored position information to make the routing decision to select the next hop, it tends to select the next-hop node at the border of the communication range. As the node selected as next hop may leave the neighborhood of the forwarder due to its mobility, it has been found that it is prone to link interruption and significantly increases the packet loss rate. Furthermore, additional retransmissions lead to increased delay. Once the GF fails to forward the message, the PF is adopted as the recovery mode. Because of the dynamic topology and uneven distribution of nodes, it is very complicated to construct and traverse the planner graph in VANETs, and the node far away from the destination may be selected, resulting in an increase in delay.

Therefore, we propose an improved algorithm based on GPSR, which introduces the routing decision zone whose size can be calculated based on forward neighbor information of the forwarder along message forwarding direction, including maximum speed, minimum speed and one-hop transmission delay. Our proposed algorithm adaptively adjusts the size of the routing decision zone for each hop to adapt to mobility, and selects the next hop in the shrunk routing decision zone at the cost of communication overhead and hop count. In such network, although the selected hop may not be closer to the destination, it can avoid the selected node driving away from the neighborhood of the forwarder during communication, thereby reducing the packet loss rate and delay. In addition, our proposed algorithm adopts carry-forwarding scheme as the recovery mode instead of perimeter mode. The forwarder encountering a local optimal problem carries the packet until there is a neighbor node that could make a progress toward the destination.

2 Related Work

Several efficient techniques and approaches have been proposed in the literature to enhance the performance of GPSR in vehicular environments. WF-GPSR [4] protocol took into account link reliability, distance and movement direction angle to formulate the weighted function of a next-hop candidate node. DGF-ETX [5] protocol integrated the link quality estimation metric ETX into a multi-metric that considered the distance and direction of the candidate forwarders. DAPBR [6] protocol applied the restricted greedy forwarding approach to select the next hop by considering neighborhood vehicles having a sufficiently dense neighborhood and the least velocity variance compared to its own neighboring vehicles. LAT-GPSR [7] protocol introduced the link available time prediction into the next hop selection of GPSR, instead of simply using the GF algorithm. MAGF [8] protocol presented the concept of motion potential by combining node mobility patterns with node position information for forwarding decisions.

However, these aforementioned protocols required additional complex calculations to get the next hop after obtaining the relevant neighborhood information. To simplify the complexity of the next hop selection and improve the performance of GPSR, our proposed algorithm selects the next hop based on the position information and the routing decision zone, and obtains the corresponding information of the next hop directly from the Hello packet. Our proposed approach dynamically recalculates the size of the routing decision zone in each hop, overcoming the problem that the node selected as next hop may leave the neighborhood of the forwarder due to mobility. Furthermore, we adopt carry-forwarding as a recovery strategy.

3 Proposed Method

3.1 Network Scenario

We consider a pure V2V network scenario without any infrastructure in a straight unidirectional and uninterrupted one-way vehicle traffic highway, as shown in Fig. 1. The highway has one entrance and exit in the opposite direction. While sending a message from the source vehicle to the destination vehicle, a routing path needs to be established in a multi-hop manner.

Certain assumptions are made for our proposed algorithm, all vehicle nodes use R as their effective communication range and are equipped with GPS for positioning. The destination node is known by default. The direction of the message forwarding is assumed to be the same as the direction in which the vehicle moves. It is assumed that each vehicle has independently assigned a random speed based on the Uniform distribution, and each vehicle maintains its randomly assigned speed while it is on the highway. Besides this, any vehicle n_i can generate self-related information, including vehicle ID_i, speed v_i and position coordinates (x_i, y_i).

We take the position of node n_i as a reference, along the direction of vehicle movement, the one-hop neighbor in front of it is called the forward neighbor, contrarily termed as the backward neighbor. During message forwarding process, the forwarder usually makes routing decisions based on the forward neighbor information of each hop to achieve efficient and fast forwarding. $S.\ neighbour$ is used to distinguish between forward and backward neighbors as follows:

$$S.\ neighbour = \begin{cases} 1, & forward\ neighbour \\ 0, & backward\ neighbour \end{cases}$$

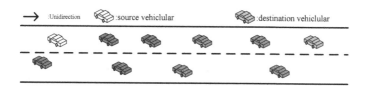

Fig. 1. Network scenario

4 Algorithm Design

In this paper, the one-hop communication range along message forwarding direction is defined as the routing query zone, whereas the forward neighbor spatial distribution range of the auxiliary routing decision is defined as the routing decision zone, and the duration of a packet received from one node to another is defined as one-hop delay. The main purpose of the proposed algorithm is to select a relatively stable next hop by reducing the occurrence of the selected node moving out of the neighborhood of the forwarder, aiming to design a novel forwarding scheme for next hop selection in VANETs.

The idea of proposed algorithm is illustrated in Fig. 2. Where n_0 represents the number of forward neighbors, n' represents the number of vehicles located inside the boundary of communication range, depending on the number of lanes. In addition to that, v_{f_max} and v_{f_min} respectively represent the maximum speed and minimum speed of the forward neighbor node, and t_{delay} is one-hop delay of the packet. $S.\ existence = 1$ indicates that the forward neighbor node exists in the routing decision zone, and the recovery mode basically adopts carry-forwarding mechanism.

Firstly, an information query is initiated by the forwarding node at the beginning of each hop, and then one-hop neighbor list is established, and the routing decision zone is adaptively set based on forward neighborhood information, including v_{f_max}, v_{f_min} and t_{delay}. Secondly, the next hop is selected in the routing decision zone. Thirdly, the packet forwarding process is performed on the selected next hop.

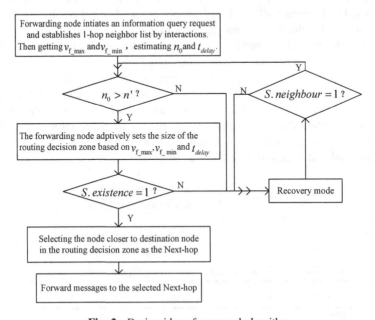

Fig. 2. Design idea of proposed algorithm

4.1   Implementation Steps of Proposed Algorithm

Getting 1-Hop Neighbor Information. Every vehicle periodically sends a Hello packet to its neighboring vehicles, which carry following information such as position, vehicle ID, velocity of itself, and the timestamp attached to packet sending time. The purpose of the Hello packet is to get routing information of neighbor nodes, and its format is presented in Fig. 3.

When the forwarder initiates an information query request, the *query* packet is sent by the broadcast method within its communication range. The node that receives the *query* packet for the first time sends a *reply* packet to the forwarder, otherwise the *query* packet is dropped. The format of the *query* packet and the *reply* packet have been shown in Fig. 3. Thus, the forwarder can establish a one-hop neighbor list based on these received *reply* packets, and then get complete information about its one-hop neighbor nodes. The forwarder can extract information about the forward neighbor from the list, including n_0, v_{f_max}, v_{f_min} and one-hop delay.

| Node ID | Node Position | Node Speed | Time stamp |

Fig. 3. The format of a packet

Setting the Size of the Routing Decision Zone. In the GF scheme, the size of the routing query zone is the same as the size of the routing decision zone, as shown in Fig. 4(a). It is prone to select the boundary node as the next hop based on the expired routing information. The selected next hop is invalid if it satisfies the following two conditions: (1) After the boundary node successfully sends the *reply* packet to the forwarder, it instantly moves out of the communication range of the forwarder. (2) When the forwarder sends a message to the selected next hop, the selected node moves outside the communication range of the forwarder during that period.

It can be noted that an invalid relay node can easily interrupt the communication link between the forwarder and the selected next hop, resulting in packet loss, which tends to increase delay due to its retransmission. To avoid selecting an invalid relay node, the forwarder should appropriately shrink the size of the routing decision zone in the routing query zone along the data forwarding direction, as shown in Fig. 4(b). Although selecting the next hop in the shrunk routing decision zone may increase the hop count, it can avoid the selected node moving away from the neighborhood of the current forwarder and greatly enhance the probability of receiving messages.

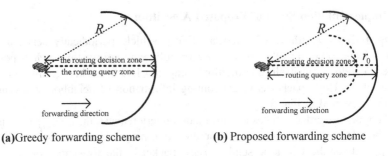

(a) Greedy forwarding scheme   **(b)** Proposed forwarding scheme

Fig. 4. Comparison of two forwarding schemes

Moreover, if the size of the routing decision zone is set extensively large, once the selected relay node moves too fast, it may head out of the communication range of the forwarder during communication, and cannot efficiently ensure reliable message forwarding. Correspondingly, if the size of the routing decision zone is set extensively small, although the message can be reliably forwarded, it will result in an excessive increase in the number of hops, which may increase the probability of packet loss. Therefore, it is obligatory to dynamically set the size of the routing decision zone based on the dynamic forward information.

To reasonably set the size of the routing decision zone in each hop, we define a critical zone r_0 ($r_0 \in [r_{min}, r_{max}]$) for the forwarding node at the edge of R, as shown in Fig. 4(b). From the instant when a node is selected as the next hop to the instant when the selected relay successfully receives the message, the selected relay node just arrives at the communication boundary of the forwarding node. In this section, we also investigated and calculated r_0 value, as shown below.

We assume that a forwarding node n_f has a neighbor node n_x moving to the communication boundary, and node n_f and node n_x happen to be in a critical connection state within Δt time. In this case, node n_x becomes the critical node, and the displacement that node n_x covered at speed v_x in Δt time is taken as the critical value r_0, and $r_0 = \Delta t \cdot v_x$. The time when node n_x receives the *query* packet from node n_f is denoted as t_1, and the time when node n_x receives the message from node n_f is denoted as t_2. Therefore, we get $\Delta t = t_2 - t_1$.

The duration from node n_x receiving the *query* packet to node n_f receiving the *reply* packet is the one-hop delay of the reply, which is denoted as $t_{one-hop\_reply}$. Node n_f receives the *reply* packet from node n_x, and then node n_f sends a message to node n_x until node n_x receives the message from node n_f. The duration of the process is called the one-hop delay of the message, which is denoted as $t_{one-hop\_message}$.

Based on the above analysis, we obtain:

$$\Delta t = t_2 - t_1$$
$$= t_{one-hop\_reply} + t_{one-hop\_message} \tag{1}$$

Then:

$$r_0 = \Delta t \cdot v_x = \left(t_{one-hop\_reply} + t_{one-hop\_message}\right) \cdot v_x, \; v_x \in \left[v_{f\_max}, \; v_{f\_min}\right] \quad (2)$$

When $v_x = v_{f\_max}$, $r_{max} = \left(t_{one-hop\_reply} + t_{one-hop\_message}\right) \cdot v_{f\_max}$

When $v_x = v_{f\_min}$, $r_{min} = \left(t_{one-hop\_reply} + t_{one-hop\_message}\right) \cdot v_{f\_min}$

We can get the range of the size of the routing decision zone $(R - r_0)$ as follows:

$$(R - r_0)_{max} = R - r_{min} \quad (3)$$

$$(R - r_0)_{min} = R - r_{max} \quad (4)$$

In each hop, the size of the routing decision zone is adaptively estimated as follows:

$$R - r_0 = (R - r_0)_{min} + (1 - \beta) \cdot \left[(R - r_0)_{max} - (R - r_0)_{min}\right] \quad (5)$$

Where the value of β determines the size of the routing decision zone. If the value is 0, $(R - r_0)$ will be $(R - r_0)_{max}$. If the value is 1, $(R - r_0)$ will be $(R - r_0)_{min}$. However, when all vehicles on this segment make a uniform linear at the same speed v_x, we can get $(R - r_0)_{max} = (R - r_0)_{min}$. Thus, the size of the routing decision zone is as follows:

$$R - r_0 = R - \Delta t \cdot v_x \quad (6)$$

After determining the size of the routing decision zone, and making a reasonable trade-off between the communication overhead and the hop count, the forwarder selects the node closest to the destination as the next hop in the shrunk routing decision zone based on the position information, which can avoid the selected node driving away from the communication range of the forwarder during communication. If the forwarder cannot find a node closer to the destination than itself, it carries the packet until it encounters a neighbor node that may be heading towards the destination. Then repeat the process until the message is successfully forwarded to the destination node.

5 Analysis and Simulation Results

In this paper, we use MATLAB R2014a as a simulation platform to evaluate the performance of our proposed algorithm. We compare the performance of our proposed protocol with the GPSR in the same simulation environment. The setting of the simulation scenario is described in Table 1. Furthermore, we set β as 0.5, one hop delay as 0.1 s, and here we ignore the delay of electromagnetic wave propagation in the air. However, variations in vehicle speed results in frequent link changes and error-prone transmissions, which may increase delay. To analyze the impact of vehicle speed on the end-to-end delay, we assume all vehicles on this segment make a uniform linear at the same speed, and the number of vehicles is set to moderately 55.

The simulation results in Fig. 5 depict that the delay of our proposed algorithm remains relatively stable, while the delay of GPSR increases with the increase of vehicle density. However, the increase in vehicular density ultimately increases the chance of packet collisions, and GPSR has a higher probability to select an invalid relay, which may result in an increase in packet loss and delay. Furthermore the proposed algorithm selects the next hop in the shrunk routing decision zone can avoid selection of an invalid relay as much as possible, hence the average delay is expected to reduce compared to GPSR. When the vehicle density is small, the forwarder in the proposed algorithm cannot find any node as the next hop to deliver data packet in the shrunk routing decision zone, which will result in much bigger delay. Therefore, the average end-to-end delay of our proposed algorithm is better than GPSR in a relatively dense vehicle scene.

Table 1. Parameter setting in simulations

Parameter	Value
Lane length/km	1
Lane width/m	5
Number of lanes	2
Number of traffic direction	1
Transmission range/m	150
Velocity distribution model	Uniform
Vehicle maximal speed	108 kmph
Vehicle minimal speed	72 kmph
GPSR Hello interval/s	5
Vehicle density/(veh/km)	40–70

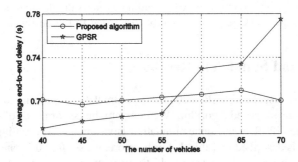

Fig. 5. Impact of the number of vehicles on the average end-to-end delay.

The simulation results in Fig. 6 show that the delay of our proposed algorithm is smaller than GPSR. On one hand, the increase of vehicle speed increases the probability of occurrence of the void region, which may increase the average delay. On the other hand, with the increase of vehicle speed, the next-hop selected in GPSR has a higher probability of leaving the communication range of the forwarder because of its high speed, thus it certainly increases chances of packet loss and extra delay of retransmissions. The proposed algorithm selects the next hop in the shrunk routing decision zone, which can reduce the probability of occurrence of this phenomenon. However, the proposed algorithm sacrifices the hop-count as a compromise to gain stable links, and the delay increases with the increase of vehicle speed.

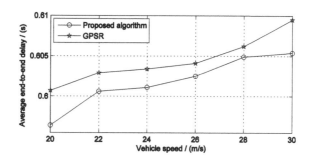

Fig. 6. Impact of vehicle speed on the average end-to-end delay

6   Conclusion

In this paper, a novel next-hop selection scheme based on GPSR is proposed, which adaptively sets the size of the routing decision zone based on the information of the dynamic forward neighbor in each hop, including speed and one-hop delay. Then, we consider position information and the routing decision zone to select the next hop, which reduces the chance of the selected node moving out of the effective communication range of the forwarder during communication. Simulation results indicate that the performance of our proposed protocol performs better than GPSR in some cases.

References

1. Devangavi, A.D., Gupta, R.: Routing protocols in VANET—a survey. In: 2017 International Conference on Smart Technologies for Smart Nation (SmartTechCon), pp. 163–167. IEEE (2017)
2. Fonseca, A., Vazão, T.: Applicability of position-based routing for VANET in highways and urban environment. J. Netw. Comput. Appl. **36**(3), 961–973 (2013)
3. Karp, B., Kung, H.T.: GPSR: greedy perimeter stateless routing for wireless networks. In: International Conference on Mobile Computing and Networking, pp. 243–254. ACM (2000)
4. Cui, Z., Li, D., Zhang, G., et al.: The next-hop node selection based GPSR in vehicular Ad Hoc networks. J. Comput. Commun. **04**(10), 44–56 (2016)
5. Alzamzami, O., Mahgoub, I.: An enhanced directional greedy forwarding for VANETs using link quality estimation. In: Wireless Communications and Networking Conference. IEEE (2016)

6. Ram, A., Mishra, M.K.: Density aware position based routing (DAPBR) protocol for VANET. In: Sixth International Symposium on Embedded Computing and System Design, pp. 142–146. IEEE (2017)
7. Yang, H., Yu, M., Zeng, X.: Link available time prediction based GPSR for vehicular ad hoc networks. In: IEEE International Conference on Networking, Sensing and Control, pp. 293–298. IEEE (2017)
8. Li, J., Shatz, S.M.: Toward using node mobility to enhance greedy forwarding in geographic routing for mobile Ad Hoc networks. In: Proceedings of MODUS (2008)

Learning-to-Rank Based Strategy for Caching in Wireless Small Cell Networks

Chenxi Zhang[1,2], Pinyi Ren[1,2(✉)], and Qinghe Du[1,2]

[1] Department of Information and Communications Engineering,
Xi'an Jiaotong University, Xi'an, China
`zcx110708@hotmail.com`, {`pyren,duqinghe`}`@mail.xjtu.edu.cn`
[2] Shaanxi Smart Networks and Ubiquitous Access Research Center, Xi'an, China

Abstract. Caching in wireless network is an effective method to reduce the load of backhaul link. In this paper, we studied the problem of wireless small cell network caching when the content popularity is unknown. We consider the wireless small cell network caching problem as a ranking problem and propose a learning-to-rank based caching strategy. In this strategy, we use the historical request records to learn the rank of content popularity and decide what to cache. First, we use historical request records to cluster the small base stations (SBS) through the k-means algorithm. Then the loss function is set up in each cluster, the gradient descent algorithm is used to minimize the loss function. Finally we can get the ranking order of the content popularity for each SBS, and the files are cached to the SBS in sequence according to the order. From Simulation results we can see that our strategy can effectively learn the ranking of content popularity, and obtain higher cache hit rate compared to the reference strategies.

Keywords: Wireless networks · Caching · Learning-to-rank

1 Introduction

As smart phones and mobile network develop, the main business of wireless network is transferred from traditional communication services to mobile data services, this makes the mobile network face heavier and heavier load pressure. In 5G networks, deploying small base stations in hotspots to relieve the load pressure of wireless networks is a common method, but there are still lots of problems to be solved. Especially the limited capacity of backhaul links has attracted the attention of many scholars. One solution to handle this problem is to deploy caching devices in wireless network edge devices to cache popular

The research reported in this paper is supported by the Key Research and Development Program of Shaanxi Province under Grant No. 2017ZDXM-G-Y-012 and the Fundamental Research Funds for the Central Universities.

content, such as base stations or mobile terminals, so that users can get those content directly from the caching devices when they request. This method can not only reduce the backhaul link's load pressure, but also reduce the file transfer delay and improve the user satisfaction [1,2,10].

In the wireless network caching technology, how to determine the popular content is a very important issue. Most of the research uses content popularity to determine the popular content. However, In the actual system, content popularity is constantly changing and difficult to get the accurate value in advance. So how to determine the content of caching without knowing of content popularity has become the focus of research. In [3], the author consider the cache problem as a multi-armed bandit problem, and propose a learning based caching algorithm. Then author in [8] take the context information into consideration. Author in [12] use the collaborative filtering algorithm in mobile social network, and propose a learning based caching strategy. But most of these works are focus on predict the value of content popularity.

When content popularity is already known, many researches modeled the caching problem as a 0–1 integer programming problem, and many researchers use greedy algorithm to handle this problem, such as [5,6,9]. Besides, the most popular caching strategy [8] which always caching the files with highest content popularity. We can find that many caching strategies are caching the most popular file or the file have highest gain to the system. So in this paper, we model the wireless small cell network caching problem as a ranking problem, no longer focusing on the value of content popularity, but is the ranking relationship between the content popularity. And we solve this problem by a learning-to-rank method [11].

The main contribution of this paper are as follows. We model the caching problem in wireless small cell network as a sorting problem when the system have no knowledge of content popularity. And we give a learning-to-rank based algorithm to solve this problem. In the simulation results, we can find that our strategy can get higher cache hit rate than reference strategies and close to the optimal strategy.

Rest of this paper is organized as follows: In Sect. 2 we give the system model and the optimization problem. Section 3 we describe the details of the learning-to-rank based caching strategy. Section 4 we simulate the propose strategy and compared with some existing strategies. Finally, we make a summary of this paper.

2   System Model

We consider a wireless small cell network with M SBSs which are distributed in one area, assuming that the coverage of these SBSs does not overlap, and each one of them is equipped with a caching device, which can cache popular files and connect to the core network through the backhaul link. When a user in a SBS's service range requests a file, the system first look for the file on the SBS's cache device, if found, it will be sent directly to the user by the SBS. If the file is not cached, the SBS will request the file from the core network, then the core network will send the file to SBS through the backhaul link and send it to the user.

All users request files from the file library $F = \{f_1, f_2 \ldots, f_N\}$ with a total number of N, each file have a size of $S_f = \{s_1, s_2 \ldots, s_N\}$. All the SBS equipped a cache device with cache memory of S Mb. We assume that the small base stations in the system are distributed in different areas, such as schools, factories, office buildings, etc. Therefore, users of different SBS may have different content preferences, so the content popularity in different SBS is not the same. The probability that the user is in the service range of each SBS is equal. We use $P \in \mathbb{R}^{M \times N}$ to represent the content popularity of all SBSs, which the element $p_{m,i}$ represents the probability of the user in i-th SBS's coverage request for i-th file. And we use the matrix X to represent the cache matrix, the element $x_{m,i} = 1$ represents the m-th SBS cached the i-th file, otherwise is 0.

We aim to maximize the cache hit rate in our system, which is the probability of finding the corresponding file on the SBS's cache device when the user requests a file. So the optimization problem for our system is as follows:

$$
\max_{X} \quad CHR = \frac{1}{M} \sum_{m=1}^{M} \sum_{i=1}^{N} p_{m,i} x_{m,i}
$$

$$
s.t. \quad \sum_{i=1}^{N} s_i x_{m,i} \leq S, m-1, 2, \cdots, M, i-1, 2, \cdots, N \tag{1}
$$

$$
x_{m,i} \in \{0,1\}, m = 1, 2, \cdots, M, i = 1, 2, \cdots, N
$$

We can see that the optimization variable of this optimization problem is a binary variable. With known of the content popularity P, the optimization problem is a 0–1 integer programming problem. The 0–1 integer programming problem is a typical NP problem [7], there are many algorithm can solve this problem, such as greedy algorithm, branch and bound. But in our system, content popularity P is not a known parameter, so solving this problem becomes very difficult. In next section we propose a strategy based on a learning-to-rank to solve this problem.

3   Learning-to-Rank Based Caching Strategy

In this section, we propose a cache strategy based on learning-to-rank. Learning-to-rank is a kind of machine learning method that used for information retrieval [4]. We use the historical request record to learn the rank of content popularity and decide what to cache.

The caching strategy proposed in this section has two steps. The first step is to cluster the SBSs, which replaces the process of feature extraction in learning-to-rank [4]. Next step is to determine the caching files of each SBS based on the historical request record in each cluster by learning-to-rank method.

3.1 k-means Based Small Base Station Clustering

We cluster the SBSs according to the historical request record $R \in \mathbb{R}^{M \times N}$, $r_{i,j}$ represents the number of requests to j-th file received by i-th SBS, and R_i represent the i-th row of the historical request record which is the historical request record of the i-th SBS. We normalize the historical request record matrix R, so as to improve the clustering accuracy, and get a new matrix R^*,. Then we use k-means algorithm to cluster the SBSs, the process is as follows:

(1) We initialize k clusters $C_1, C_2 \ldots C_k$, and randomly select k rows from R^* as the initial centroid $\mu_1^{(1)}, \mu_2^{(2)} \ldots \mu_k^{(k)}$ of these clusters.
(2) We allocate each SBS into the cluster which have the smallest Euclidean distance between the SBS's historical request record and the centroid.

$$C_m = \{R_i^* : \|R_i^* - \mu_m^{(t)}\|_2 \leq \|R_i^* - \mu_n^{(t)}\|_2, 1 \leq n \leq k\} \tag{2}$$

(3) According to the clustering results, we can get the new centroid of each cluster which is calculated by the follow formula:

$$\mu_m^{(t+1)} = \frac{1}{|C_m|} \sum_{R_i^* \in C_m} R_i^* \tag{3}$$

Repeat the step 2 and step 3, until the centroid of all the clusters is unchanged. Then we can get the cluster results and each cluster's historical request record $H_1, H_2 \ldots, H_k$. And the SBSs in the same cluster with similar content popularity.

3.2 Learning-to-Rank Based Caching Algorithm

In this step, we use the learning-to-rank method to enable the base station to learn the ranking of content popularity and decide what file to cache for each cluster.

We assume there is a final ranking matrix $H_m^* = U^T V$ for each cluster. And we use the top one probability [4] to calculate the probability of the highest number of requests for each file, the formula is as follows:

$$P_{H_m(i)}(h_{i,j}^m) = \frac{\varphi(h_{i,j}^m)}{\sum_{n=1}^N \varphi(h_{i,n}^m)} \tag{4}$$

The $\varphi(x)$ represents a monotone increasing and constant positive function, in this paper we define this function as an exponential function.

Then the cross entropy loss function in our system can be obtained as follows:

$$L(U,V) = \sum_{i=1}^{M_m} \{-\sum_{j=1}^{N} P_{H_m(i)}(h_{i,j}^m) \log P_{H_m(i)}(g(U_i^T V_j))\} + \frac{\lambda}{2}(\|U\|_F^2 + \|V\|_F^2)$$

$$= \sum_{i=1}^{M_m} \{-\sum_{j=1}^{N} I_{ij} \frac{exp(h_{i,j}^m)}{\sum_{n=1}^{N} I_{in} exp(h_{i,n}^m)} \log \frac{exp(g(U_i^T V_j))}{\sum_{n=1}^{N} I_{in} exp(g(U_i^T V_n))}\}$$

$$+ \frac{\lambda}{2}(\|U\|_F^2 + \|V\|_F^2)$$

$$(5)$$

where $g(x)$ represents the sigmoid function, which is $g(x) = 1/(1 + e^{-x})$, and λ represents the regularization coefficient. The value of the assumed probability distribution H_m^* can be obtained by minimizing this loss function. In order to We use the gradient descent algorithm to minimize the loss function, the gradient is shown in the following form:

$$\frac{\partial L(U,V)}{\partial U_i} =$$

$$\sum_{j=1}^{N} I_{ij} \left(\frac{exp(g(U_i^T V_j))}{\sum_{n=1}^{N} I_{in} exp(g(U_i^T V_n))} - \frac{exp(h_{i,j}^m)}{\sum_{n=1}^{N} I_{in} exp(h_{i,n}^m)} \right) g'(U_i^T V_j)V_j + \lambda U_i$$

$$\frac{\partial L(U,V)}{\partial V_j} =$$

$$(6)$$

$$\sum_{i=1}^{M_m} I_{ij} \left(\frac{exp(g(U_i^T V_j))}{\sum_{n=1}^{N} I_{in} exp(g(U_i^T V_n))} - \frac{exp(h_{i,j}^m)}{\sum_{n=1}^{N} I_{in} exp(h_{i,n}^m)} \right) g'(U_i^T V_j)U_i + \lambda V_j$$

We can update the value of U and V until the loss function is convergence, and the updating formula are as follows:

$$U_i = U_i - \alpha \frac{\partial L(U,V)}{\partial U_i}$$

$$V_j = V_j - \alpha \frac{\partial L(U,V)}{\partial V_j}$$

$$(7)$$

Finally, we can get the value of U and V, then we can calculate $P_i = U_i^T V$ to get the ranking of the files for each SBS by sort the value of each element of P_i in descending order. According the ranking of the files for each SBS, we can cache the files in descending unit the cache memory is filled. The whole process of our algorithm are given in Algorithm 1.

Algorithm 1. Learning-to-Rank Based Caching Algorithm

1: **Inputs:** historical request record R,cluster number k
2: Cluster the SBSs by k-means algorithm
3: Get historical request record $H_1, H_2 \ldots, H_k$
4: **for** $m = 1, 2 \cdots, k$ **do**
5:    initialize the loss function by (5)
6:    **while** Loss function is not convergence **do**
7:       Calculate the gradient by (6)
8:       Update U and V by (7)
9:    **end while**
10:    **for** $i = 1, 2 \cdots, M_m$ **do**
11:       Get the ranking of the files by (8)
12:       **while** $S_M < S$ **do**
13:          Cache file to the SBS
14:       **end while**
15:    **end for**
16: **end for**
17: **Output:** cache matrix X

---

4   Simulation Results

In this section, we simulate the performance of the proposed strategy and compared it with some existing algorithms.

We assume that there are 100 SBSs which are divided into four categories, each with a different Zipf distribution file order. Each SBS has different Zipf distribution parameters, which follow a uniform distribution with a mean of 0.5. The learning rate of gradient descent algorithm is 0.5, and the regularization coefficient $\lambda = 0.8$. There is 100 files in the file library, and all files with size of 1 Mb. The capacity of the cache device in each SBS is 20 Mb. And each SBS receive 200 times request from the user to set up the historical request record.

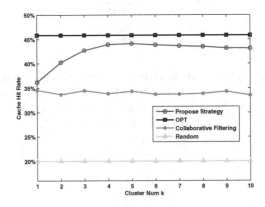

Fig. 1. Cache hit rate varies with cluster number k.

To better observe the simulation results of the proposed strategy, we compared our strategy with three existing strategies. First is the OPT strategy, which let the SBSs known the content popularity in advance and cache the optimal files. Next is the random cache strategy, in this way the SBSs will cache some files by random. Finally is the collaborative filtering based cache strategy [12], this strategy uses collaborative filtering algorithm to decide which file should cache. The idea of the collaborative filtering algorithm is the SBSs have similar request records will request the same files in the future. All simulation results are averaged after repeating 1000 times.

In Fig. 1 we simulate the cache hit rate in different number of cluster. From the Fig. 1, we can see that when the number of clusters k = 1, the SBSs is not clustered, the cache hit rate of the system is 36% when our strategy is used, slightly higher than the collaborative filtering strategy. With increase of the cluster number, the cache hit rate of propose strategy is gradually rising, and it reaches the maximum of about 44.5% when the number of clusters is between 4 and 5. This is also in accordance with our simulation settings. When the cluster number is higher than 5, the cache hit rate has decreased, this is because the number of SBSs in each cluster is decreased. The cluster numbers can be decided by the elbow method.

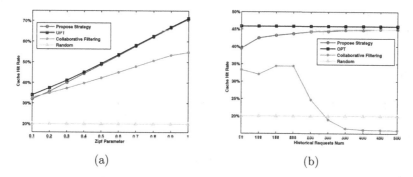

(a)   (b)

Fig. 2. (a) Cache hit rate varies with Zipf parameter (b) Cache hit rate varies with historical requests numbers

Next, we analyze the influence of Zipf distribution parameters on the system performance. From Fig. 2(a) we can see that, when the Zipf parameter is less than 0.2, the cache hit rate in propose strategy is slightly less than the strategy based on collaborative filtering. This is because when the Zipf parameter is very small the request probability of each file is almost same, so it is difficult to learn the rank of content popularity. With the increase of Zipf parameters, the cache hit rate obtained in propose strategy is more than the strategy based on collaborative filtering and close to the optimal strategy. Finally, we present the simulation results of the cache hit rate varies with the historical request numbers in Fig. 2(b). In this picture, it can be seen that when the number of requests is only 50 times, the proposed strategy can get a cache hit rate of close to 40%. As the number of requests increase, the cache hit rate of the proposed strategy is also increasing, and the cache hit rate is about 45% when the number of requests is 500 times. On the other hand, the performance of strategy based on collaborative filtering is much lower than random cache strategy, when the numbers of requests is higher than 300 times, this is because the collaborative filtering strategy only considers whether the file was requested rather than the numbers of requests. So we can get the conclusion that the strategy proposed in our paper has stronger learning ability and can make the system performance close to the performance of the best strategy by learning.

5   Conclusions

In this paper, we propose a learning-to-rank based caching strategy for wireless small cell networks. This strategy consider the wireless small cell network caching problem as a sorting problem, and the history request record is used to learn the rank of content popularity. Then the files are cached to the small base station in sequence according to the order. Simulation shows that this strategy can effectively learn content popularity ranking, thus caching the appropriate content, and the performance is close to the optimal cache strategy. Compared to the reference caching strategy this strategy can adapt to more scenes and obtain higher cache hit rate.

References

1. Bastug, E., Bennis, M., Debbah, M.: Living on the edge: the role of proactive caching in 5G wireless networks. IEEE Commun. Mag. **52**(8), 82–89 (2014)
2. Bastug, E., et al.: Big data meets telcos: a proactive caching perspective. J. Commun. Netw. **17**(6), 549–557 (2016)
3. Blasco, P., Gunduz, D.: Learning-based optimization of cache content in a small cell base station. In: IEEE International Conference on Communications, pp. 1897–1903 (2014)
4. Cao, Z., Qin, T., Liu, T.Y., Tsai, M.F., Li, H.: Learning to rank: from pairwise approach to listwise approach. In: International Conference on Machine Learning, pp. 129–136 (2007)
5. Jiang, D., Cui, Y.: Partition-based caching in large-scale SIC-enabled wireless networks. IEEE Trans. Wirel. Commun. **PP**(99), 1 (2017)
6. Keshavarzian, I., Zeinalpour-Yazdi, Z., Tadaion, A.: A clustered caching placement in heterogeneous small cell networks with user mobility. In: IEEE International Symposium on Signal Processing and Information Technology, pp. 421–426 (2016)
7. Korbut, A.A., Sigal, I.K.: Exact and greedy solutions of the knapsack problem: the ratio of values of objective functions. J. Comput. Syst. Sci. Int. **49**(5), 757–764 (2010)
8. Muller, S., Atan, O., Schaar, M.V.D., Klein, A.: Context-aware proactive content caching with service differentiation in wireless networks. IEEE Trans. Wirel. Commun. **16**(2), 1024–1036 (2017)

9. Sermpezis, P., Spyropoulos, T., Vigneri, L., Giannakas, T.: Femto-caching with soft cache hits: improving performance with related content recommendation. In: 2017 IEEE Global Communications Conference, pp. 1–7 (2018)
10. Shanmugam, K., Golrezaei, N., Dimakis, A.G., Molisch, A.F., Caire, G.: Femto-caching: wireless content delivery through distributed caching helpers. IEEE Trans. Inf. Theory **59**(12), 8402–8413 (2011)
11. Shi, Y., Larson, M., Hanjalic, A.: List-wise learning to rank with matrix factorization for collaborative filtering. In: ACM Conference on Recommender Systems, Recsys 2010, September, Barcelona, Spain, pp. 269–272 (2010)
12. Wang, Y., Ding, M., Chen, Z., Luo, L.: Caching placement with recommendation systems for cache-enabled mobile social networks. IEEE Commun. Lett. **PP**(99), 1 (2017)

Multi-network Communication Gateway of IoT for Complex Environment

Jinhong Li[1], Xiaoyan Pang[1(✉)], and Zhou Hu[2]

[1] Northwestern Polytechnical University, Xi'an, China
xypang@nwpu.edu.cn
[2] China Electronic Technology Group Corporation No. 20 Institute,
Xi'an, China

Abstract. In the paper a multi-network communication gateway of Internet of Things (IoT) is designed for complex environments with poor mobile network. This design is developed based on ARM processor and Android operating system, which can forward the data of IoT to the Internet through two mobile networks (from different mobile operators) and a satellite network. Our experimental tests show that the system runs stably and can forward data of IoT accurately and instantly. This design of gateway can be applied to special scenarios, such as the tunnel construction and oil exploration.

Keywords: IoT · Multi-network · Mobile network · Satellite network

1 Introduction

With the development of the Internet era of big data, the concept of the Internet of Things (IoT) has been proposed and has attracted much attention [1]. The IoT is defined as a network that according to a specific protocol connects any item with the Internet and exchanges information with each other to realize intelligent identification, location, tracking, monitoring and management. It is a network that extends and expands on the Internet [2].

In most cases, the data collected in IoT need to be transmitted through a gateway device to the server in the Internet [3]. Usually, the gateway only needs to connect to mobile network in many applications like the agricultural standardization production monitoring, smart city, medical management [4], etc. However, in some complex environments, for instance the tunnel construction [5] and oil exploration [6], the mobile infrastructure is not constructed well, and the mobile network even does not exist. Considering the IoT applications in these special scenarios, in this paper we design a multi-network communication gateway of IoT and examine its performance. This gateway is developed based on ARM processor and Android operating system, which can forward the data of IoT to the Internet through two mobile networks (from different mobile operators) and a satellite network. Our proposal thus can work efficiently in the scenarios with poor mobile network and be applied to the complex environments.

B. Li et al. (Eds.): IoTaaS 2018, LNICST 271, pp. 120–126, 2019.
https://doi.org/10.1007/978-3-030-14657-3_13

2 Gateway Design

The main framework of this communication gateway is divided into four layers: hardware layer, operation systems layer, driver layer and application layer (see Fig. 1).

The hardware layer mainly includes Ethernet, two mobile modules, one satellite module and each hardware module is connected to the ARM mini PC. The operating system layer adopts the Android operating system which is based on a modified version of the Linux kernel and other open source software and designed initially for mobile devices. Hence this operating system can support ARM processor, flexible driver development and easily system tailoring. The driver layer is mainly responsible for driving the interface of each network hardware. The driver of the mobile network module is mainly related to Radio interface layer (RIL), which is mainly divided into two parts, RILJ and RILC. The RILJ runs in the framework layer, and the RILC runs in the hardware abstraction layer. The interaction between RILJ, RILC and Modem is based on the data interaction mode of network socket. The satellite module is connected to the ARM as a serial port device and is controlled by AT command. The Application layer is the place where the main functions are implemented by the communication gateway of IoT. It mainly includes: a data encryption, a SIM card status detection, a mobile network switching algorithm, and a satellite communication switching algorithm.

The configuring modules: mobile network driver, mobile switching algorithm, data forwarding configuration and data encryption are crucial parts in this gateway and in the following we will discuss these parts explicitly.

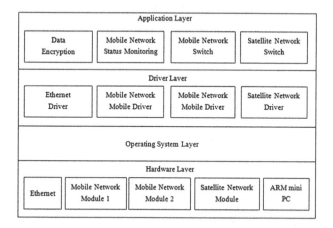

Fig. 1. System block diagram

2.1 Mobile Network Driver

Here the version of the both two mobile modules used in this multi network communication gateway is EC20. The mobile module is connected to the ARM mini PC via a Mini-PCIE interface which is essentially a USB interface, thus the 4G module can be connected to ARM as a USB device. Based on USB to serial port drive, each mobile device is mapped to several virtual serial port devices, and the Android/Linux kernel sends AT commands and receive/send data to the mobile module through these virtual serial port devices.

In order to develop the mobile network driver, first we need to add the 'USB to serial port driver' and the 'PPP protocol support' into the Linux kernel, then also add the mobile module ID in the USB to serial port driver, as is shown in Fig. 2.

```
Linux Kernel Configure by "make menuconfig"
1.1 USB Driver Configure
Device Drivers  --->
    [*] USB support  --->
          <*> USB Serial Converter support
               [*] USB Generic Serail Driver
               [*] USB Driver for GSM and CMDA modems

EC20's Vender ID and Product ID in option_ids[] of drivers/usb/serial/option.c
    static const struct usb_device_id option_ids[] = {
                    { USB_DEVICE(0x05C6, 0x9215) },    //for EC20

PPP Configure
    Device Drivers  --->
        [*] Network device support  --->
              <*> PPP (point- to- point protocol) support
              <*> PPP support for async serial ports
              <*> PPP support for sync tty ports
              <*> PPP Deflate compression
```

Fig. 2. Mobile module driver configuration

Secondly, we modify and add the source code provided by EC20 to the Android source code, and modify the Android launch configuration file to start the RILD service in, as is shown in Fig. 3. From this figure, it can be seen that the two mobile networks are set as two individual services and two different ttyUSB ports are assigned to these services, which is the basis for achieving the Dual SIM Dual Standby (DSDS).

```
service ril-daemon /system/bin/rild -l /system/lib/libreference-ril.so -c 0 -- -d /dev/ttyUSB2
class main
socket rild stream 660 root radio
socket rild-debug stream 660 radio system
user root
group radio cache inet misc audio log

service ril-daemon2 /system/bin/rild -l /system/lib/libreference-ril2.so -c 2 -- -d /dev/ttyUSB7
class main
socket rild2 stream 660 root radio
socket rild-debug2 stream 660 radio system
user root
group radio cache inet misc audio log
```

Fig. 3. RILD service configuration

2.2 Mobile Switching Algorithm

Since these two mobile networks of the communication gateway come from two different mobile operators and in the complex environments their signal strengths and qualities are also quite different from each other, it needs to monitor the signals from these two mobile networks and select the better one for data forwarding.

The TelephonyManager.listen function of the Android operating system is used to monitor the signal strength of the default mobile network. When there are two mobile networks, the 'subId' of mobile network has to be determined. However, the 'subId' of the PhoneStateListener class is a hidden parameter that cannot be set or got directly by the application programs. Therefore, we have to use the java "reflection" method to make the 'subId' be accessible, and then the signal strength of the both mobile network can be monitored. After that the average historical signal strength of the two mobile networks can be calculated according to the following equation,

$$ave = 1.0 \times sigsum/signum \tag{1}$$

here sigsum represents the sum of the signal strengths and signum is the number of times of recording the signal strength. Then the mobile network with better average historical signal strength in a time window is chosen as the main mobile communication network, and the other is used as a standby mobile network.

2.3 Data Forwarding Configuration

The data from IoT are usually sent to the gateway through Ethernet and will be forwarded to the server through mobile network or satellite network. However, when the amount of data is too large or the data packet is very big, using the traditional Data forwarding configuration will lead to losing some packets. In order to avoid this lost, here another method–using iptables to forward data directly from the driver layer is adopted. Iptables is a packet filtering management tool based on Netfilter architecture. The most important function is to build firewall or transparent proxy. To use iptables, we first change the firewall to a forwarding mode, and then set the relevant parameters according to its command rule, as is shown in Fig. 4.

```
echo 1 > /proc/sys/net/ipv4/ip_forward
iptables -F
iptables -t nat -F
iptables -t mangle -F
iptables -X
iptables -t nat -X
iptables -t mangle -X
iptables -A INPUT -i eth0 -j ACCEPT
iptables -A INPUT -i ppp0 -j ACCEPT
iptables -A INPUT -i ppp1 -j ACCEPT
iptables -A OUTPUT -o eth0 -j ACCEPT
iptables -A OUTPUT -o ppp0 -j ACCEPT
iptables -A OUTPUT -o ppp1 -j ACCEPT
iptables -A FORWARD -i eth0 -j ACCEPT
iptables -A FORWARD -i ppp0 -j ACCEPT
iptables -A FORWARD -i ppp1 -j ACCEPT
iptables -P INPUT ACCEPT
iptables -P OUTPUT ACCEPT
iptables -P FORWARD ACCEPT
iptables -t nat -A PREROUTING -p udp --dport PORT -i eth0 -j DNAT --to-destination REMOTE_MOBILE_IP:REMOTE_PORT
iptables -t nat -A POSTROUTING -d REMOTE_MOBILE_IP -j SNAT --to LOCAL_MOBILE_IP
```

Fig. 4. Iptables parameter settings

In Fig. 4, PORT is the port where the gateway receives data of IoT; REMOTE_MOBILE_IP and REMOTE_PORT are the IP address and port of the server, and LOCAL_MOBILE_IP is the port of the local mobile network. With these commands, iptables can transmit packets directly to the server through the driver layer.

2.4 Data Encryption

In the complex environment the data of IoT need to be encrypted. In the design of the gateway, we use end-to-end encryption, that is, the packet is automatically encrypted by the gateway, while it is decrypted at the receiving end. The encryption algorithm used here is RSA encryption algorithm which is the first algorithm that can be used both for data encryption and digital signature. RSA is a kind of asymmetric encryption algorithm with a changeable key length. The principle of key generation is as follows:

1. Randomly selecting two large prime numbers p and q, where p is not equal to q, then to calculate $N = p * q$;
2. Selecting a natural number e which is greater than 1 less than N, and e must be compatible with $(p - 1) * (q - 1)$.
3. Using the Euclid algorithm to calculate the decryption key d and also to make it satisfies $e * d = 1[\mod(p - 1) * (q - 1)]$ (where n, d are also mutually prime).
4. Getting the public key which is (n, e), and the private key which is (n, d).

In the process of encryption (or decryption) of information, the message m (or c) which is going to be encrypted (or decrypted) is first divided into equal length data blocks before they are encrypted (or decrypted). The corresponding formula are shown as follows:

$$c = m^e (\mod N) \tag{2}$$

$$m = c^d (\mod N) \tag{3}$$

3 Results and Discussions

We implement the communication gateway of IoT in an ARM mini PC and test its system performance by using Android Studio software. In the following we show some main results from the debug window of the Android Studio.

3.1 Mobile Network Test Results

As shown in Fig. 5, after the communication gateway is powered and both the mobile modules work normally, the system can obtain the IP address of both mobile network (denoted as ppp0 and ppp1), and get the signal strength and the operator's information of the mobile network. This illuminates that the mobile networks of the communication gateway work well.

```
D/MainService: SIM card status changes, SIM card 1 is true, SIM card 2 is true
D/IPUtils: getpppOIPAddress() enter...
D/IPUtils: getpppOIPAddress() success 10.172.199.87
D/IPUtils: getppp1IPAddress() enter...
D/IPUtils: getppp1IPAddress() success 10.36.82.252
D/MainService: The 4G network IP address is normal, create a mobile network send thread for: 10.172.199.87
D/MainService: Started 4G network UDP thread
D/MainService: The signal strength of the SIM card changes: Card: SIM_CARD_2 Signal strength: 4
V/MainService: Mobile operator 1: ChinaMobile
V/MainService: Mobile operator 2: CHN-CT
V/MainService: Data network type 1:13
V/MainService: Data network type 2:13
```

Fig. 5. Mobile networks test results

3.2   Mobile Networks Switching Test Results

After the system is powered, the mobile network 1 (denoted by SIM card 1 in Fig. 6) is selected as the main mobile communication network, and the average historical signal strength of the two mobile networks is monitored at the same time. If the signal strength of the mobile network 2 (denoted by SIM card 2 in Fig. 6) is better than the he mobile network 1 (denoted by SIM card 1) in a time window, the mobile network 2 (denoted by SIM card 2) is automatically switched as the main mobile communication network. In addition, manual switching is also possible. The debugging information is shown in Fig. 6.

```
D/MainService: The current network is SIM card 1, signal strength: 2
D/MainService: SIM card 2 signal strength changes, signal strength: 4
D/MainService: Automatically switch to SIM card 2
D/MainService: Switched to SIM card 2 successfully
```

Fig. 6. Mobile networks switching test results

3.3   Data Forwarding Test Results

After the data forwarding configuration is done, the data packets from IoT in the Ethernet interface are directly forwarded to the server through mobile network or satellite network. In Fig. 7 the data forwarding process in the debug window of Android Studio is given.

```
D/EthUdpServerNio: Received data, length: 25
D/DataBroadcastReceiver: Data received from Ethernet, command word: text message
I/Mobile UDP client process: Send packets to the mobile server
```

Fig. 7. Data forwarding test results

3.4   Data Encryption Test Results

The data encrypting process is shown in Fig. 8, from which we can see that the original data is 'Message from IoT' and the encrypted data cannot be read directly.

```
I/Mobile UDP client process: Unencrypted data: Message from IoT
I/Mobile UDP client process: Encrypted data: Wad2qs[]syw˜˜
```

Fig. 8. Data encryption test results

4   Conclusion

For some complex scenarios of IoT applications lacking of good mobile networks, a multi-network communication gateway is proposed and tested. The gateway is developed based on ARM processor and Android operating system and it can forward the data of IoT to the Internet through two mobile networks (from different mobile operators) and one satellite network. The working principle and performance of four crucial configuring modules including the mobile network driver, the mobile switching algorithm, the data forwarding configuration and the data encryption in this gateway have been discussed separately in detail. Our results have proved that this gateway system runs steadily and can forward the data of IoT accurately and safely. The combination of the two mobile networks and one satellite network ensures the communication in a complex environment effectively.

References

1. Breur, T.: Big data and the internet of things. J. Mark. Analytics **3**(1), 1–4 (2015)
2. Al-Fuqaha, A., Guizani, M., Mohammadi, M.: Internet of Things: a survey on enabling technologies, protocols, and applications. IEEE Commun. Surv. Tutorials **17**(4), 2347–2376 (2015)
3. Saxena, N., Roy, A., Sahu, B.J.R.: Efficient IoT gateway over 5G wireless: a new design with prototype and implementation results. IEEE Commun. Mag. **55**(2), 97–205 (2017)
4. Zanella, A., Bui, N., Castellani, A.: Intenet of things for smart cities. IEEE IoT J. **1**(1), 22–32 (2014)
5. Cheng, H., Wu, N., Lian, J.: The management and monitor system of tunnel construction based on internet of things. In: Wang, W. (ed.) Proceedings of the Second International Conference on Mechatronics and Automatic Control. LNEE, vol. 334, pp. 1019–1026. Springer, Cham (2015). https://doi.org/10.1007/978-3-319-13707-0_112
6. Khan, W.Z., Aalsalem, M.Y., Khan, M.K., et al.: A reliable Internet of Things based architecture for oil and gas industry. In: International Conference on Advanced Communication Technology, pp. 705–710 (2017)

An Integrated Architecture for IoT Malware Analysis and Detection

Zhongjin Liu[1]($\boxtimes$), Le Zhang[2], Qiuying Ni[2], Juntai Chen[2], Ru Wang[2],
Ye Li[3], and Yueying He[1]

[1] National Computer Network Emergency Response Technical
Team/Coordination Center of China, Beijing, China
lzj@cert.org.cn
[2] Beijing University of Posts and Telecommunications, Beijing, China
[3] Beihang University, Beijing, China

Abstract. Along with the rapid development of the IoT, the security issue of the IoT devices has also been greatly challenged. The variants of the IoT malware are constantly emerging. However, there is lacking of an IoT malware analysis architecture to extract and detect the malware behaviors. This paper addresses the problem and propose an IoT behavior analysis and detection architecture. We integrate the static and dynamic behavior analysis and network traffic analysis to understand and evaluate the IoT malware's behaviors and spread range. The experiment on Mirai malware and several variants shows that the architecture is comprehensive and effective for the IoT malware behavior analysis as well as spread range monitoring.

Keywords: IoT malware · Behavior analysis · Mirai · Architecture

1 Introduction

With the rapid development of the Internet of things (IoT) technology, the number of smart devices has increased greatly. According to analysis reports, the number of smart devices (excluding smart phones, tablets and computers) will grow to 28.1 billion by 2020. The IoT devices have been used in a wide range of applications, such as smart grid, intelligent transportation, intelligent home, etc.

Along with the rapid development of the IoT, the security issue of the IoT devices has also been greatly challenged [1]. The vulnerability of IoT devices have drawn hackers' attention, especially for those who are interested in DDoS attacks. Considering the significant number of devices, IoT has been gradually becoming one of the weakest part of the computer network. The spread and evolution of the IoT malware, such as botany, worm, and malicious software, are both speeding up.

The most famous IoT malware "Mirai" has been used in some of the largest and most disruptive distributed denial of service (DDoS) attacks [2]. Since the source code of Mirai was published on the Internet, it has become an architecture for building new malware. The variants of the Mirai are constantly emerging.

However, there is lacking of an IoT malware analysis architecture to extract and detect the malware behaviors, which mainly lies in three aspects:

B. Li et al. (Eds.): IoTaaS 2018, LNICST 271, pp. 127–137, 2019.
https://doi.org/10.1007/978-3-030-14657-3_14

(1) Network security researchers have carried out extensive and in-depth research on PC and mobile malware analysis. Since IoT devices run embedded systems on different hardware architectures and platforms, such as ARM, PowerPC, etc. There is lacking of a cross-platform architecture for analyzing the malicious code.

(2) The analysis of malware is usually carried out case by case. These work are independent and short of association analysis. Therefore, there is lacking of an architecture for comparative analysis of different malware variants.

(3) Malware network behavior monitoring is an effective method for malware detection. How to extract network behaviors of IoT malware and detect them on wide area of Internet still needs to study.

A survey [3] show that The detection patterns used in static analysis include string signature, byte-sequence n-grams, syntactic library call, control flow graph and opcode (operational code) frequency distribution etc. In view of this problem that the extracted opcode sequences cannot represent the true behaviors of an executable. Ding presents a control flow-based method to extract executable opcode behaviors [4]. But they don't have much concern about the key behavior in malicious code, however the key behavior can reflect malicious behavior.

Researchers have less dynamic analysis of IoT malware and more dynamic analysis of mobile mal-ware. Because dynamic analysis tools require intensive computation power, which are inadaptable to IoT devices due to the resource-constraint problem [5]. Moreover, most advanced analysis techniques are highly dependent on the underlying system platform. Building these analysis techniques require ad-hoc development for different platforms in the diversified IoT environments [6, 7]. Besides it is difficult to make dynamic analysis due to the trouble in applying it in an actual environment and be-cause of the overhead of tracking data flow to a low level [8].

In [9, 10], techniques to cluster network traffic patterns associated with botnets are presented. The characteristics observed include the flow patterns between hosts, such as the number of connections and amount of data exchanged. Similarly, using machine learning and network traffic features, [11] presents an approach to detect malware related traffic.

Researchers have proposed a few approaches [12, 13] to detect the existence of botnets in monitored networks. Almost all of these approaches are designed for detecting botnets that use IRC or HTTP based C&C, but these solutions lack correlation analysis of traffic from two dimensions of time and space.

This paper proposes an integrated analysis architecture for the IoT malware. The architecture integrates static analysis, dynamic analysis, variant evolution analysis and network behavior detection to achieve the goal of extracting and detecting IoT malware in the large-scale network.

2   Architecture Design

To solve the problem, we propose an architecture to analyze the IoT malware behaviors in static and dynamic manner. Based on the behavior, we focus on the network behavior analysis and intend to detect IoT malware in large-scale network (Fig. 1).

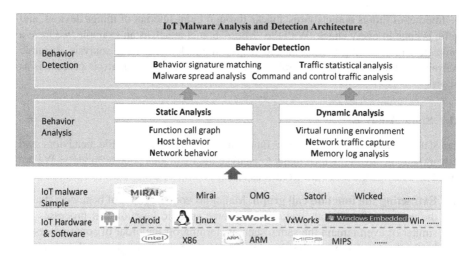

Fig. 1. IoT malware analysis and detection architecture

The architecture consists of two layer: Behavior Analysis layer and Behavior Detection layer.

The Behavior Analysis layer consists of two module, the static analysis module extracts function call graph, host behavior and network behavior by code reverse analysis; the dynamic analysis module captures network traffic and analyze memory log by running code in the virtual environment.

The Behavior detection layer matches the malware traffic with behavior signature. Based the traffic, the module is able to analysis the malware spreading and traffic pattern between bots and Command and control server (C&C).

3 Design Details

Based on the proposed architecture, we propose the method to derive and detect the static and dynamic behavior of different IoT malware. The analysis process is shown in Fig. 2.

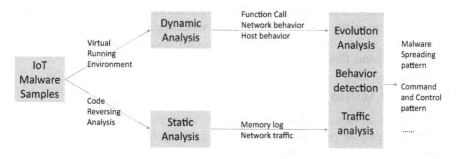

Fig. 2. The malware analysis process

Internet of things botnets virus attacks through more Internet of things devices, and traffic is huge in every attack. With the "Mirai" botnet attack as an example, we have a complete understanding of the embedded system attack on its static analysis, dynamic analysis, evolution analysis and traffic analysis, and our overall analysis process is as follows:

3.1   Static Analysis

The static analysis can extract the function call diagram, host-side behaviors and network-side behaviors, which helps to analyze the malicious code accurately. We are intend to analysis the malware from 3 aspects:

(1) **Extracting the function call graph.** Malicious code has a lot of behaviors, but in the process of malicious code variety, some specific function call processes are similar or even remain the same, such as: turning off the order of some processes in the host, occupying the resource of the system, and sending the specified information to the network server. Through the analysis of the function call process, it can be more easily determined from the whole structure of malicious code, and can provide the basis for the process of dynamic analysis to achieve the function of common verification.

(2) **Extracting the behaviors of host.** The host may be forced to perform some behavior after the host is infected with malicious code, which may prevent the running of the normal function of the host or open the new function without affecting the normal function of the host. The form or naming of malicious code may change, but the behavior of the final implementation will not change. The analysis of the behavior of the host end can help us correctly judge the behavior similarity and make up for the shortcomings of the simple analysis of the multiplexing function. In Mirai, for example, the host's behaviors: closing the watchdog process, preventing host to clear malicious code during reboot, closing the specific port and taking up it, deleting a specific file and kill corresponding to the process.

(3) **Extracting the behaviors of network.** The attacker in order to quickly infect the IoT equipment and effectively control of the host, the host usually performs certain specific network behavior operations, these network behavior operations maintain the connection between the attacker and the host, and the spread of the malicious code. Taking Mirai as an example, the behaviors of the network: the scanner module scans other potentially infected devices, and reports the infected devices to reach the goal of expanding the botnet; regularly sending messages to the C&C server, this host has been identified and stay active; upload and download some files.

3.2   Dynamic Analysis

The sandbox is a virtual environment which is often used to execute untested or untrusted programs or code without risking harm to the host machine or operating system. The sandbox is a powerful tool for dynamic analysis of IoT malware. However,

IoT malware has compilation formats for various platforms. Current sandbox is not able to run IoT malware directly.

We intend to handle this problem in two ways: First, run malicious code based on x86 sandbox directly by source code cross-platform compilation; second, integrate virtual machine, such as qemu, into sandbox, to simulate an embedded running environment for non-x86 instruction sets.

3.3   Evolution Analysis

Based on the extracted features from static and dynamic result, we are able to carry out evolution analysis. By comparing the extracted features of different malwares, such as function calls, traffic pattern, configuration table, we are able to analyze the evolution of the IoT malwares.

From the static result, take Mirai for example, if a set of functions that exist in Mirai sample also exists in the variant samples, then the Mirai and variant may belong to the same malware family. Therefore, we compare the similarity of different samples based on the reusable library. When we get a variant sample, we calculate the similarity of the function in the reuse function library, then divide the samples into different families.

For the dynamic result, we compared the similarities and differences between IoT malware by comparing host behaviors and network behaviors.

3.4   Detection and Traffic Analysis

Based on the extracted behaviors, we match the network traffic data by source IP, destination IP, source port number, destination port number and data packet signature in large-scale network. We can further extract relevant information based on these network traffic data.

We perform statistical analysis on network traffic data, including statistics on port numbers and the geographical situation of infected bots. Further, we perform cluster analysis on the Network traffic data and measure the similarities between different data sources to find out if there are multiple different variants in the data.

We analyze the behaviors of different variants from space and time dimensions, including the scope, time distribution, propagation method, and communication behavior, and then assess the degree of harm of different variants.

4   Experiment and Results

To evaluate the architecture, we integrate many tools and develop the entire system based on these tools. For static analysis, we use IDA disassembly tools to analyzes malicious code and extract key behaviors of malicious code; for dynamic analysis, we use cuckoo sandbox to extract the characteristics of host behaviors and traffic. To make comparison of different variant of malware, we obtain and test Mirai, satori, OMG and Wicked samples in the experiment. Furthermore, we capture malware related traffic in the large-scale network based on the static and dynamic signatures. The result is shown below.

4.1   Static Analysis

(1) Function flow chart analysis

The function control flow chart can be obtained by static disassembly of the Mirai code through IDA and other disassembler tools. As shown in the Fig. 3, we obtain the following modules.

Bot main module: Firstly, anti-debugging, disabling watchdog function to prevent the device restart remove malicious files, then making sure that an independent example is running in the device, then opening the killer module, attack module, scanner module, finally, keeping links between BOT and CNC server, accepting and sending necessary information.

Kill module: Closing the specific ports and occupying, deleting specific files and killing corresponding processes to achieve the function of occupying resources alone.

Attack module: Parsing attacked command and launching Dos attack.

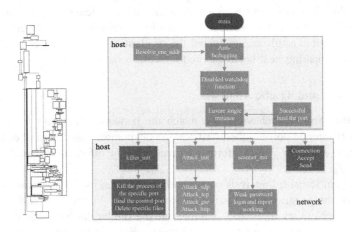

Fig. 3. Functions of the Mirai code

Scanner module: When bot scan other potentially infected devices, if bot can log in to a new bot with a weak password, the successful result will be reported to CNC server. The botnet expands rapidly through this pattern.

Connect, accept, and send module: Realizing links between BOT and CNC server, accepting and sending necessary information.

(2) Key behavioral operations

Each function represents the different functions, the key functions of many have the malicious behavior, and the key functions are divided into two classes: host behavior and network behavior, which can be better combined with dynamic analysis and flow analysis. The results of the classification are shown in the following Table 1:

Table 1. The results analyzed by IDA

Class	Function name	Means
Host behavior	anti_gdb_entry()	anti-debugging
	close_watchdog()	disabling watchdog
	ensure_single_instance()	ensure single instance
	killer_init()	killer module initialize
	bind()	bind the specific ports
	kill()	kill the specific process
Network behavior	attack_init()	attack module initialize
	scanner_init()	scanner module initialize
	establish_connection()	establish connection
	teardown_connection()	tear down connection
	accept()	accept messages from CNC
	send()	send messages to CNC

4.2 Dynamic Analysis

In the sandbox setting, both host and guest machine (virtual machine) use Ubuntu system. The sandbox is installed on the host machine and the recompiled IoT malware runs in the virtual machine environment. We analyzed the malicious code of several Mirai botnet variants and obtained the following results (Table 2).

Table 2. The results analyzed by sandbox

Source	Destination	Protocol	Source port	Destination port	Package number
192.168.56.101	224.0.0.251	MDNS	5353	5353	23
192.168.56.101	192.168.56.255	BJNP	8612	8612	16
192.168.56.101	192.168.56.255	BJNP	8612	8610	16

Analysis the packages of BJNP protocol (network protocol used by Canon printers and scanners), the malicious code can discover the network printers and scanners within the local area network (LAN) through send BJNP protocol discovery request package to broadcast address. Then find printers and scanners support BJNP protocol.

It is able to launch attacks to printers and scanners, such as Exhaustion of print consumables attack; read print log, access to private information; printer configuration changes (if the network printer has changed the administrator password, theoretically could firstly send the content to the attacker machine then send the content to the severs' printers and scanners by ARP cheat).

4.3 Evolution Analysis

According to the results of the static and dynamic analysis of the malware on the IoT, we analyzed the similarity and differences between Mirai and the Mirai variants (Table 3).

Table 3. The comparison between Mirai and variants

Name	Feature	For Mirai	For variants
Satori	Transmission Mode	A Telnet scanner component is downloaded in an attempt to scan to identify vulnerable devices and use the Mirai Trojan to infect after infecting an IoT device	Two embedded vulnerabilities are exploited in an attempt to infect remote devices connected to ports 37215 and 52869 in-stead of using the scanner component
	Target device	Scanning ports 2323 and 23	Connected to ports 37215 and 52869
OMG	Configuration table	Include killing processes, Telnet brute force logins, and launching DDoS attacks	Setting up a firewall to allow traffic to penetrate two random ports
IoTroop	Exploitation	Using the default credentials of the IoT device	Exploiting a wider range of vulnerabilities to target a wider range of products
	DDoS loader	Placing a Mirai-style DDoS engine on the device	Placing a loader that constantly communicates with the C2 server
Wicked	Persistence	Cannot persist and keep on device after restart	Downloading payloads on demand from C&C servers, and adding code to home router firmware to make malware lasting

4.4   Network Traffic Analysis

We collected traffic data of two IoT malwares. Therefore, we consider the traffic can be divided into two IoT events. Each IoT event is not related to each other. We compare the different port scanning behavior and evaluate the spread area of two malwares.

4.4.1   Port Number Statistical Analysis

We perform statistical analysis on the destination port numbers in the two events to obtain the following information (Fig. 4):

From the figure we can see that the Event 1 mainly scan port 7547, using the remote command execution vulnerability rather than the weak password Trojans are significantly different from the behavior of the previous Mirai [14]. From our statistical data, this variant is in an extremely active state, and the daily record of the active scan source is in the million level. The port number scanned in event 2 is much smaller than the port number in Event 1, and may also be related to the amount of data. Event 2 has the highest proportion of port number 23.

We can figure out that malware 1 prefer 7547 port to spread itself. However, malware 2 use many ports to spread itself, the ports include some widely used ports, such as 23, 22, 2000, 445, 80, 81, 3389, 8545, 2323.

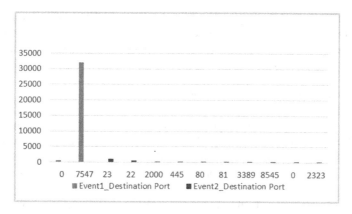

Fig. 4. Destination port numbers of different IoT malwares' interest

4.4.2 Distribution of Infected Areas

The color depth in the picture shows the severity of the infection. Focusing on the variant of scanning port number 7547, we can find that the infection of bots in China is the most serious (Fig. 5).

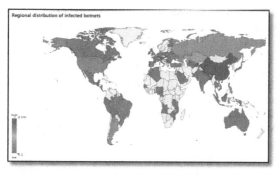

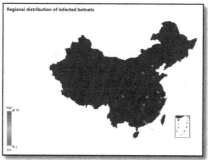

Fig. 5. Regional distribution of the infected botnets

After preliminary analysis, we analyze the distribution of infections in various cities in China. The distribution of data is not evenly distributed, but the focus is on variants of the scanning port number 23. The most serious infection area is in east area of China.

5 Conclusion and Future Work

In this paper, we propose an integrated architecture for IoT malware analysis and detection. By integrating IDA disassembly tool and cuckoo sandbox, we successfully extract the behaviors of the Mirai malware and several variants. We are also able to analysis the evolution and spread of IoT malwares through large-scale network traffic analysis.

Our future work is to make this architecture adapt to more novel variants of the IoT malware. We will apply existing static and dynamic analysis methods to different embedded systems. Simultaneously, we can perform cluster analysis on network traffic data to distinguish different traffic patterns of different variants.

References

1. Jing, Q., et al.: Security of the IoT: perspectives and challenges. Wirel. Netw. **20**(8), 2481–2501 (2014)
2. Kolias, C., et al.: DDoS in the IoT: Mirai and other botnets. Computer **50**(7), 80–84 (2017)
3. Gandotra, E., Bansal, D., Sofat, S.: Malware analysis and classification: a survey. J. Inf. Secur. **5**(02), 56 (2014)
4. Ding, Y., et al.: Control flow-based opcode behavior analysis for Malware detection. Comput. Secur. **44**, 65–74 (2014)
5. Zhang, Z.K., Cho, M.C.Y., Wang, C.W., et al.: IoT security: ongoing challenges and research opportunities. In: IEEE International Conference on Service-Oriented Computing and Applications, pp. 230–234. IEEE (2014)
6. Davidson, D., Moench, B., Jha, S., et al.: FIE on firmware: finding vulnerabilities in embedded systems using symbolic execution. In: Usenix Conference on Security, pp. 463–478. USENIX Association (2013)
7. Zaddach, J., Bruno, L., Francillon, A., et al.: Avatar: a framework to support dynamic security analysis of embedded systems' firmwares. In: Network and Distributed System Security Symposium (2014)
8. Ham, H.S., Kim, H.H., Kim, M.S., et al.: Linear SVM-based android malware detection for reliable IoT services. J. Appl. Math. **2014**(4), 1–10 (2014)
9. Gu, G., Perdisci, R., Zhang, J., Lee, W.: BotMiner: clustering analysis of network traffic for protocol-and structure-independent botnet detection. In: Proceedings of USENIX Security Symposium (2008)
10. Strayer, W.T., Lapsely, D., Walsh, R., Livadas, C.: Botnet detection based on network behavior. In: Lee, W., Wang, C., Dagon, D. (eds.) Botnet detection. Advances in Information Security, vol. 36, pp. 1–24. Springer, Boston (2008). https://doi.org/10.1007/978-0-387-68768-1_1

11. Bekerman, D., Shapira, B., Rokach, L., Bar, A.: Unknown malware detection using network traffic classification. In: Proceedings of IEEE Conference on Communications and Network Security (CNS) (2015)
12. Binkley, J.R., Singh, S.: An algorithm for anomaly-based botnet detection. In: Proceedings of USENIX SRUTI 2006, pp. 43–48, July 2006
13. Edwards, S., Profetis, I.: Hajime: analysis of a decentralized internet worm for IoT devices. Rapidity Netw. (2016)
14. Li, F.: Blog. https://blog.netlab.360.com/a-few-observations-of-the-new-mirai-variant-on-port-7547/

Ant Colony Optimization Based Delay-Sensitive Routing Protocol in Vehicular Ad Hoc Networks

Zhihao Ding[1,2], Pinyi Ren[1,2(✉)], and Qinghe Du[1,2]

[1] School of Electronic and Information Engineering, Xi'an Jiaotong University, Xi'an 710049, China
`dingzhihao@stu.xjtu.edu.cn`, {`pyren,duqinghe`}`@mail.xjtu.edu.cn`
[2] Shaanxi Smart Networks and Ubiquitous Access Research Center, Xi'an, China

Abstract. Vehicular Ad Hoc Network (VANET) is a multi-hop autonomous system that consists of vehicular nodes. VANETs aim to perform an efficient wireless communication in vehicular environments, and vehicular communication scenario is one of the typical high reliability and low delay scenarios in 5G networks. However, the special situations in VANETs like frequent link failure, unstable network topology and random change of vehicle mobility pose a number of challenges in routing protocol design. In this paper, we propose a delay sensitive routing protocol for VANETs to address these serious problems by using ant colony optimization (ACO) and we aim to find a path with a low average end-to-end delay from source to destination. We transform the next hop selection into a probability problem according to ACO concept. There are two mechanisms applied in routing discovery process which utilize pheromone information of transmission delay and heuristic information of vehicles. Two Mathematical models are proposed in pheromone deposit and evaporation prodecure to estimates transmission delay. Performance analysis and simulation results show that the proposed scheme has better performance.

Keywords: VANET · Ant colony optimization · Routing protocol · GPSR

1 Introduction

With the development of wireless communication technology and the popularization of vehicular electronic equipments, the study of vehicular communication network has become a trend. Vehicular Ad Hoc Network (VANET) is a emerging network which enables communications among vehicles (Vehicle-to-Vehicle,

The research work reported in this paper (corresponding author: Pinyi Ren) was supported by Key Research and Development Program of Shannxi Province under Grant 2017ZDXM-GY-012. the Fundamental Research Funds for the Central Universities.

© ICST Institute for Computer Sciences, Social Informatics and Telecommunications Engineering 2019
Published by Springer Nature Switzerland AG 2019. All Rights Reserved
B. Li et al. (Eds.): IoTaaS 2018, LNICST 271, pp. 138–148, 2019.
https://doi.org/10.1007/978-3-030-14657-3_15

V2V), between vehicles and roadside infrastructures (Vehicle-to-Infrastructure, V2I) and between vehicles and pedestrians (Vehicle-to-Pedestrian, V2P) [1–3]. Unlike from Mobile Ad Hoc Network (MANET), VANETs show some different features compared with MANETs to some extent. The most commonly difference in VANETs is the high speed and unstable network topology which makes the link situation more complex and difficult to control. In addition, the best effort delivery in traditional networks can not meet the applications in VANETs, since most of the emergency applications in VANETs are delay-sensitive which guarantees traffic safety. Usually, we prefer a lower transmission delay rather than a higher transmission rate as far as the emergency messages. The fundamental architecture of VANETs is shown in Fig. 1. In VANETs environment, communication range of vehicle is small compared to cellular networks. Besides, channel conditions are relatively poor due to the obstruction of the roadside buildings and the low altitude of the vehicle antenna. Therefore, the typical route from source vehicle to destination vehicle consists of multiple hops and middle node on the path acts as a forwarding node. Consider the important role of multi hop communications and the special characteristics in VANETs, its necessary to develop efficient routing protocol for VANETs. A variety of studies have been done on MANETs in order to propose efficient routing protocols since last century. Such as AODVGPSR and DSR [4,5]. However, these famous routing protocols are dedicated to solve basic requirements in MANETs and can hardly meet the harsh requirements in VANETs. Researches have focused on routing design to improve routing performance over VANETs. The authors in [6] proposed a geographical routing protocol which takes advantage of the road conditions to improve QoS performance. Each forward node selection guarantees delay, bandwidth usage and error bit while satisfying the Qos constraints. In [7], authors developed a propagation strategy and a delay analytical framework in routing algorithm using bidirectional vehicles to forward messages. The authors in [8] designed an algorithm for cluster head and gateway node selection which is combined with AODV routing protocol to provide efficient and secure communications among vehicles by grouping vehicles into different clusters. In this paper, the well-known theory of ant colony optimization (ACO) is adopted to optimize transmission delay in VANETs. The newly research is done on the basis of our previous work in [9]. We in [9] analyse the access mechanism of MAC layer and propose a mathematical model for estimating the delay consumed in the MAC layer. Consider that only the MAC delay is optimized which only is a part of end-to-end delay in our previous work. Therefore, we propose an improved scheme in this work to optimize not only MAC delay but also transmission delay by using ACO. What is expected is that the performance of the new scheme has been improved. The theory of ant colony optimization originates from the study of real ant colony behavior in nature [10–12]. Due to the strong robustness of ant colony algorithm and the convenience to combine with other algorithms, ant colony algorithm is widely applied to address all kinds of NP hard problems. We notice that the transmission process of packets is similar to the process that ants seek paths from nest to food. Hence, as similar as ants lay pheromone along the path, data packets will also lay information during transmission. Commonly, packet delay of each transmission can be recorded as pheromone information. Moreover, this information can be fed back to the transmitter. So we combine the heuristic knowledge of end-to-end delay in the packet transmission process with the characteristics of vehicles to optimize routing algorithm. The remainder of this paper is organized as follows. In Sect. 2, the principle of ACO is introduced. Section 3 presents the proposed adaptive delay-sensitive routing protocol based on ACO concept. Performance evaluations and simulation results are presented in Sect. 4. In the end, the conclusions are derived in Sect. 5.

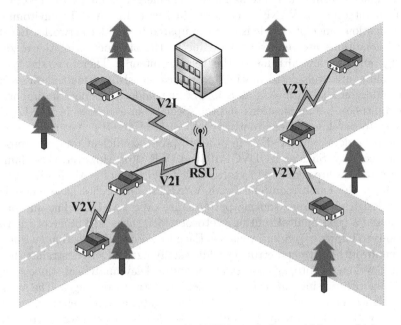

Fig. 1. The fundamental architecture of VANETs, including vehicle to vehicle communications and vehicle to infrastructure communications.

2   Ant Colony Optimization

In the early 1990s, Goss et al. proposed the concept of ant colony optimization algorithms after a long period of study on the behavior of real ants [13]. ACO is a branch of swarm intelligence (SI) and has been proved to be able to solve the complex combinational optimization problems. Figure 2 presents the behavior of ants when searching for food. As shown in Fig. 2(a), ants randomly select a path when in the beginning of finding food from their nest. After a period of time, ants are distributed on the every path from nest to food as shown in Fig. 2(b). Finally, almost all ants go through the shortest path after a long research which is shown in Fig. 2(c). Unlike human, ants lay a chemical substance called pheromone when searching the route. When ants reach a intersection, they tend to select the way with more pheromone. When a plenty of ants searching the food on different paths, ants that pass the shortest path will be the most after a period of time because shortest path takes shortest time. Therefore, the amount of pheromone on the shortest path will be the most, so all the ants will select the shortest path in the end. The characteristics of real ants will provide an idea for solving actual optimization problems.

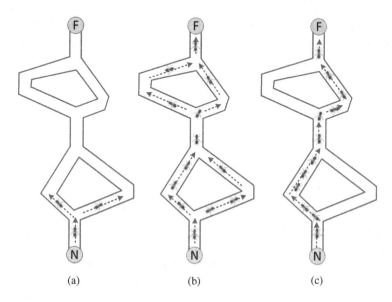

(a)   (b)   (c)

Fig. 2. The behavior of ant colony from nest N to food F. (a) The initial state of ant colony; (b) The intermediate state of ant colony; (c) The final state of ant colony.

3   The Proposed Routing Protocol

In this section, we introduce the proposed routing protocol in detail. The kernel concept of routing is to measure the reasonableness of each forward node and then choose the most reasonable node. Hence, the routing problem can be transformed into the design of probability model. Recall the process of ants searching for food, ants follow the pheromone rule to explore the network. We take the pheromone concept into routing algorithm and combined with the characteristics of vehicles in VANETs. Commonly, the probability model can be represented as follows [14]:

Where τ_{ij} is the density of pheromone from vehicle node i to vehicle node j; η_{ij} is the heuristic information between vehicle i and vehicle j, including link expiration time, distance and MAC backoff times. N_i is the set of neighbors belong to vehicle i; α is the weight parameter of pheromone and β is the weight parameter of heuristic information; P_{ij} represents the probability that vehicle i choose vehicle j as forward node. Therefore, the density of pheromone and the heuristic information of vehicles are the two main mechanisms in our proposed routing protocol. They will be introduced in the following subsections.

3.1 Pheromone Deposit and Evaporation

Consider that most applications in VANETs require a low end-to-end delay, so we can transform packet delay as pheromone information. In the packet transmission process, each packet will experience a delay from source node to destination node. We assume vehicle i transmits n packets to vehicle j in time interval Δt. Then we can get the average delay of a single packet as pheromone information from vehicle i to vehicle j as follows:

$$\tau_{ij}^{D}(\Delta t) = \frac{1}{n} \sum_{k=1}^{n} D_{ij}^{k} \tag{1}$$

where D_{ij}^{k} represents the delay of k?th packet from vehicle i to vehicle j. If the transmitter adds transmission time to the packet, the receiver can get the delay of the packet based on the receive time. According to the transmission control protocol, the transmitter will wait a acknowledge (ACK) message after send a packet. If the ACK message isnt received in a certain time, the transmitter will resend the packet. Therefore, the transmitter can collect the delay of each packet which send to the receiver according to the ACK message. Just like the pheromone information laied by real ants, it will expirence a evaporation process as time passed. In VANETs environment, the network topology will change with time passed. Therefore, the pheromone information of delay between two vehicles will gradually become valueless. In general, we assume the evaporation rate of pheromone is a constant. Then the evaporation process of pheromone can be represented as follows:

$$\tau_{ij}^{E}(\Delta t) = (1 - \rho) \cdot \tau_{ij}(t) \tag{2}$$

where ρ is the constant evaporation rate of pheromone; $\tau_{ij}(t)$ is the level of pheromone between vehicle i and vehicle j at time t. Consequently, the pheromone information of delay collected between vehicle i and vehicle j at time $t + \Delta t$ is as follows:

$$\tau_{ij}(t + \Delta t) = \tau_{ij}^{D}(\Delta t) + \tau_{ij}^{E}(\Delta t) = \frac{1}{n} \sum_{k=1}^{n} D_{ij}^{k} + (1 - \rho) \cdot \tau_{ij}(t) \tag{3}$$

As we prefer a low end-to-end delay, we normalize the result as follows:

$$\tau_{ij} = \frac{\tau_{max} - \tau_{ij}(t + \Delta t)}{\tau_{max}} = \frac{\tau_{max} - \frac{1}{n} \sum_{k=1}^{n} D_{ij}^{k} + (1 - \rho) \cdot \tau_{ij}(t)}{\tau_{max}} \tag{4}$$

where τ_{max} is the maximum value of pheromone between vehicle i and its neighbors.

3.2 Heuristic Information of Vehicles

With the utilization of modern vehicular electrical equipments, the mobility information can be collected by these devices, such as speed, location, move direction and so on. We assume that vehicles can get the mobility information of neighbors and destination under the assistance of Global Positioning System (GPS). These information has an important influence on the performance of routing protocol to a large extent. For example, the well-known GPSR routing protocol selects the forwarding node with the forwarding regulation of shortest distance. Hence, we take distance into consideration as heuristic information. In addition, the number of MAC backoff times and link expiration time can be considered as well.

1. Link Expiration Time: Link expiration time (LET) measures the lifetime of a link, and wireless links with a long LET will maintain a longer communication time. If the locations of vehicle i and vehicle j are (x_i, y_i) and (x_j, y_j), the velocity of vehicle i and vehicle j are v_i and v_j respectively. Then we can calculate the distance and angel between vehicle i and vehicle j. According to communication range R and locations, we can get Δd_{ij} as the relative distance which maintenances the wireless link. In addition, the relative velocity Δv_{ij} between vehicle i and vehicle j can be calculated by speed. As a result, the LET between vehicle i and vehicle j can be simply written as follows:

$$LET_{ij} = \frac{\Delta d_{ij}}{\Delta v_{ij}} \tag{5}$$

2. MAC Backoff Times: According to the MAC layer protocol of VANETs, vehicles must experience a contention process before transmission in a cluster. Thus, the delay in MAC contention process is a part of end-to-end delay. Its essential to reduce MAC delay because it influence the ultima performance of delay. Regard that the basic backoff algorithm in MAC layer is the Binary Exponential Backoff (BEB) algorithm, therefore, node has a low MAC delay if the number of backoff times is less. First, we analysis the procedure of MAC contention. In MAC layer contention process, nodes will start a backoff process before transmission. The backoff time is generated randomly. Only the backoff ends, node can start a transmission. The problem is how to estimate the successful probability in a contention process. We assume the packet arrival rate follow the Poisson distribution and the probability that vehicle i has n packets to be transmitted during a time interval t is written as follows:

$$P_i(t, n) = \frac{(\lambda t)^n}{n!} e^{(-\lambda t)} \tag{6}$$

where λ is the packet arrive rate of a vehicle. Then we can get the successful probability $P_i(t, 0)$ and failure probability $1 - P_i(t, 0)$ of each contention.

Consequently, the average number of backoff times following the above analysis is calculated as follows:

$$\bar{N}_i = \lim_{n \to -\infty} \sum_{k=1}^{n} k \times (1 - P_i(t, 0))^{k-1} \times P_i(t, 0) = \frac{1}{e^{-\lambda(j) \times t}} \tag{7}$$

where $\lambda(j)$ is the overall packet arrival rate in cluster j. We normalize the results of heuristic information (distance, LET and extimated MAC backoff times) between vehicle i and vehicle j and we get the weight value of heuristic information is as follows:

$$\eta_{ij} = w_1 \frac{\bar{N}_{max} - \bar{N}_j}{\bar{N}_{max}} + w_2 \frac{LET_{ij}}{LET_{max}} + w_3 \frac{D_{max} - D_i}{D_{max}} \tag{8}$$

where $\bar{N}_{max}$, LET_{max} and d_{max} are the maximum value of backoff times, LET and distance; D_j is distance between vehicle j and destination; w_1, w_2 and w_3 are weight value and $w_1 + w_2 + w_3 = 1$. Finally, we combine pheromone information with heuristic information and apply Eq. (5), Eq. (9) to Eq. (1). Then we get the mathematical model of forward probability between two vehicles. Following the analysis, we develop the routing algorithm by solving the following problem:

$$\max_{j \in N_i} P_{ij} = \max_{j \in N_i} \frac{[\tau_{ij}^\alpha] \cdot [\eta_{ij}^\beta]}{\sum_{l \in N_i} [\tau_{il}^\alpha] \cdot [\eta_{il}^\beta]} \tag{9}$$

The procedure of the proposed ACO based delay-sensitive Routing Protocol can be summarised as follows. Firstly, measure the delay of pheromone information by flooding RREQ packets. Secondly, measure the heuristic information including LET, backoff times and distance. Thirdly, combine these two mechanisms by the method of ACO using Eq. (1). The detailed ACO based delay-sensitive routing algorithm is shown in Algorithm 1.

Algorithm 1. The Proposed Routing Algorithm

1: Collect neighbor vehicles of transmit vehicle in set M;
2: Calculate and update pheromone information of delay τ_{ij} by Eq.(5);
3: Calculate heuristic information (link expiration time, MAC backoff times and distance) η_{ij} by Eq. (9);
4: Apply Eq.(5) and Eq. (9) to calculate forward probability P_{ij};
5: Select the vehicle with maximum value of P_{ij} in neighbor vehicle set M;

---

4   Performance Evaluations

The performance evaluation of the proposed routing protocol has been investigated and compared with GPSR routing protocol and AODV routing protocol. In addition, we compare the proposed scheme with the scheme in [9] as well. The simulation scenario is a 1000 m 25 m highway area. Vehicles move with a average velocity 50 km/h. The number of vehicles is variable from 12 to 60. The communication range of vehicle is 250 m. Table 1 shows the detailed simulation parameters. Figure 3 presents the average throughput among different routing protocols versus the number of vehicles. In this figure, the average throughput increases as the number of vehicles increases, which is expected because numerous nodes provide connectivity as the number of vehicles increases. Hence, the probability that packets transmitted to the destination increases which leads to a higher average throughput. Also, Fig. 3 shows that the average throughput of our proposed routing protocol outperforms the scheme in [9], GPSR routing protocol and AODV routing protocol, which can be explained according to the following reasons. In our proposed routing protocol, we select each forward node with better delay and connectivity performance as we estimated the transmission delay and MAC delay (MAC backoff times represents the delay in MAC layer), and consider the influence of vehicle mobility. However, GPSR only select the forward node according to distance which results in a terrible link connectivity. AODV selects the path with minimum hops which causes network congestion easily, result a bad performance. The above observations demonstrate that our routing protocol has a better performance than GPSR and AODV.

Table 1. Simulation Parameters.

Parameters	value
Scenario layout	Highway
Terrain size	$1000\,\mathrm{m} * 25\,\mathrm{m}$
Packet size	512 bytes
Transmission power	$23\,\mathrm{dBm}$
ρ	0.1
(α, β)	$(1/2, 1/2)$
(w_1, w_2, w_3)	$(1/3, 1/3, 1/3)$
Speed of vehicles	$10\text{--}30\,\mathrm{km/h}$
Vehicle density	Up to 60
Transmission range	$250\,\mathrm{m}$
Bandwidth	$10\,\mathrm{MHz}$
Comparison schemes	The Proposed, Scheme in [9], GPSR, AODV
Simulation time	500TTI

Figure 4 compares average end-to-end delay of different routing protocols against the number of vehicles. As shown in Fig. 4, average end-to-end delay increases both in all protocols with the increase of vehicle density. This is because MAC layer contention becomes intensely competitive, which determined by the fact vehicles in a cluster use channel resource through competition according to the 802.11 DCF (Distributed Coordination Function) mechanism. In addition, we can see that delay of our proposed routing protocol is lower than scheme in [9], GPSR and AODV, which demonstrates our routing protocol has better

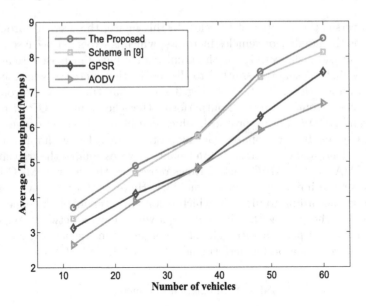

Fig. 3. Average throughput vs vehicle density.

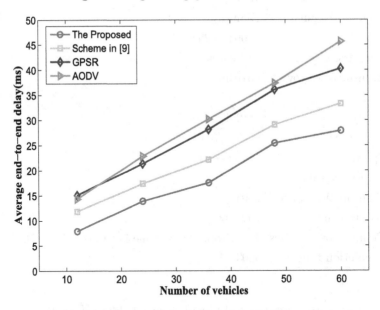

Fig. 4. Average end-to-end delay vs vehicle density.

performance again. Two reasons explain the result. Firstly, recall the pheromone information applied in our routing protocol, this is a positive feedback which guides us to select a forward node with a lowest delay. Secondly, we estimated the backoff times in MAC layer according to the vehicle density in a cluster, which further optimize the delay.

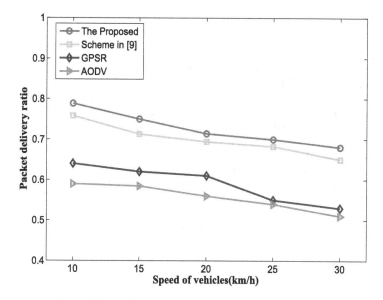

Fig. 5. Packet delivery rate vs speed of vehicles.

Figure 5 depicts the packet delivery rate of different routing protocol versus the speed of vehicles, and once more, our proposed routing protocol performs better than other three schemes. As mentioned previously, we apply link expiration time as heuristic information which influenced by vehicle speed and direction. Therefore, the path in our proposed routing protocol has better connectivity and less broken than GPSR and AODV. On the other hand, from this figure the packet delivery rate of all these protocols decreases while vehicle speed increases from Fig. 5 This is because link reliability and stability decreases when the speed of vehicles increases which causes link interruption more easily.

5   Conclusions

In this paper, we propose an ant colony optimization based delay-sensitive routing protocol in VANETs. We investigate two mechanisms of ACO to be applied to the routing protocol. For pheromone information, we take the estimated transmission delay between two nodes as pheromone and propose the pheromone deposit and evaporation process. For heuristic information, we develop the scheme combining the link expiration time, MAC backoff times and distance to estimate the connectivity and reliability of a node. The abundant and religious simulation results verify that the proposed routing protocol performs much better than GPSR and AODV and can efficiently support the routing requirements in VANETs.

References

1. Zheng, K., Zheng, Q., Chatzimisios, P., Xiang, W., Zhou, Y.: Hetero-geneous vehicular networking: a survey on architecture, challenges, and solutions. IEEE Commun. Surv. Tutor. **17**(4), 2377–2396 (2015)
2. Chen, S., Hu, J., Shi, Y.: Vehicle-to-everything (v2x) services supported by LTE-based systems and 5G. IEEE Commun. Stand. Mag. **1**(2), 70–76 (2017)
3. Zhao, J., Cao, G.: VADD: vehicle-assisted data delivery in vehicular ad hoc networks. IEEE Trans. Veh. Technol. **57**(3), 1910–1922 (2008)
4. Karp, B., Kung, H.: GPSR: greedy perimeter stateless routing for wireless networks. In: International Conference on Mobile Computing and Networking, pp. 243–254 (2000)
5. Karp, B., Kung, H.: Ad hoc on demand vector (AODV) Routing. RFC **6**(7) (2003)
6. Saleet, H., Langar, R., Naik, K.: Intersection-based geographical routing protocol for VANETs: a proposal and analysis. IEEE Trans. Veh. Technol. **60**(9), 4560–4574 (2011)
7. He, J., Cai, L., Pan, J., Cheng, P.: Delay analysis and routing for two-dimensional VANETs using carry-and-forward mechanism. IEEE Trans. Mob. Comput. **16**(7), 1830–1841 (2017)
8. He, J., Cai, L., Pan, J., Cheng, P.: Clustering in vehicular ad hoc network for efficient communication. Int. J. Comput. Appl. **115**(11), 15–18 (2015)
9. Ding, Z., Ren, P., Du, Q.: DownloadURL. http://gr.xjtu.edu.cn/upload/2497558/mobility.pdf
10. Khan, M.S., Sharma, V.: Ant colony optimization routing in mobile adhoc networks? A survey paper. In: International Conference on Computing, Communication and Automation, pp. 529–533 (2017)
11. Martens, D., Backer, M., Haesen, R.: Classification with ant colony optimization. IEEE Trans. Evol. Comput. **11**(5), 651–665 (2007)
12. Duan, H., Wang, D.: Development on ant colony algorithm theory and its applications. Control Decis. **19**(12), 1320–1321 (2004)
13. Goss, S., Aron, S., Deneubourg, J., Pasteels, J.: Self-organized shortcuts in the argentine ant. Naturwissenschaften **76**(12), 579–581 (1989)
14. Dorigo, M., Birattari, M., Stutzle, T.: Ant colony optimization. IEEE Comput. Intell. Mag. **1**(4), 28–39 (2016)

Reinforcement Learning-Based Cooperative Spectrum Sensing

Wenli Ning[(⊠)], Xiaoyan Huang, Fan Wu, Supeng Leng,
and Lixiang Ma

School of Information and Communication Engineering,
University of Electronic Science and Technology of China,
Chengdu 611731, China
WenliNing@126.com,
{xyhuang,wufan,spleng,lixiangma}@uestc.edu.cn

Abstract. In cognitive radio (CR) networks, the detection result of a single user is susceptible due to shadowing and multipath fading. In order to find an idle channel, the secondary user (SU) should detect channels in sequence, while the sequential detection may cause excessive overhead and access delay. In this paper, a reinforcement learning (RL) based cooperative sensing scheme is proposed to help SU determine the detection order of channels and select the cooperative sensing partner, so as to reduce the overhead and access delay as well improve the detection efficiency in spectrum sensing. By applying Q-Learning, each SU forms a dynamic priority list of the channels based on neighbors' sensing results and recent act-observation. When a call arrives at a SU, the SU scans the channel in list order. To improve the detection efficiency, the SU can select a neighbor with potential highest detection probability as cooperative partner using multi-armed bandit (MAB) algorithm. Simulation results show that the proposed scheme can significantly reduce the scanning overhead and access delay, and improve the detection efficiency.

Keywords: Spectrum sensing · Reinforcement learning ·
Cooperative sensing · Q-Learning · Multi-armed bandit

1 Introduction

In wireless networks, inefficient and fixed spectrum usage mode is the main reason for low utilization of spectrum resources. CR technology is envisaged to solve this problem by exploiting the existing wireless spectrum opportunistically [1, 2]. In CR networks, SU can opportunistically transmit in the vacant portions of the spectrum already assigned to licensed primary users (PUs). The goal of spectrum sensing is to find idle spectrum for SUs to occupy while reducing the interference to PUs.

There are two main problems in spectrum sensing. Firstly, due to the detection errors caused by fading and shadowing, the local detection result of a single user on a channel is susceptible [3]. Secondly, we usually use energy detection in local detection. But when there is a demand, SU needs to detect the licensed channels in sequence until it finds an available channel, which can cause excessive overhead and access delay.

© ICST Institute for Computer Sciences, Social Informatics and Telecommunications Engineering 2019
Published by Springer Nature Switzerland AG 2019. All Rights Reserved
B. Li et al. (Eds.): IoTaaS 2018, LNICST 271, pp. 149–161, 2019.
https://doi.org/10.1007/978-3-030-14657-3_16

Hence, selecting the most likely idle channel to sense can reduce the scanning overhead and access delay.

The problems above cause serious access delay, overhead and inefficiency in spectrum sensing. Cooperative spectrum sensing technology [4, 5] has been used in CR network to improve the detection efficiency. Authors in [6] proposed that when a SU's detection ability is higher than the other nodes, taking the local decision of this node as the final decision can obtain better performance than cooperation. So in this case, SU hopes the neighbor with highest detection ability can help him detect the channel. Reinforcement Learning [7] techniques are often applied in dynamic environment to maximum rewards, Q-Learning [8] and multi-armed bandit [9, 10] are two of the RL algorithms. In order to alleviate scanning overhead and access delay in spectrum access, authors in [11] use Q-Learning technique to estimate channels states based on the past history of channel usage. In [12], authors use Q-Learning to select independent users under correlated shadowing for cooperation to improve detection efficiency. In [13], authors formulate the online sequential channel sensing and accessing problem as a sequencing multi-armed bandit problem to improve the throughput.

To address the issues of access delay, scanning overhead and inefficiency in spectrum sensing, a novel cooperative sensing scheme based on RL is designed in this paper. Reinforcement learning is an online learning algorithm. The action-taking agent interacts with the external environment through reward mechanisms, and then adjusts its action according to the reward values. The aim of the agent is to learn the optimal action to maximize the reward. In our scheme, each SU is an agent who needs to learn the behaviors of channels and neighbors, and then takes action to improve the spectrum sensing performance.

Our contributions can be summarized as follows:

- We propose a channel status prediction algorithm based on Q-Learning for SUs to determine the detection order of channels. Specifically, each SU learns the channel patterns by neighbors and detection results. A dynamic priority list of the channels is formed accordingly during the learning procedure. Whenever there is a demand, an SU probes the channels in list order.
- We propose a cooperative partner selection algorithm based on MAB for SUs. Each SU estimates the detection probabilities of its neighbors by MAB algorithm. When detecting, the SU can select a neighbor with potential highest detection probability to help it sense the spectrum.
- Simulation results show that the proposed RL-based cooperative sensing scheme can greatly improve the performance in terms of the access delay, scanning overhead, and detection efficiency.

The remainder of this paper is organized as follows: Sect. 2 describes the system model. Section 3 elaborates the proposed RL-based cooperative sensing scheme. Section 4 evaluates the performance of the proposed scheme. Finally Sect. 5 concludes the paper.

2 System Model

We consider a CR network as shown in Fig. 1. We assume there are N SUs randomly distributed in the network. Each SU can communicate control packets with its neighbors over a channel of the ISM band, which is known to every node. PU network has L licensed channels. PUs may appear in a set of licensed channels. Due to the random distribution of SUs, the effects of fading and shadowing between each SU and PU are different. Thus the detection probability of each node is different.

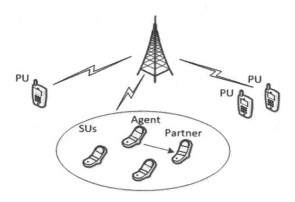

Fig. 1. CR network

To improve channel utilization, SUs attempt to find available spectrum to access by spectrum sensing. When there is a demand, SU needs to scan the licensed channels in sequence until it finds an available channel. In order to find the idle channel quickly, SUs use Q-Learning technique to predict the availability of a channel in our scheme. Q-value in Q-Learning technique represents the probability of each channel being idle. A dynamic priority list of channels is formed according to Q-values of all channels. When there is a demand at a SU, the SU takes an action by scanning a channel in the order of priority list, and then calculates the reward based on neighbors' sensing results and local detection. Then the SU uses the reward to update Q-value of this channel. Finally, the priority list is updated based on the updated Q-value. Whenever there is an update of channel status, the SU shares it with its neighbors.

However, because the multipath fading and shadowing effects in wireless channels can lead to detection errors, the sensing result obtained by a single SU is susceptible. Cooperative spectrum sensing can effectively combat shadowing and multipath fading. When a node cooperates with the partner with lower detection probability, the partner likely degrades the detection performance [7]. So in our scheme, SU would select the neighbor with highest detection ability as its cooperative partner. The selected cooperative neighbor will perform local energy detection, and then send its local binary decisions 1/0 to the SU. 1 and 0 indicate the absence and presence of the PU on the detected channel respectively.

3   RL-Based Cooperative Sensing Scheme

3.1   Q-Learning Based Channel Status Prediction Algorithm

When SU needs access to the channel, the SU hopes to choose the most likely idle channel to detect. Q-learning technique is applied to predict the statuses of channels and form a priority list of channels accordingly. When there is a demand, the SU can detect the channel using action select strategy according to the list, so as to reduce access delay and scanning overhead. Specially, when detecting, each SU select a channel according to the list, and then computes reward of the channel based on neighbors' sensing results and local detection. Then, SU uses reward to update Q-value of the channel. Q-value represents the estimated probability of the channel being idle. The dynamic priority list is updated according to Q-values of all channels.

Q-Learning. Q-Learning is a RL algorithm that includes two entities: agent and environment. An agent in a state s interacts with the environment by taking an action $a \in A$, and then the agent receives a reward $r(s, a)$. So the agent uses $r(s, a)$ to update $Q(s, a)$ and goes in state s'. The agent learns from the state-action-reward history. $Q(s, a)$ is updated in every iteration using the following formula:

$$Q(s,a) = (1 - \alpha)Q(s,a) + \alpha \left\{ r(s,a) + \beta \max_{b \in A}[Q(s',b)] \right\} \tag{1}$$

Here a is the learning rate, $0 \leq \alpha \leq 1$. With α closer to 0, the agent learns less from instant rewards and concentrates more on the history. β is the discount factor, $0 \leq \beta \leq 1$, which notes the attenuation of rewards in the future.

In this paper, agent is each SU. SU_k represents the secondary user k, $1 \leq k \leq N$. State indicates the occupancy statuses of all channels in the primary network. When a channel status turns to busy from idle or turns to idle from busy, the state changes. So the state changes dynamically as the PUs occupy the channel or not. An action is a decision an agent makes in a state. That SU_k chooses an action $a = c_i$, $1 \leq i \leq L$ indicates SU_k selects the c_i as the channel to be detected. How to choose actions depends on the action selection strategy. The choice of current action is evaluated by the reward. Reward function $r(s, a)$ maps the state-action transition to a real-valued reward. Considering the uncertainty of the result detected by a single user, thus SU_k calculates $r(s, a)$ based on both neighbors' sensing results and local detection.

Action Selection Strategy. When a call arrives at SU_k, SU_k selects c_i as the channel to be detected using ε-greedy strategy. That is, the channel with the highest priority according to the priority list will be selected with the probability of $1 - \varepsilon$, which is called exploitation. The channel will be randomly selected with the probability of ε, which is called exploration. ε controls the degree of exploration versus exploitation. For large ε, SU_k concentrates more on exploring the statuses of more channels, so as to help find potentially idle channels. For small ε, SU_k concentrates more on exploiting the current knowledge to perform current best selections, so as to reduce scanning overhead. The ε - greedy strategy helps SUs adapt to the channels' dynamic statuses.

The ε - greedy strategy is an improvement of the algorithms in [11], which helps SUs adapt to the dynamic statuses of channels to choose an idle channel.

Reward Calculation. The reward is used for evaluating the choice of current action. Suppose at time t, SU_k chooses the channel c_i according to the action selection strategy, and then performs local detection on the selected channel c_i. If the detection result is idle, SU_k attempts to access the channel and then obtains the access result. Considering the uncertainty of the result detected by a single user, SU_k calculates the $r^k(s_t, c_i)$ combining local result and detection results of its neighbors:

$$
r^k(s_t, c_i) = \begin{cases} 1 - \sum_{j=1}^{N} \frac{(1-s^j(c_i)) * W_t^j(c_i)}{N}, & if \ s^j(c_i) = 1 \\ -\sum_{j=1}^{N} \frac{(1-s^j(c_i)) * W_t^j(c_i)}{N}, & if \ s^j(c_i) = 0 \end{cases} \tag{2}
$$

Here $s^j(c_i)$ is the sensing result of c_i obtained by SU_j, if the detection result is idle and SU_k accesses c_i successfully, $s^j(c_i) = 1$, otherwise $s^j(c_i) = 0$. $W_t^j(c_i)$ is the detection weight of SU_j about c_i at time t, it represents the estimated value of SU_j's detection probability, which would be obtained by cooperative partner selection algorithm. We'll discuss it in the next section.

After computing the reward point, SU_k updates the corresponding Q-value of channel c_i as:

$$
Q^k(s_{t+1}, c_i) = (1 - \alpha) \cdot Q^k(s_t, c_i) + \alpha \cdot \left\{ r^k(s_t, c_i) - \beta(exp^{-\tau \cdot m}) \right\} \tag{3}
$$

Here α is the learning rate, $0 \leq \alpha \leq 1$. β is the discount factor, $0 \leq \beta \leq 1$. τ, $0 \leq \tau \leq 1$ is a constant. m represents that it's the mth attempt for SU_k to find an idle channel.

The Q-learning based channel status prediction algorithm is applied for SUs to determine the detection order of channels. Each node maintains a Q table which consists of the Q-values of all channels. Q table is initialized to zero at start. When a call arrives at SU_k, SU_k selects a channel by ε-greedy strategy according to the priority list. Then SU_k detects the selected channel and attempts to access it if the detection result is idle. After that, SU_k calculates the reward and uses it to update Q-value of this channel. Based on the updated Q-value, SU_k prepares the new dynamic priority list of the channels for the next round. Cycling until SU_k accesses an idle channel or reaches the maximum number of attempts. The updated channel's status and corresponding weight are broadcasted in each round.

3.2 Multi-armed Bandit Based Cooperative Partner Selection Algorithm

In this paper, we assume that the detection probability of each SU is fixed. In order to increase the detection efficiency, when a call arrives at SU_k, SU_k hopes select a neighbor with highest detection probability to help it sense the spectrum. Since the detection probabilities of SUs are unknown, MAB technique is applied for SUs to estimates the detection probabilities of its neighbors. It also can enable users to learn a strategy to select cooperative partner to maximize detection probability.

MAB. The MAB problem is the model of a gambler (agent) who is playing a slot machine (arm). At time t, the agent gets a reward R_t by pulling/selecting arm $a \in A$. R_t has an independent and appropriate distribution. MAB has two stage of exploration and exploitation, the agent collects information of arms in exploration stage and exploits it in exploitation stage. The purpose of agent is to maximize the total rewards during pulling the arms.

In this paper, the agent is each SU. Arms are its neighbors which are the potential partners of the SU. Let f represent the arm, $1 \leq f \leq N$. $a = f$ means SU selects SU_f as its cooperative partner. Reward R_t represents the detection result of SU_f is right or not, $R_t = 1$ means SU_f detects correctly, or $R_t = 0$ means that SU_f makes a wrong conclusion of the channel status. The expected reward $p(f)$ represents the detection probability of SU_f, $p(f) = E[R|a = f]$, we call it the true value. $\hat{p}_t(f)$ represents the estimation of $p(f)$ at time t. $\hat{p}_t(f)$ is calculated by the obtained information, we call it the estimated value.

MAB has two stage of exploration and exploitation. In exploration stage, agent can obtain more information of arms for selecting better arms, in exploitation stage agent can use the obtained information to maximize its current reward. But when the algorithm focuses more on exploitation, it produces regrets. When the algorithm focuses more on exploitation, it can't find better arms. This is the exploration versus exploitation dilemma. Bandit algorithms look for a balance between exploration and exploitation.

Sample Mean Method. Because the detection probability of SU_f doesn't change with time in our scheme, so it's reasonable to choose the mean of the samples as $\hat{p}_t(f)$, $\hat{p}_t(f)$ is calculated by formula (4):

$$\hat{p}_t(f) = \frac{1}{t} \sum_{i=1}^{t} R_i \tag{4}$$

Action Selection Strategy. Upper confidence bound (UCB) algorithm takes into account both estimated value and selection times of each action to explore and exploit. The aim of UCB algorithm is to choose the most potential user to achieve a balance between exploration and exploitation. At each time t, the action is selected by following formula:

$$a_t = arg_f max \left[\hat{p}_t(f) + c\sqrt{\frac{\log t}{N_t^f}} \right] \tag{5}$$

Here c controls the degree of exploration versus exploitation. N_t^f represents the times that SU_f has been selected as partner until time t, if $N_t^f = 0$, SU_f will be chosen firstly.

The MAB based algorithm is applied for each SU to select a neighbor for cooperation. If there is a demand at SU_k. Firstly, SU_k selects a cooperative partner SU_f using formula (5). And then, SU_k selects a channel c_i using Q-learning algorithm. SU_f performs local energy detection and sends its local binary decisions 1/0 to SU_k. SU_k attempts to access c_i if SU_f's decisions is 1. According to the detection and access result, SU_k updates the estimated value $\hat{p}_t(f)$ of SU_f. The updated $\hat{p}_t(f)$ is also the detection weight $W_t^k(c_i)$ of SU_k in channel status prediction algorithm. Then, SU_k continues to performs Q-Learning based algorithm.

3.3   RL-Based Cooperative Sensing Scheme

In summary, the proposed RL-based cooperative sensing scheme consists of the aforementioned Q-learning based channel status prediction algorithm and MAB based cooperative partner selection algorithm, as presented in Table 1.

Each node maintains a Q table which consists of the Q-values of all channels. Q table is initialized to zero at start. When a call arrives at a node SU_k, the main flow of the proposed scheme is as follows:

(1) SU_k selects a channel c_i using ε-greedy according the priority list, and then selects a cooperator SU_f using UCB.
(2) Update the $\hat{p}_t(f)$ or $W_t^k(c_i)$ by MAB algorithm according to the detection result.
(3) Update the priority list of channels by Q-Learning algorithm according to the sensing result.
(4) Broadcast the updated channel's status, the corresponding weights $W_t^k(c_i)$ and $\hat{p}_t(f)$ to its neighbors. Loop 1–3 until SU_k accesses an idle channel or reaches the maximum number of attempts.

The time complexity of channel selection strategy and cooperative partner selection strategy are $O(L)$ and $O(M)$ respectively. Here, M is the maximum number of attempts for each call before declaring a call block. L is the total number of channels in primary network. It can be seen from Algorithm 1, if we consider the worst case, the time complexity of each SU for one call is $O(M(L+N))$. N is the total number of neighbors of one SU.

Table 1. Pseudo code of the proposed algorithm.

Algorithm 1: RL-based Cooperative Sensing Scheme

Input: the set of SUs, $W_{t-1}^k(c_i)$ and $Q^k(s_{t-1}, c_i)$ of each SU_k
for all c_i, $\hat{p}_{t-1}(f)$ of each SU_k.

Output: $s^k(c_i)$, $W_t^k(c_i)$, $\hat{p}_t(f)$

for each SU_k **do**

  if (a demand appears) **then**

    success =0; attempt=0;

    repeat

      Select a channel c_i using ε-greedy;

      Select a cooperator SU_f using UCB;

      if (SU_f detects c_i correctly) **then**

$$R_t = 1;$$

      else

$$R_t = 0;$$

      end

      Update $\hat{p}_t(f)$;

$$W_t^k(c_i) = \hat{p}_t(f);$$

      if (SU_k access c_i successfully) **then**

$$s^k(c_i) = 1$$

$$r^k(s_t, c_i) = 1 - \sum_{j=1}^{N}\left(1 - s^j(c_i)\right) * W_{t-1}^j(c_i)\Big/N ;$$

      success = 1;

      else

$$s^k(c_i) = 0$$

$$r^k(s_t, c_i) = -\sum_{j=1}^{N}\left(1 - s^j(c_i)\right) * W_{t-1}^j(c_i)\Big/N ;$$

      end

      Update $Q^k(s_t, c_i)$;

      ++attempt;

    until success =1 || attempt=M;

    if(success = 0)

      Declare call dropped.

    end

    Broadcast $s^k(c_i)$, $W_t^k(c_i)$ and $\hat{p}_t(f)$;

  end

end

---

4 Performance Evaluation

4.1 Simulation Setup

In this section, we evaluate the performance of the proposed scheme. In this paper, it is assumed that time is discrete with fixed time unit. In CR network, each SU has 4 neighbors and inquires whether there is a demand at each time unit. The arrival of call request follows Poisson process with $\lambda = 0.5/time\ unit$. There are 10 potential available channels, and PUs' usage rate of channels varies from 40% to 90% [11]. It is assumed that the maximum number of attempts of one call for each SU is 5, if the SU fails to access a channel for 5 times, the call is abandoned and announced blocked.

4.2 Effect of System Parameters

The parameter in the proposed scheme needs to be set according to the specific situations. c is the control parameter of MAB algorithm, which controls the degree of exploration versus exploitation. If c is too large or too small, the probability estimation of neighbors will be inaccurate, which will lead to inefficient cooperation. We can use the average detection probability to evaluate c of different values.

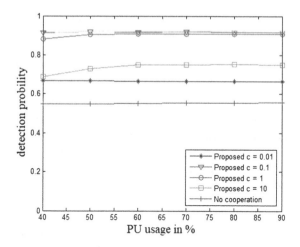

Fig. 2. Average detection probability versus PU usage for different parameter

Figure 2 shows the average detection probability versus PU usage. It can be seen from Fig. 2, cooperation can significantly improve the detection probability of SUs. The exploration coefficient c of MAB has a great influence on partner selection. When the value of c is set about 0.1, the algorithm achieves the balance of exploration and exploitation, so SU can select partner with high detection probability to cooperate. It reflects that if the exploration coefficient c is set properly, SU in our proposed scheme can indeed select partner with high detection probability.

4.3   Comparison with Other Algorithms

To evaluate the performance of the proposed scheme, we considered other two algorithms. The algorithm proposed in [11] (denoted as QLNC) estimates the status of
channels based on the Q-Learning. The other algorithm uses Q-Learning approach to
estimate the status of channels and then uses K/N rule to cooperate (QLKN). Figures 3,
4 and 5 compare the performance of our proposed scheme with other two algorithms.

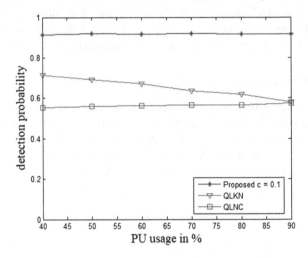

Fig. 3. Average detection probability versus PU usage for different algorithms

Figure 3 shows the average detection probability versus PU usage. It can be seen
from Fig. 3, our proposed scheme performs much better than other two algorithms.
This is because that when the parameter c is set to 0.1 in our simulation scene, the D-
UCB algorithm can learn the dynamic detection probabilities of its neighbors well, thus
SU can select the potential best neighbor to cooperate to improve the detection efficiency. So when the discount factor c is set properly in a specific dynamic situation, our
proposed scheme can significantly improve the detection efficiency.

Figure 4 shows the average number of attempts for a successful access versus PU
usage. It can be seen from Fig. 4, our proposed algorithm has the least average attempts
in all the cases, and the average attempts increase with the PU usage in all the algorithms. This stems from the fact that Q-Learning technique forms a priory list of
channels according to their statues, thus SU in our scheme just needs fewer times of
detection to find an idle channel. With PU usage increasing, there are less opportunities
for SUs to explore available channels in Q-Leaning based algorithm. So that the
priority list can't be updated accurately. Average attempts reflect the scanning overhead
and access delay, hence our proposed scheme indeed improves the scanning overhead
and access delay.

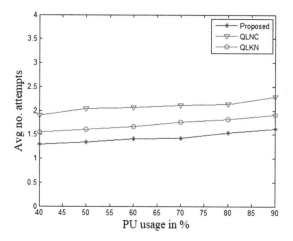

Fig. 4. Average attempts versus PU usage for different algorithms

Figure 5 shows average block rate versus PU usage. It can be seen from Fig. 5 that our proposed algorithm has the least average block rate in all the cases, the average block rate increases with the PU usage in all the algorithms. This stems from the fact that MAB algorithm can help SU learn the detection probabilities of its neighbors, thus SU can select the potential best neighbor to cooperate to improve the detection efficiency. With PU usage increasing, the decrease of the number of available channels leads to more exploration errors. Block rate reflects the quality of service provided to users, hence our proposed scheme performs better than the other two algorithms in terms of communication quality.

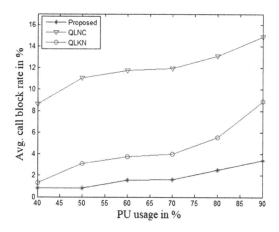

Fig. 5. Average block rates versus PU usage

4.4   Advantages and Disadvantages of the Proposed Scheme

According to the above simulation and analysis, the main advantages can be summarized as follows: the Q-Learning based channel status prediction algorithm can help SUs form a priority list of channels. When there is a demand, the SU can scan the channels in the list order, which helps reduce scanning overhead and access delay. MAB based cooperative partner selection algorithm can help SUs select a partner with high detection probability to cooperative. It improves the average detection probability. The proposed scheme also has some disadvantages: when we apply this scheme to a specific scenario, it takes some time to adjust the parameters. Also the scheme has a poor performance in scenes where the detection probabilities of SUs change dynamically.

5   Conclusion

In this paper, we proposed a RL based cooperative sensing scheme including the Q-Learning based channel status prediction algorithm and the MAB based cooperative partner selection algorithm. The Q-Learning based channel status prediction algorithm is applied for SUs to determine the detection order of channels. MAB based cooperative partner selection algorithm can help SUs select the neighbor with potential highest detection probability to cooperate. Simulation results demonstrate that compared to the existing algorithms (e.g., QLNC in [11] and QLKN), the proposed RL-based scheme has less scanning overhead, less access delay, and higher detection efficiency. In the future, effective learning strategies for mobile SUs will be studied.

Acknowledgement. This work is supported by the National Natural Science Foundation of China under Grant No. 61601083, and the Fundamental Research Funds for the Central Universities, China, No. ZYGX2016J001.

References

1. Wang, B., Liu, K.J.R.: Advances in cognitive radio networks: a survey. IEEE J. Sel. Topics Sig. Process. **5**(1), 5–23 (2011)
2. Haykin, S., Thomson, D.J., Reed, J.H.: Spectrum sensing for cognitive radio. In: Proceedings of the IEEE, pp. 849–877. IEEE (2009)
3. Uchiyama, H., Umebayashi, K., Kamiya, Y.: Study on cooperative sensing in cognitive radio based AD-HOC network. In: IEEE, International Symposium on Personal, Indoor and Mobile Radio Communications, pp. 1–5. IEEE, Athens (2007)
4. Akyildiz, I.F., Lo, B.F., Balakrishnan, R.: Cooperative spectrum sensing in cognitive radio networks: a survey. Phy. Commun. **4**(1), 40–62 (2011)
5. Mishra, S.M., Sahai, A., Brodersen, R.W.: Cooperative sensing among cognitive radios. In: IEEE International Conference on Communications, pp. 1658–1663. IEEE, Istanbul (2006)
6. Zheng, Y., Xie, X., Yang, L.: Cooperative spectrum sensing based on SNR comparison in fusion center for cognitive radio. In: International Conference on Advanced Computer Control, pp. 212–216. IEEE, Singapore (2009)

7. Gosavi, A.: reinforcement learning: a tutorial survey and recent advances. Inf. J. Comput. **21** (2), 178–192 (2009)
8. Watkins, C.J.C.H., Dayan, P.: Machine Learning. Kluwer Academic Publishers, Dordrecht (1992)
9. Kang, S, Joo, C.: Combinatorial multi-armed bandits in cognitive radio networks: a brief overview. In: International Conference on Information and Communication Technology Convergence. IEEE, Jeju (2017)
10. Niimi, M., Ito, T.: Budget-limited multi-armed bandit problem with dynamic rewards and proposed algorithms. In: 4th International Congress on Advanced Applied Informatics, pp. 540–545. IEEE, Okayama (2015)
11. Das, A., Ghosh, S.C., Das, N.: Q-learning based cooperative spectrum mobility in cognitive radio networks. In: IEEE 42nd Conference on Local Computer Networks, pp. 502–505. IEEE, Singapore (2017)
12. Lo, B.F., Akyildiz, I.F.: Reinforcement learning-based cooperative sensing in cognitive radio ad hoc networks. In: 21st Annual IEEE International Symposium on Personal, Indoor and Mobile Radio Communications, pp. 2244–2249. IEEE, Instanbul (2010)
13. Li, B., Yang, P., Wang, J.: Almost optimal dynamically-ordered channel sensing and accessing for cognitive networks. IEEE Trans. Mob. Comput. **13**(10), 1 (2014)

Path Planning Algorithm for UAV Sensing Data Collection Based on the Efficacy Function

Siqi Tao[1](✉), Jianhua He[1], Yiting Zhang[1], Wensheng Ji[1], and Libin Chen[2]

[1] School of Electronics and Information, Northwestern Polytechnical University, Xi'an, China
873692005@qq.com
[2] China State Shipbuilding Corporation, Beijing 100094, China

Abstract. Data collection is one of the most fundamental tasks of wireless sensor networks. At present, the information collection methods of the sensor network mainly include static information collection methods and information collection methods based on mobile sink nodes. Static information collection methods have "energy void problem". However, in another method, the movement of sink nodes will be limited by the environmental terrain. Therefore, these two methods are difficult to effectively collect information in many application scenarios for a long time. In order to solve the above problems, we use the UAV to collect sensing data from the sensor network. It can also choose the order of collecting information based on the importance and the demand of information. In order to solve the problem of unmanned aerial vehicle's energy limitation and time delay of data collection in the real environment, an efficiency function is constructed which considered data value, energy consumption, time and risk. An improved A* path planning algorithm based on efficiency function is proposed for planning the flying path between SDG nodes. We also propose a bee colony path planning algorithm for solving the problem of SDG node allocation and access order.

Keywords: UAV · Wireless sensor network data collection · Path planning

1 Introduction

With the development of Internet of Things technology, the Internet of Things is widely used in intelligent transportation [1], military [2], agricultural production [3], emergency transaction processing [4], disaster relief [5] and environmental monitoring. All aspects of application are inseparable from the collection of data. The sensor network, as a carrier of the Internet of Things, plays an important role in the information acquisition of the Internet of Things. The data collected by the sensor nodes from the monitoring scene is a key part of the Internet of Things application. It is the link between the sensing layer and the application layer. If no data is sensed, the IoT network will lose its application value. Therefore, a reasonable data collection method is essential.

© ICST Institute for Computer Sciences, Social Informatics and Telecommunications Engineering 2019
Published by Springer Nature Switzerland AG 2019. All Rights Reserved
B. Li et al. (Eds.): IoTaaS 2018, LNICST 271, pp. 162–178, 2019.
https://doi.org/10.1007/978-3-030-14657-3_17

At present, sensor networks mainly use sink nodes to collect data. Sensor nodes transmit data to sink nodes through one or more hops. Therefore, the node near the sink node sends more data packets than the remote node. This causes the energy of the nearby sink node to be quickly depleted, eventually causing the entire network to break and the remote node cannot send data to the sink node [6]. In [6], this phenomenon is called "energy void problem". In [7], a data collection method based on a mobile sink node is proposed. The electric trolley equipped with the sink node is moved throughout the monitoring area to collect the data.

Although the "energy void problem" has been solved, when the sensor network is deployed in the wild terrain or in dangerous areas such as cliffs and landslides, the sink node cannot move within the monitoring area. When the sensor network is deployed in a fragile ecological environment protection area, the movement of the sink node may cause damage to the environment, so it is not feasible to collect data through the mobile sink node. Therefore, UAV (unmanned aircraft vehicle) can be used to collect data for the special application scenarios. It can not only solve the "energy void problem" but also be free from environmental terrain restrictions.

The sink node is mounted on the UAV. The UAV navigates the monitoring area and collects data. In order to reduce the flying distance, reduce energy consumption and reduce data delays. In [14], the vertical distance between the UAV and the ground sensor node is equal to or slightly less than the maximum effective communication distance, so as to reduce the flying distance of the UAV and optimize flying path. In [15], according to the value of data and the similarity of data to select key nodes with large data value. The UAV only collect the data of key nodes. In this way, UAV data collection tasks are greatly reduced. In [16], in order to minimizes the maximum energy consumption of all sensor nodes while ensuring that the UAV can collect the required data from each sensor node. The author jointly optimizing the wakeup schedule of the sensor nodes and the trajectory of the UAV. In [17], by deploying cooperative relays, instead of being limited to cluster head nodes, the waypoints for the UAV flying can be selected more freely. Data collection can be more efficient and energy consumption can be reduced.

The above references mainly use a single UAV to collect data, but when the scope of the monitoring area is large and the number of sensor nodes is large, a single UAV has been unable to collect data for all sensor nodes within the maximum data delay and energy limit. We need to use multiple UAVs for data collection and plan the flying path so that each UAV can complete the data collection task within the time limit and energy limit.

In this paper, we have established an efficacy function to evaluate the quality of the path. The efficacy function takes into account some factors, such as data value, energy consumption, time and risk. And then we use the improved A-star algorithm and the bee colony algorithm to plan the flying path when the UAV collects data. The A-star algorithm is used to plan the flying path of the UAV from one SDG (Sensing Data Gather) node to another SDG node. Using the bee colony algorithm to assign each SDG node to each UAV and determine the order of data collection.

The rest of this paper is organized as follows. The efficacy function is described in Sect. 2. In Sect. 3, the A-star algorithm is improved and is used to plan the flying path between SDG nodes. In Sect. 4, UAV flying path planning based on Bee Colony Algorithm. The performance evaluation with simulation results is revealed in Sect. 5. Conclusions are drawn in Sect. 6.

2   Efficacy Function

When planning flying path for multiple UAVs, it is necessary to establish corresponding evaluation index for evaluating the quality of the planned flying path. This article uses the efficacy function as an evaluation index. The efficacy function takes into account some factors, such as data value, energy consumption, time and risk.

Multiple UAVs collect data from the wireless sensor network in the monitoring area R. There are several SDG (Sensing Data Gather) nodes in the monitoring area R for transmission of sensing data to the UAV. Before the UAV collecting data, the location of the SDG nodes is known, and the type, quantity, and data value of the sensors in each SDG node are known. The UAV stays in the SDG node area during data collection and communicates with ground sensors to acquire all sensed data. UAVs cannot fly beyond the no-fly zone.

2.1   Data Value

The value of the data that the UAV obtain from each SDG node depends on the number of various types of sensors in the SDG node and the related data value. v^y is the data value of the data sensed by the y type sensor. Therefore, the value of the data collected by the UAV at the SDG node i is:

$$D_i^s = \sum_{y=1}^{J} n_i^y v^y \tag{1}$$

n_i^y is the number of y sensor nodes in the SDG node.

Therefore, the value of the data that the UAV fly from the SDG node i to the SDG node j is:

$$D_{ij} = \frac{1}{2}(D_i^s + D_j^s) \tag{2}$$

2.2   Energy Consumption

Because the energy of UAV is limited. Therefore, we must ensure that the UAV completes its mission and returns safely to its destination before it runs out of energy. The energy of UAV is mainly used for flying and hovering.

d_{ij} is the distance from SDG node i to SDG node j. The energy consumption of each point on the flying path of the UAV is f^e, and f^e is related to the terrain. Therefore, the flying energy consumption from SDG node i to SDG node j is:

$$E_{ij}^f = -\int_0^{dij} f^e dl \tag{3}$$

When the UAV collects data, it needs to hover over SDG nodes for a period of time. The energy consumption of the UAV in the hovering process is called hover energy consumption. We assume that the energy consumption of the UAV hovering per unit time is h^e. The hover energy consumption of the UAV at SDG node i is [8]:

$$E_i^h = -h^e t_s \sum_{y=1}^{J} n_i^y n_y^d \tag{4}$$

n_i^y is the number of y sensor nodes in the SDG node i. t_s is the exchange time of a single data packet. n_y^d is the number of data packets in the y sensor.

Therefore, the energy consumption that the UAV fly from the SDG node i to the SDG node j is:

$$E_{ij} = E_{ij}^f + \frac{1}{2}(E_i^h + E_j^h) \tag{5}$$

2.3 Time

The UAV mainly spends time in flying and hovering. $\bar{v}_{ij}^f$ is the average flying speed of the UAV from SDG node i to SDG node j, so the flying time from SDG node i to SDG node j is:

$$T_{ij}^f = -\frac{dij}{\bar{v}_{ij}^f} \tag{6}$$

The time for the UAV to hover over the SDG node i is:

$$T_i^h = -t_s \sum_{y=1}^{J} n_i^y n_y^d \tag{7}$$

Therefore, the time that the UAV fly from the SDG node i to the SDG node j is:

$$T_{ij} = T_{ij}^f + \frac{1}{2}(T_i^h + T_j^h) \tag{8}$$

2.4 Risk

When the UAV flies and hover, it may be at risk. The risk that the UAV encounters while flying, we call it flying risk. The flying risk of each point on the flying path of the UAV is r^f, so the flying risk from SDG node i to SDG node j is:

$$R_{ij}^f = -\int_0^{dij} r^f dl \tag{9}$$

The risk of the UAV hovering over the SDG node to collect data, we call it the hovering risk, Hovering risk for SDG nodes is:

$$R_i^s = -r_i^h t_s \sum_{y=1}^{J} n_i^y n_y^d \tag{10}$$

r_i^h is the risk of hovering within unit time.
So the risk that the UAV fly from the SDG node i to the SDG node j is:

$$R_{ij} = R_{ij}^f + \frac{1}{2}(R_i^s + R_j^s) \tag{11}$$

2.5 Efficacy Function and Restrictions

M is the number of UAVs for data collection, numbering multiple UAVs. Expressed as $F = \{f_1, f_2, \ldots, f_M\}$. Therefore, the flying path for the m-th UAV data collection is $P_m = \{S_0, S_i, S_j \ldots, S_0\}$. In order to define the efficacy function and describe constraints, define 0-1 decision variables as follows:

$$x_{ijm} = \begin{cases} 1 & \text{The i−th and j−th SDG nodes are on the m−th UAV data collection path} \\ 0 & \text{others} \end{cases}$$

The relevant parameters of the m-th UAV are as follows:
Data value:

$$D_m = \sum_{i=0}^{N} \sum_{j=0}^{N} D_{ij} x_{ijm} \tag{12}$$

Energy consumption:

$$E_m = \sum_{i=0}^{N} \sum_{j=0}^{N} E_{ij} x_{ijm} \tag{13}$$

Time:

$$T_m = \sum_{i=0}^{N} \sum_{j=0}^{N} T_{ij} x_{ijm} \tag{14}$$

Risk:

$$R_m = \sum_{i=0}^{N} \sum_{j=0}^{N} R_{ij} x_{ijm} \tag{15}$$

All UAV related parameters are as follows:
Total data value:

$$D_{ma} = \sum_{m=1}^{M} D_m \tag{16}$$

Total energy consumption:

$$E_{ma} = \sum_{m=1}^{M} E_m \tag{17}$$

Total time:

$$T_{ma} = \sum_{m=1}^{M} T_m \tag{18}$$

Total risk:

$$R_{ma} = \sum_{m=1}^{M} R_m \tag{19}$$

The efficacy function of multi-UAV data collection is as follows:

$$U_m = a \times \frac{D_{ma}}{D_{max}} + b \times \frac{T_{ma}}{T_{max}} + c \times \frac{E_{ma}}{E_{max}} + d \times \frac{R_{ma}}{R_{max}} \tag{20}$$

$D_{max}, T_{max}, E_{max}, R_{max}$ are the maximum value of each parameter which are set in advance.

The restrictions of multi-UAV data collection is as follows:

Ensure that each UAV can complete data collection for all SDG nodes before the limited time and energy are exhausted. So energy and time constraints:

$$\begin{cases} max(E_m) \leq E_{max} \\ max(T_m) \leq T_{max} \end{cases} \tag{21}$$

Ensure that each SDG node's data is collected:

$$\begin{cases} \sum_{m=1}^{M} \sum_{j=0}^{N} x_{ijm} = 1 \quad i \in S \\ \sum_{m=1}^{M} \sum_{i=0}^{N} x_{ijm} = 1 \quad j \in S \end{cases} \tag{22}$$

Each UAV is guaranteed to start from the starting point and eventually return to the end.

$$\begin{cases} P_m(0) = S_0 \\ P_m(end) = S_0 \end{cases} \tag{23}$$

Ensure that each UAV will collect data. No UAVs will be idle.

$$\sum_{i=1}^{N} \sum_{j=1}^{N} x_{ijm} \neq 0 \quad m \in M \tag{24}$$

3   UAV Flying Path Planning Between SDG Nodes Based on A-Star Algorithm

In this section, we use the A-star algorithm to plan the flying path of the UAV from one SDG node to another SDG node.

First, the monitoring area is rasterized, and then an open list is created to record the neighborhood of the evaluated area. A close list is used to record the areas that have already been evaluated, and the estimated distances from the "starting point" to the "target point" is calculated, the closed list holds all the nodes that have been explored or evaluated. In the process of path finding, the nodes are expanded according to the evaluation function. The nodes in the open list and the close list are changed at any time, and the same node may appear repeatedly in the open list and the close list. According to the evaluation function to find the target point, and then through the backtracking way to get the final path from the starting point to the end point.

3.1 Evaluation Function

$$f(n) = k \times (g(n) + h(n)) + \frac{1-k}{3} cost(n) \tag{25}$$

$g(n)$ is the distance from the starting point to the current point, when the node n is in the vertical or horizontal direction of the node n−1, use the formula (26) to calculate, and when the node n is in the diagonal direction of the node n-1, use formula (27) to calculate:

$$g(n) = g(n-1) + 1 \tag{26}$$

$$g(n) = g(n-1) + 1.4 \tag{27}$$

$h(n)$ is a heuristic function that represents the estimated distance from the current node to the end point, the A* algorithm usually uses Euclidean distance to represent $h(n)$:

$$h(n) = \sqrt{(x_n - x_{end})^2 + (y_n - y_{end})^2} \tag{28}$$

$cost(n)$ is the flying cost of the current node:

$$cost(n) = \frac{UAV_{risk}(n)}{UAV_{risk\_max}} + \frac{UAV_{energy}(n)}{UAV_{energy\_max}} + \frac{1}{UAV_{velocity}(n)} \times \frac{1}{t_{max}} \tag{29}$$

3.2 Determination of k-Value in Evaluation Function

The k value indicates the weight value of flying cost in the evaluation function. The flying path planned by the A-star algorithm is different with different k value.

In order to assess the UAV flying path planned by the A-Star algorithm is bad or good, we define the flying consumption from SDG node to SDG node j as:

$$fly_{con} = \frac{E_{ij}}{E_{ijmax}} + \frac{T_{ij}}{T_{ijmax}} + \frac{R_{ij}}{R_{ijmax}} \tag{30}$$

In order to more fully demonstrate the impact of the k value on the planned path, we have shown the time, energy consumption, risk, and flying consumption in Fig. 1.

As can be seen from Fig. 1, when k is larger than 0.6, energy consumption and time tend to decrease as k increases. Because the larger the value of k is, the greater the weight of the distance in the evaluation function is, the length of the planned path will be shortened, and the time and energy consumption of the UAV will be reduced. The risk decreases firstly and then increases, because the larger the value of k, the smaller the weight value of the risk in the evaluation function, and the smaller the impact on the

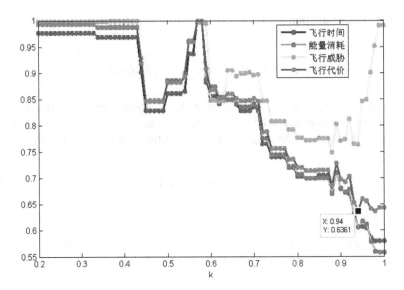

Fig. 1. k value vs. flight consumption graph

planned route, the lower the search distance and the higher the risk value. The area will therefore cause the flight risk to increase.

The node marked in Figs. 3, 4, 5, 6, 7, 8, 9 and 10 is the minimum value of the flying consumption. At this time, the flying consumption is 0.6361, corresponding to the value of k is 0.94. By changing the position of the starting point and the ending point, we found that when the flying consumption is the minimum value, the corresponding k value is mainly distributed within the range of 0.92−0.96. Therefore, we determine the value of k is 0.94.

3.3   Comparison Between A-Star Algorithm and Dijkstra Algorithm

In order to prove the performance of the improved A-star algorithm, we compare the A-star algorithm with the Dijkstra algorithm. We select a starting point and an ending point in the area R, and draw the search range map of the A star algorithm and the Dijkstra algorithm from the start point to the end point respectively.

It can be seen from Figs. 2 and 3 that the search range of the improved a-star algorithm is much smaller than the search range of the Dijkstra algorithm, so that the optimal path can be obtained in a shorter time. And the path of the improved a star algorithm is straighter. Because the UAV is difficult to make a large range of maneuvers, the UVA cannot fly according to the path planned by the Dijkstra algorithm, so the path planned by the improved a star algorithm is more reasonable.

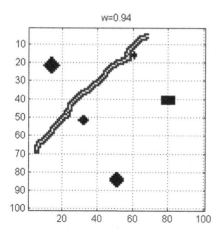

Fig. 2. A star algorithm search range map

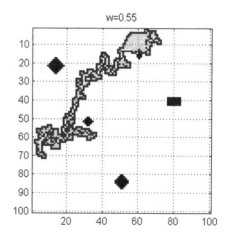

Fig. 3. Dijkstra algorithm search range map

4 UAV Flying Path Planning Based on Bee Colony Algorithm

We use the bee colony algorithm to assign each SDG node to each UAV. The UAVs exchange SDG nodes and change the order of collecting SDG node data, so that the total efficacy value of all paths planned is maximized.

In the bee colony algorithm, bees are divided into three categories: Leader, Follower and Scouter. Each Leader corresponds to a honey source (a feasible solution). The Leaders generate new honey sources according to the neighborhood strategy, evaluate the performance values of the new and old honey sources, use greedy strategies to select, and share this information with others with a certain probability. Follower to select leader following a certain probability value according to the efficacy value of the honey source, and seek for other honey sources in the neighborhood, and try to change the honey source of the leader to be followed to make the efficacy value larger. When a certain honey source cannot be improved after a limited search time (the honey source has been fully utilized), the honey source will be discarded, and the corresponding honey bee will be converted into a scout bee, and the scout bee will randomly generate a new one within the search scope.

4.1 Generation of Initial Solution

We use three UAVs and ten SDG nodes as examples to generate the initial solution. The UAVs are numbered 1, 2 and 3. The SDG nodes are numbered 1, 2, 3, ... 9 and 10. In each group of initial solution, each UAV randomly selects one SDG node as the first data collection node. Each SDG node can only be selected once, and then according to the previous SDG node of each UAV, the node with the largest value of the efficacy function of the previous SDG node is found among the remaining unselected nodes. In this way, n groups of initial populations are generated (Table 1).

Table 1. Generation of initial solution

The first UAV	0-1-6-8-0
The second UAV	0-5-7-9-10-0
The third UAV	0-3-2-4-0

4.2 Neighborhood Search Strategy

The neighborhood search strategy is to change the solution by some operations based on the solutions already generated, so as to obtain a new solution. In this paper, we use the reverse strategy, the nearest strategy, and the cross strategy to search neighborhood.

a. Reverse strategy

SDG nodes with two different random positions in the UAV path are reversed.

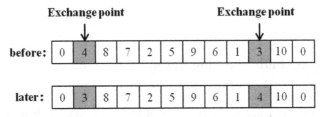

Fig. 4. Reverse strategy

b. Nearest strategy

A SDG node is randomly selected in the path string, we called it as PSDG and then select a SDG which has the highest efficacy value with PSDG and insert it behind the PSDG.

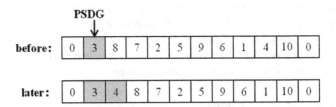

Fig. 5. Nearest strategy

c. Cross strategy

We randomly select one section of the two UAV paths from the same group of solutions to exchange.

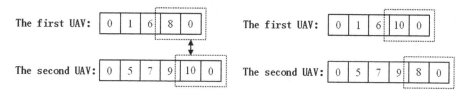

Fig. 6. Cross strategy

5 Simulation Analysis

5.1 The Initial Parameters are Set as Follows

See (Table 2).

Table 2. Initial parameters

Parameters	Symbols	Values
Number of y-sensor in the i-th SDG node	n_i^y	0–5
The data value of y-sensor	v^y	1–3
Velocity	v	8–13 m/s
Energy Consumption	f^e, h^e	10-20
Risk	μ^f	0–1
Time limit	T_{limit}	30 min
Energy limit	E_{limit}	4000

5.2 Simulation Steps

Step 1: Enter initial parameters.
Step 2: Using the A-Star algorithm to plan the path between SDG nodes.
Step 3: Using bee colony to assign SDG nodes to each UAV and determining the order in which UAVs collect data.
Step 4: Smoothing the planning path.
Step 5: Draw data collection path map and output simulation results.

5.3 Simulation Result

This section we will show the path planning results when we use multiple UAVs to collect data form SDG nodes. Figures 7, 8, 9, 10, 11, 12, 13 and 14 shows the path planning results of using different numbers of UAVs to collect data form different amounts of SDG nodes. Figure 13 shows the path planning results when the starting node is not in the center of the target area. Figure 14 shows the path planning results when the SDG nodes are relatively concentrated. The red point is the start point (end point), the green point is the SDG node, and the black area is the no-fly area. Different colored lines are the flying paths of different UAVs.

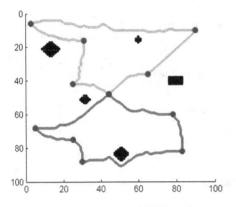

Fig. 7. Two UAVs 10 SDG nodes

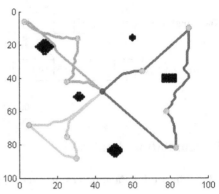

Fig. 8. Three UAVs 10 SDG nodes

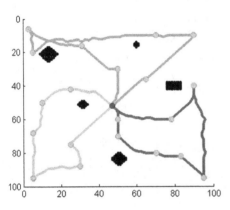

Fig. 9. Three UAVs 20 SDG nodes

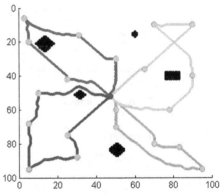

Fig. 10. Four UAVs 20 SDG nodes

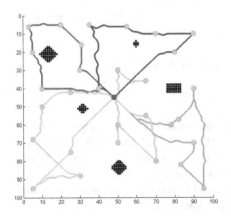

Fig. 11. Five UAVs 30 SDG nodes

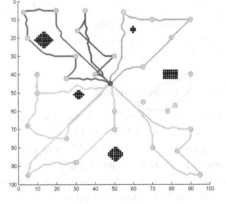

Fig. 12. Seven UAVs 30 SDG nodes

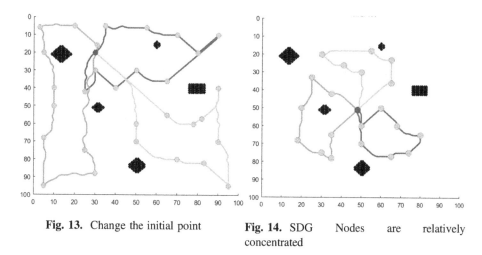

Fig. 13. Change the initial point

Fig. 14. SDG   Nodes   are   relatively concentrated

In order to further analyze the effect of using different numbers of UAVs on the value of the efficacy function, we use 1–7 UAVs to collect data from 20 SDG nodes and get the total energy consumption, total Time, and total risk. As shown in Tables 3 and 4.

In order to comparative analysis, each parameter is presented in Figs. 15 and 16.

Table 3. Efficacy function simulation result parameters (the initial point is at the center of the target area)

UAV number	Total energy consumption	Total risk	Total time(s)
1	5829.4	119.32	4279.1
2	6340.6	130.43	2595.3
3	7062.2	146.59	1501.4
4	7340.6	153.43	1430.3
5	8243.4	184.72	1199.7
6	9120.2	195.40	1351.0
7	9694.0	210.77	347.9

Table 4. Efficacy function simulation result parameters (the initial point isn't at the center of the target area)

UAV number	Total energy consumption	Total risk	Total time(s)
1	5691.6	118.54	4212.3
2	6803.8	133.75	2389.6
3	7791.2	154.50	1913.8
4	8481.8	171.55	1720.7
5	9495.2	186.90	1123.3
6	10593.6	208.74	1096.3
7	11371.0	220.48	754.5

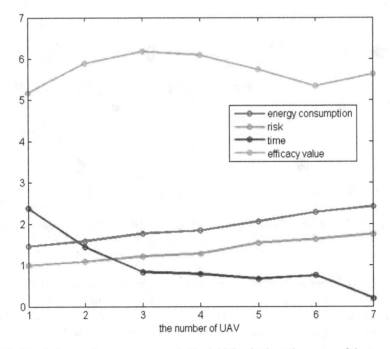

Fig. 15. Simulation result parameters graph (the initial point is at the center of the target area)

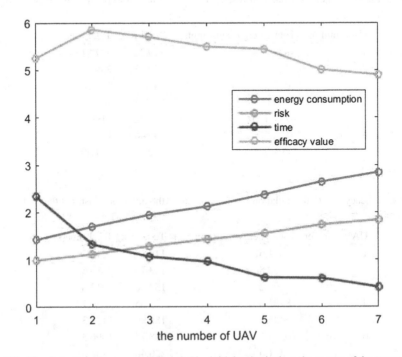

Fig. 16. Simulation result parameters graph (the initial point isn't at the center of the target area)

From Fig. 15, the total energy consumption increases with the increase in the number of UAVs, because each UAV needs to start from the start point and finally return to the start point, so the total distance will increase, and the total energy consumption of UAVs will also increase. Risk values are similar to the energy consumption. As the increase of flying distances, the possibility of risk to UAVs during flight is also increasing. So the risk value increases with the increase in the number of UAVs. Due to the increase in the number of UAVs, the number of SDG nodes that each UAV needs to collect is reduced, so each UAV has a shorter flying distance and less time. Because each UAV departs from the start point at the same time, the flying time of the last UAV that completes the task and return to the start point is the time value of this data collection.

From Fig. 15, it can be seen that the value of the efficacy increases first and then decreases, and reaches the maximum value when the number of UAVs is three. When only one UAV is used for data collection, the energy consumption and risk value are the smallest, but the maximum flying time and maximum energy consumption of a single UAV have been exceeded. The efficacy function value is lower. When seven UAVs collect information, the time required is the shortest, but due to the large number of UAVs, a large amount of energy is required, and the risk value is also large, so the value of the efficacy function is also low. When using three UAVs to collect data, although all parameters are not the minimum value, the time is greatly shortened, and the increase in energy consumption and risk value is small, so the efficacy value is maximum.

It can be seen from Fig. 16 that when the initial node is not at the center of the target area, the trend of energy consumption, risk and time does not change as shown in Fig. 8, but the efficacy value of data collection is the largest when using two unmanned aerial vehicles. Because when the initial node is not in the center of the target area, the UAV needs to fly a long distance to reach the area with the SDG nodes. So the more the number of unmanned aerial vehicles used, the longer the total path length of the UAV flying without the SDG node. Therefore, the UAV consumes more energy and time.

From the above diagrams and analysis, when the initial node is in the center of the target area, it is best to use 3 UAVs collect information from 20 SDG nodes. When the initial node is not in the center of the target area, it is best to use 2 UAVs collect information from 20 SDG nodes.

6   Conclusion

In this paper, we propose an algorithm for UAV data collection in wide IoT sensor networks. The efficacy function is established to evaluate the quality of the path. The efficacy function takes into account some factors, such as data value, energy consumption, time and risk. The A-star algorithm is used to plan the flying path of the UAV from one SDG (Sensing Data Gather) node to another SDG node. Using the bee colony algorithm to assign each SDG node to each UAV and determine the order of data collection. Finally, we obtained the path planning results of data collection by using multiple drones. The simulation results show that our method can optimize the flying path of UAV and reduce energy consumption, also provide basis for multiple UAV data collection.

References

1. Hirankitti, V., Krohkaew, J., Hogger, C.: A multi-agent approach for intelligent traffic-light control. In: World Congress on Engineering, London, UK, 2–4 July, WCE 2007, pp. 116–121 DBLP (2009)
2. Chaimowicz, L., Kumar, V.: Aerial shepherds: coordination among UAVs and swarms of robots. In: Proceedings of Dars', pp. 243–252 (2007)
3. Minbo, L., Zhu, Z., Guangyu, C.: Information service system of agriculture IoT. Automatika 54(4), 415–426 (2013)
4. Xiao, Q.F., Wang, Y., Wang, Y.: Research on emergency disposal platform based on multi-agent with a cooperative model. Adv. Mater. Res. 712–715, 3106–3111 (2013)
5. Domnori, E., Cabri, G., Leonardi, L.: Multi-agent approach for disaster management. In: International Conference on P2p, Parallel, Grid, Cloud and Internet Computing IEEE Computer Society, pp. 311–316 (2011)
6. Jung, J.W., Ingram, M.A.: Residual-energy-activated cooperative transmission (REACT) to avoid the energy hole. In: IEEE International Conference on Communications Workshops IEEE, pp. 1–5 (2010)
7. Di Francesco, M., Das, S., Anastasi, G.: Data collection in wireless sensor networks with mobile elements: a survey. ACM Trans. Sens. Netw. (2011)
8. Yoo, S.J., et al.: Flying path optimization in UAV-assisted IoT sensor networks. ICT Express 2(3), 140–144 (2016)
9. Wang, C., et al.: Approximate data collection for wireless sensor networks. In: IEEE, International Conference on Parallel and Distributed Systems IEEE Computer Society, pp. 164–171 (2010)
10. Chakrabarti, A., Sabharwal, A., Aazhang, B.: Using predictable observer mobility for power efficient design of sensor networks. In: International Conference on Information Processing in Sensor Networks, pp. 129–145. Springer (2003)
11. Ma, M., Yang, Y.: SenCar: an energy-efficient data gathering mechanism for large-scale multihop sensor networks. IEEE Trans. Parallel Distrib. Syst. 18(10), 1476–1488 (2007)
12. Kim, D., et al.: Minimizing data collection latency in wireless sensor network with multiple mobile elements. In: 2012 Proceedings IEEE INFOCOM, pp. 504–512. IEEE (2012)
13. Gu, Z., et al.: Reducing information gathering latency through mobile aerial sensor network. In: IEEE INFOCOM, pp. 656–664. IEEE (2013)
14. Kashuba, S.V., et al.: Optimization of UAV path for wireless sensor network data gathering. In: Actual Problems of Unmanned Aerial Vehicles Developments, pp. 280–283. IEEE (2015)
15. Li, W.: Research on data collection method based on UAV in wireless sensor network (2016)
16. Zhan, C., Zeng, Y., Zhang, R.: Energy-efficient data collection in UAV enabled wireless sensor network. IEEE Wirel. Commun. Lett. 7, 328–331 (2017)
17. Ho, D.T., et al.: Performance evaluation of cooperative relay and particle swarm optimization path planning for UAV and wireless sensor network. In: GLOBECOM Workshops IEEE, pp. 1403–1408 (2013)

A Distributed Algorithm for Constructing Underwater Strong k-Barrier Coverage

Weiqiang Shen[1], Chuanlin Zhang[1]([✉]), Jinglun Shi[2], Xiaona Zhang[3], and Zhihong Wang[1]

[1] College of Information Science and Technology, Jinan University, Guangzhou 510632, China
weiqshen@gmail.com, tclzhang@jnu.edu.cn, horace@139.com
[2] School of Electronic and Information Engineering, South China University of Technology, Guangzhou 510641, China
shijl@scut.edu.cn
[3] School of Computer Science and Engineering, South China University of Technology, Guangzhou 510641, China
zxn0304@qq.com

Abstract. Sensor barrier coverage has been recognized as an appropriate coverage model for intrusion detection, and many achievements have been obtained in two-dimensional (2D) terrestrial wireless sensor networks. However, the achievements based on 2D assumption cannot be directly applied in three-dimensional (3D) application scenarios, e.g., underwater wireless sensor networks. In this paper, we aim to devise a distributed algorithm for constructing maximum level underwater strong k-barrier coverage with mobile sensors in 3D underwater environment. Considering that an underwater strong k-barrier coverage is constituted with k underwater strong 1-barrier coverage which is referred to as layer in this work, we first derive the optimal positions of the sensors in each layer, then we propose a distributed algorithm for constructing maximum level underwater strong k-barrier coverage with available mobile sensors layer by layer from left to right in 3D underwater environment. Simulation results show that the proposed algorithm outperforms the optimal centralized approach (i.e., Hungarian algorithm) in terms of duration and achieves performance close to Hungarian algorithm with respect to several performance metrics.

Keywords: Wireless sensor networks ·
Underwater wireless sensor networks ·
Underwater sensor barrier coverage ·
Three-dimensional underwater strong k-barrier coverage

1 Introduction

Wireless sensor networks (WSNs) have many real life applications in environmental monitoring, battlefield surveillance and intrusion detection, etc. As an

B. Li et al. (Eds.): IoTaaS 2018, LNICST 271, pp. 179–194, 2019.
https://doi.org/10.1007/978-3-030-14657-3_18

important problem in WSNs, barrier coverage is garnering more and more attention in recent years [1,17–19]. Compared with the area coverage problem, barrier coverage does not necessarily cover every point of the monitored region, but rather only needs to detect intruders that cross the border [12]. Therefore, it is more cost-efficient for large-scale deployment of wireless sensors, and has been widely employed in practical security related applications, e.g., international border surveillance, intrusion detection and critical infrastructure protection.

Most existing works on barrier coverage assume that sensors are deployed in 2D long thin belt region, where a barrier is a chain of sensors from one end of the deployment region to the other end with overlapping sensing zones of adjacent sensors. This assumption is reasonable in 2D terrestrial WSNs where the height of the network is usually negligible as compared to its length and width. However, the 2D assumption may not be appropriate when considering WSNs in 3D application scenarios, e.g., underwater wireless sensor networks (UWSNs), where the sensors are finally distributed over 3D underwater environment. As technology advances, efforts are currently underway to extend sensor barrier coverage to underwater application scenarios. For example, underwater sensor barrier has been considered for detecting submarine intrusion in marine environment [3].

In 3D underwater application scenarios[1], the sensors composing UWSNs are distributed at different depths in underwater environment. In this case, a sensor barrier is not a chain of sensors from one end of the deployment region to the other end with overlapping sensing zones of adjacent sensors any more. Instead, a sensor barrier deployed in underwater environment should be a set of sensors with overlapping sensing zones of adjacent sensors that covers an entire (curly) surface that cuts across the 3D underwater space [3]. In practical underwater environment, as the existence of sudden sensor failures and water current which may led to that a sensor deviates from its desired position, a single underwater sensor barrier usually fails to provide adequate service quality. Hence, many real life underwater applications require k-barrier coverage to guarantee their service quality. The notion of k-barrier coverage is first defined in [11], the authors introduced two types of k-barrier coverage including weak barrier coverage, which guarantees to detect intruders moving along congruent paths, and strong barrier coverage, which guarantees to detect intruders no matter what crossing paths they choose.

In this paper, we aim to devise a distributed algorithm for constructing maximum level underwater strong k-barrier coverage with available mobile sensors in underwater environment, to thwart the intruders crossing the monitored 3D underwater environment. Considering that an underwater strong k-barrier coverage is constituted with k underwater strong 1-barrier coverage which is referred to as layer in this work, we first derive the optimal positions of the sensors in each layer, then we propose a distributed algorithm for constructing maximum level underwater strong k-barrier coverage with available mobile sensors layer by layer from left to right in 3D underwater environment. Simulation results show

---

[1] In this paper, we only consider 3D underwater application scenarios where the sensors are deployed in 3D underwater environment.

that the proposed algorithm outperforms the optimal centralized approach (i.e., Hungarian algorithm) in terms of duration and achieves performance close to Hungarian algorithm with respect to several performance metrics.

The rest of the paper is organized as follows. Next section reviews the related work. In Sect. 3, we explain the network model and provide the problem statement. In Sect. 4, we propose a distributed deployment algorithm for constructing maximum level underwater strong k-barrier coverage with available mobile sensors in underwater environment. Section 5 evaluates the performance of the proposed algorithm through extensive simulations, and finally, Sect. 6 concludes the paper.

2 Related Work

As an important problem in WSNs, barrier coverage has been extensively studied in the past decades. Most existing works consider 2D barrier coverage in terrestrial WSNs, while only recently 3D barrier coverage in UWSNs has been studied. In the following, we review the works on 2D and 3D sensor barrier coverage.

2.1 2D Sensor Barrier Coverage

The concept of barrier coverage was first appeared in [7] in the context of many-robot systems. In [11], Kumar et al. developed theoretical foundations for laying barriers of wireless sensors. They defined two types of barrier coverage including weak barrier coverage, which guarantees to detect intruders moving along congruent paths, and strong barrier coverage, which guarantees to detect intruders no matter what crossing paths they choose. Chen et al. [4] introduced the concept of "quality of barrier coverage" and proposed an effective way to measure it. Fan et al. [6] studied the coverage of a line interval with a set of wireless sensors with adjustable coverage ranges. Liu et al. [14] studied the strong barrier coverage of a randomly-deployed sensor network on a long irregular strip region. They showed that in a rectangular area of width ω and length ℓ with the relation $\omega = \Omega(\log \ell)$, if the sensor density reaches a certain value, then there exist, with high probability, multiple disjoint sensor barriers across the entire length of the area such that intruders cannot cross the area undetected; On the other hand, if $\omega = o(\log \ell)$, then with high probability there is a crossing path not covered by any sensor regardless of the sensor density. He et al. [9] presented a condition under which line-based deployment is suboptimal, and proposed a new deployment approach named curve-based deployment. Wang et al. [18] explored the effects of location errors on barrier coverage on a 2D plane by considering two scenarios (i.e. only stationary nodes have location errors, stationary and mobile nodes both have location errors), and proposed a fault-tolerant weighted barrier graph to model the barrier coverage formation problem.

With the advances of technology, sensor mobility has been incorporated into sensor deployment framework [20], which offers more flexibility for designing more efficient sensor deployment strategies to solve coverage problem in WSNs.

Li and Shen [13] studied the 2D MinMax barrier coverage problem of moving n sensors in a 2D plane to form a barrier coverage while minimizing the maximum sensor movement for the sake of balancing battery power consumption. Dobrev et al. [5] studied three optimization problems related to the movement of sensors to achieve weak barrier coverage, i.e., minimizing the number of sensors moved, minimizing the average distance moved by the sensors, and minimizing the maximum distance moved by the sensors. Silvestri and Goss [17] proposed an original algorithm called MobiBar, which has the capability of constructing k-barrier coverage in WSNs, self-reconfiguration and self-healing. Ban et al. [2] considered k-barrier coverage problem in 2D wireless sensor networks, and devised an approximation algorithm called AHGB to construct 1-barrier efficiently. Furthermore, based on AHGB, a Divide-and Conquer algorithm was proposed to construct k-barrier coverage for large scale WSNs. Saipulla et al. [16] explored the fundamental limits of sensor mobility on barrier coverage, and presented a sensor mobility scheme that constructs the maximum number of sensor barriers with the minimum sensor moving distance. Li and Shen [12] studied the 2D MinMax problem of barrier coverage in which the barrier is a line segment in a 2D plane and the sensors are initially resided on this plane.

2.2   3D Sensor Barrier Coverage

There are only a handful of works that have considered 3D sensor barrier coverage. The most related works to ours was presented in [3], the authors considered constructing underwater sensor barriers to thwart illegal intrusion of submarines, and devised a centralized approach and a decentralized approach to achieve this goal, respectively. However, both the centralized and decentralized approaches require one or several sensor nodes as leader to collect and propagate sensor positions. In this case, single point failure problem may not be avoid. In this work, we aim to devise a fully distributed deployment algorithm, which does not require any leader nodes to collect and propagate sensor positions, for constructing maximum level underwater strong k-barrier coverage with available mobile sensors in underwater environment, and hope to provide insights into further researches in 3D wireless sensor networks.

3   Network Model and Problem Statement

3.1   Network Model

We model the underwater deployment region as a cuboid of size $l \times w \times h$, where l, w, and h denote the length, the width, and the height of the cuboid, respectively. For the sake of clearness, we assume that the cuboid is located in a 3D Cartesian coordinate system with origin at the center point of the cuboid. Initially, n sensors are deployed in the cuboid where the sensor positions can follow any type of distribution, such as uniform distribution, Poisson distribution and normal distribution, as shown in Fig. 1. We assume that all the deployed sensors have the same sensing radius r_s and communication radius r_c, where $r_c \geq 2r_s$. Similarly to most previous works in underwater sensor deployment, we assume that all the deployed sensors are able to identify their current locations, and each sensor is able to relocate itself from its initial position to another specified position at a maximum speed of V_{max} m/s in underwater environment. For simplicity, we assume an ideal $0/1$ sphere sensing model that an object within (outside) a sensor's sensing sphere is detected by the sensor with probability one (zero).

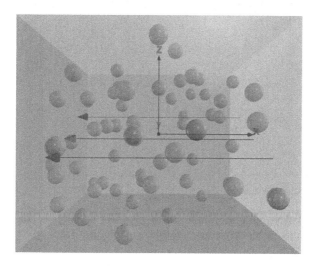

Fig. 1. Initially the sensors are deployed in the cuboid. Without loss of generality, we assume that an intruder's traversing path is a continuously moving trajectory starting at the cuboid's right face and ending at the opposite face.

Without loss of generality, we assume that the illegal intruders to be detected by underwater sensor barrier move along the direction of cuboid length, as shown in Fig. 1, the intruder's traversing path, i.e., the red lines, is a continuously moving trajectory starting at the cuboid's right face and ending at the opposite face.

3.2   Problem Statement

According to the aforementioned assumptions, the underwater deployment region is modeled as a cuboid of size $l \times w \times h$, where l, w and h denote the length, the width and the height of the cuboid, respectively. The position of a point in the cuboid is denoted by coordinates (x, y, z), where x, y, and z are the x-coordinate, the y-coordinate and the z-coordinate of this point, respectively. We suppose that n mobile sensors $S = \{s_1, s_2, ..., s_n\}$ is deployed in the cuboid with initial positions $p_1, p_2, ..., p_n$. Let (x_i, y_i, z_i) denote the coordinates of position p_i of sensor s_i.

The goal of our work is to devise a distributed deployment algorithm to drive the sensors to move to desired positions, and thus provide maximum level underwater strong k-barrier coverage, as shown in Fig. 2. Our problem is how to devise such a distributed deployment algorithm to achieve our goal.

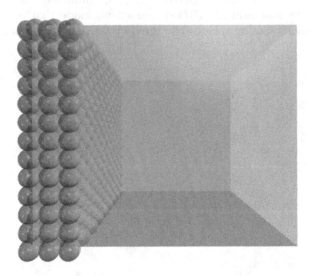

Fig. 2. In 3D UWSNs, an underwater strong k-barrier coverage is constituted with k underwater strong 1-barrier coverage (In this example, k = 3).

4   A Distributed Algorithm

In this section, we propose a distributed algorithm for constructing maximum level underwater strong k-barrier coverage with available mobile sensors.

4.1   Main Ideas

As showed in Fig. 2, an underwater strong k-barrier coverage is constituted with k underwater strong 1-barrier coverage. For simplicity, we refer to an underwater strong 1-barrier coverage as a layer in this work, and from left to right, we enumerate the sensor barriers b0, b1, b2,..., where b0 is base-layer, and the indexes of the other layers increase with the distance from b0. The main idea of the proposed algorithm is to construct underwater strong k-barrier coverage layer by layer from left to right.

For the sake of clearness, we introduce two important concepts. One is constructed-layer, on which no sensor has adjacent vacant positions resided on the same layer as itself, in other words, this layer has been constructed. The other is constructing-layer, on which there are still some vacant positions to be occupied by sensors, in other words, this layer is being constructed. Since underwater strong k-barrier coverage is to be constructed layer by layer, there is only one constructing-layer at the same time during the construction process. For example, base-layer b_0 is the constructing-layer at the beginning, and after the construction of b_0 is completed, b_0 becomes a constructed-layer and layer b_1 is the constructing-layer, and so on. Furthermore, we refer to the sensors resided on constructed-layer and constructing-layer as fixed sensors, while the sensors not already resided on constructed-layers or constructing-layer are referred to as movable sensors. A fixed sensor can not move any more, but a movable sensor can move freely in the underwater space according to its local information. In order to make each movable sensor move orderly to avoid collision as much as possible, and reduce the complexity of the proposed algorithm to ensure that the proposed algorithm performs efficiently, we set a movable sensor's movement route as follows. If a movable sensor s_i with initial position (x_s, y_s, z_s) intends to move to position (x_e, y_e, z_e), it first moves to position (x_s, y_e, z_s) by means of parallel moving, which means that a sensor moves on a plane parallel to the base-layer (i.e., the sensor's x-coordinate will not be changed in the moving process), then moves to position (x_e, y_e, z_e) by means of vertical moving, which means that a sensor moves along a straight line perpendicular to the base-layer. Nevertheless, in rare circumstances, a sensor moving by means of parallel moving may still collide with other sensors moving by means of parallel or vertical moving. Fortunately, the sensors can avoid collision by existing approaches [8], hence we will not discuss the collision avoidance among sensors in deeper in this work.

4.2   The Optimal Final Positions of Sensors in Each Layer

In order to construct maximum level underwater strong k-barrier coverage, we need to minimize the number of sensors required in each layer. Practically, as shown in Fig. 2, if we project a layer onto the cuboid's left face, the sensors on this layer will completely cover the cuboid's left face which is a rectangle of size $w \times h$, where w and h are the width and height of the cuboid, respectively. Hence, for each layer, the minimum number of sensors is equal to the minimum number of hexagons to completely cover a rectangle of size $w \times h$, as shown in Fig. 3. In the context of our work, for each layer, once we place a sensor in the center of each regular hexagon, the sensing range of all sensors will completely cover the rectangle and thus provides strong 1-barrier coverage. According to the above analysis, in each layer, the optimal final positions of sensors are the center points of regular hexagons whose x-coordinates equal to the layer's x-coordinate. In the following, we derive y, z-coordinates of each center point of regular hexagon.

As shown in Fig. 3, we first obtain the y-coordinate of each column via Eq. (1), where i denotes the i-th column.

$$f_y(w, r, i) = \begin{cases} \frac{r}{2}, & i = 0 \\ \frac{r}{2} + \frac{i \times r}{2}, & \frac{r}{2} + \frac{(f_c(w,r)-1) \times r}{2} < w \\ w, & \frac{r}{2} + \frac{(f_c(w,r)-1) \times r}{2} \geq w. \end{cases} \tag{1}$$

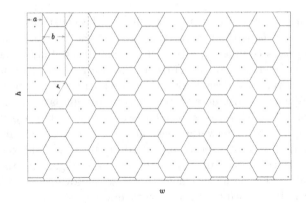

Fig. 3. A rectangle that is completely covered by minimum regular hexagons.

Then we get the z-coordinate of each point row by row. For odd-number columns, h denotes the cuboid height, r denotes the sensing radius, j denotes the j-th row, and $0 \leq j < \lceil \frac{h}{r \times \sqrt{3}} \rceil$, we have

$$f_z(w, r, j) = h - \frac{r \times \sqrt{3}}{2} - j \times r \times \sqrt{3}. \qquad (2)$$

For even-number columns, h denotes the cuboid height, r denotes sensor's sensing radius, j denotes the j-th row, and $0 \leq j < \lceil \frac{h - \frac{r \times \sqrt{3}}{2}}{r \times \sqrt{3}} \rceil + 1$, we have

$$f_z(w, r, j) = h - j \times r \times \sqrt{3}. \qquad (3)$$

Combining Eqs. (2) and (3), we have

$$f_z(w, r, j) = \begin{cases} h - \frac{r \times \sqrt{3}}{2} - j \times r \times \sqrt{3}, & \text{odd-number columns} \\ h - j \times r \times \sqrt{3}, & \text{even-number columns.} \end{cases} \qquad (4)$$

Finally, by combining the layer's x-coordinate x, we obtain the optimal final positions $(x, f_y(w, r, i), f_z(w, r, j))$ of sensors in each layer.

4.3 A Distributed Algorithm for Constructing Underwater Strong k-Barrier Coverage

The proposed algorithm provides the interleaved execution of three main activities, namely vertical movement, vacant position processing and parallel movement. Figure 4 shows the flowchart of these activities.

Before starting the execution of the three main activities, each sensor performs an initial movement aiming to move to the closest centerline, which is perpendicular to the base-layer and pass through the hexagon center point of each layer, as shown in Fig. 5.

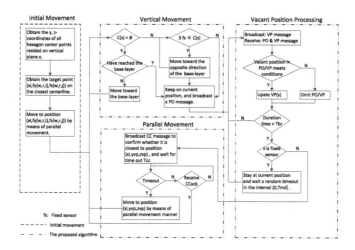

Fig. 4. Flowchart of the initial movement and the proposed algorithm executed by a sensor s with initial position (x, y, z).

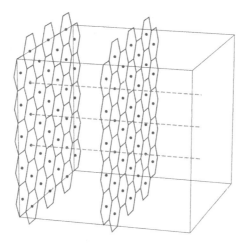

Fig. 5. The centerlines are perpendicular to the base-layer and pass through the hexagon center points in each layer. As shown in this figure, the red dash lines are centerlines. (Color figure online)

Initial Movement. In order to make each movable sensor move orderly in underwater environment, and thus ensure that the proposed algorithm performs effectively and efficiently, each sensor initially moves to the centerline closest to itself by means of parallel moving. It is worth noting that after the initial movement, all sensors are on the centerlines, and by executing the proposed algorithm, each sensor only moves along the centerline or moves to another centerline by means of parallel moving. In the following, we describe how to obtain the centerline closest to a sensor s_i.

For a sensor s_i with initial position (x_i, y_i, z_i), it first obtains the position list of the hexagon center points via Eqs. (1) and (4). Then it calculates the distance between sensor s_i and each hexagon center point $(x_i, f_y(w, r, i), f_z(w, r, j))$. Obviously, a minimum distance between sensor s_i and point $(x_i, f_y(w, r, i'), f_z(w, r, j'))$ will be obtained, the minimum distance is the distance between sensor s_i and the centerline closest to it. Thus, we obtain the centerline closest to sensor s_i:

$$\begin{cases} y = f_y(w, r, i'), \\ z = f_z(w, r, j'). \end{cases} \tag{5}$$

Vertical Movement. The vertical movement activity is twofold. On one hand, in order to increase the connectivity of the network, each movable sensor moves toward base-layer by means of vertical moving if it has no sensor in radio proximity which is resided on the same centerline and closer to base-layer than itself. The movement is stopped as soon as such a sensor is found or the base-layer is reached. On the other hand, to ensure that all sensors can finally move their desired positions, each movable sensor with a distance less than $2r_s$ from a fixed sensor moves toward the opposite direction of the base-layer until the distance not less than $2r_s$. By vertical movement, a movable sensor connects with other sensors to form a network component and thus prevent it becoming an isolated sensor. Moreover, sensors in the same network component can communicate with each others by means of a multi-hop manner. The vertical movement is based on the following described protocol.

For a sensor s_i, we refer to $C(s_i)$ as a set of sensors in the sensor's radio proximity and resided on the same centerline and closer to base-layer with respect to itself. If $C(s_i) = \emptyset$, it moves toward the base-layer along the centerline until $C(s_i) \neq \emptyset$ or it reaches the base-layer. If otherwise $C(s_i) \neq \emptyset$, it checks whether $C(s_i)$ contains fixed sensors. If $C(s_i)$ contains fixed sensors, it moves toward the opposite direction of the base-layer until $C(s_i)$ does not contain fixed sensor any more. If otherwise $C(s_i)$ does not contain fixed sensors, it keeps on its current position. It is worth noting that, moving a sensor s_i to the opposite direction of the base-layer when it finds that $C(s_i)$ contains fixed sensors, ensures that all sensors can finally move to their desired positions even if they are all initially placed on base-layer.

Finally, if sensor s_i has reached the constructing-layer, it broadcasts a PO (position occupation) message, which contains sensor ID, timestamp and its current position information, in the network to remind other sensors to remove this position from their vacant position queues, if their queue contain this position. And then, it gives start to the vacant position processing. If otherwise, it starts the vacant position processing without broadcasting a PO message.

Vacant Position Processing. There are three aspects in vacant position processing. First, to notify the other sensors of the vacant positions resided on the constructing-layer, each sensor periodically detects and broadcasts its adjacent vacant positions resided on the same layer as itself in the network. Second, each sensor receives and processes the vacant positions broadcasted by other sensors. It is worth noting that, in order to manage the vacant positions resided on the constructing-layer, each sensor maintains a vacant position queue, where the vacant positions are sorted by the distance between vacant position and the sensor in an ascending order. Finally, by leveraging the data cached in vacant position queue, a sensor makes its movement decision. The detailed protocol of the vacant position processing is described as follows.

For a sensor s_i with a vacant position queue $VP(s_i)$ (initially $VP(s_i) = \emptyset$), it first broadcasts a VP (Vacant Position) message containing sensor ID, timestamp and its adjacent vacant positions resided on the same layer as itself in the network, then it waits for the expiration of its timeout T_{rcv} to receive VP message and PO message broadcasted by other sensors. Notice that, we assume that the proposed algorithm is implemented over a communication protocol stack which handles possible transmission errors and message losses by means of timeout and retransmission mechanisms, and each sensor keeps track of the VP messages and PO messages sent so far to avoid multiple retransmissions and multiple processing by checking the unique message mark consisting of sensor ID and time stamp. When sensor s_i receives a VP message, it checks the cardinality of $VP(s_i)$. If $|VP(s_i)| = 0$, the vacant positions are put into $VP(s_i)$. If otherwise, it compares the x-coordinate of vacant position with that of one sensor in $VP(s_i)$, if larger, these vacant positions are omitted; if equal, these vacant positions are inserted into $VP(s_i)$; if smaller, it sets $VP(s_i) = \emptyset$ and puts these vacant positions into $VP(s_i)$. Similarly, when sensor s_i receives a PO message containing position p (p is a vacant position before being occupied), it removes p from its vacant position queue if p is in the queue.

By means of the above protocol, each sensor maintains a vacant position queue, where vacant positions are all resided on the constructing-layer[2] and sorted by the distance between vacant position and the sensor in an ascending order.

After the expiration of time T_{rcv}, sensor s_i determines whether it is a movable sensor or a fixed sensor by the following protocol. Sensor s_i first obtains the constructing-layer's x-coordinate by returning the first element's x-coordinate from its vacant position queue, then it compares the constructing-layer's x-coordinate with its, if larger, sensor s_i is a fixed sensor resided on constructed-layer; if equal, sensor s_i is a fixed sensor resided on the constructing-layer; if smaller, sensor s_i is a movable sensor. Then, according to this determination, sensor s_i makes its movement decision as follows.

If sensor s_i is a fixed sensor, it keeps its current position and waits a random timeout in the interval $(0, T_{md}]$ to restart the vacant position processing activity. If otherwise sensor s_i is a movable sensor, it terminates this activity and starts the parallel movement activity.

---

[2] The vacant positions resided on constructing-layer have the smallest x-coordinate, consequently they are finally cached in vacant position queue according to the above protocol.

Parallel Movement. In this activity, a movable sensor first confirms whether it is closest to the vacant position at the head of its vacant position queue. If it is the closest one, it moves to this vacant position by means of parallel moving then starts the vertical movement activity. If otherwise, it terminates this activity and gives start to the vertical movement activity. The approach to find the closest vacant position is described as follows.

For a movable sensor s_i with coordinates (x_i, y_i, z_i), it broadcasts a CC (Confirm the Closest) message containing sensor ID, timestamp, the vacant position vp with coordinates (x_{vp}, y_{vp}, z_{vp}) and the distance between sensor s_i and the vacant position vp, and then waits for the expiration of its timeout T_{cc}. When the other sensors receive the CC message, they check whether they are closer to position vp than sensor s_i, if closer, they broadcast a CCack message containing CC's original content. If otherwise, they omit this CC message. If, while waiting for the expiration of its timeout T_{cc}, sensor s_i receives a CCack message, it terminates this parallel movement activity and starts vertical movement activity. After the expiration of a timeout T_{cc}, sensor s_i moves to position (x_i, y_{vp}, z_{vp}) by means of parallel moving. Once sensor s_i reaches position (x_i, y_{vp}, z_{vp}), it gives start to the vertical movement. Similar to the vacant position processing activity, we assume that each sensor keeps track of the CC/CCack messages sent so far to avoid multiple retransmissions and multiple processing by checking the unique message mark consisting of sensor ID and time stamp.

5   Performance Evaluation

In this section, we evaluate the performance of the proposed algorithm by comparing with Hungarian algorithm [10] which obtains the optimal total movement distance of all sensors. The simulation parameters are described as follows. We conduct the experiments in a Python-based simulator, each data point of our experiment results is an average value of the data collected by running the experiments 100 times.

Considering that the average ocean depth is 3795 m [15], and some commercial magnetic sensor can detect submarines at distances of several hundred meters [3]. We assume that the underwater space where the underwater strong k-barrier coverage to be constructed is a cuboid of length $l = 4200$ m, width $w = 4000$ m, and height $h = 3800$ m, respectively. We set r_s and r_c to 200 m and 420 m, respectively.

As far as we know, the proposed algorithm is the first fully distributed algorithm for constructing maximum level underwater strong k-barrier coverage with available mobile sensors. To evaluate the performance of the proposed algorithm, we compare it to a classic centralized approach, namely Hungarian algorithm [10], which minimizes the total movement distance of all sensors. We assume that all sensors are initially randomly deployed throughout the underwater environment. The number of available sensors ranges from 100 to 800 with an increment of 100 each time. Three performance metrics are considered in the simulation experiments, namely maximum movement distance of all sensors, total movement distance of all sensors and duration of the construction of underwater strong k-barrier coverage.

Figure 6 shows the maximum movement distance of all sensors achieved by the proposed algorithm and Hungarian algorithm. Both algorithms show a gentle decreasing behavior of the maximum movement distance as the number of sensors increases. This is because more sensors locate closer to their final positions as the number of sensors increases. Hungarian algorithm achieves about 35% lower maximum movement distance due to the fact that, under Hungarian algorithm each sensor moves from its initial position to its final position along a straight line, while most of sensors in the proposed algorithm move to their final positions by means of vertical or parallel moving according to their movement decisions, which makes them move along zigzag route.

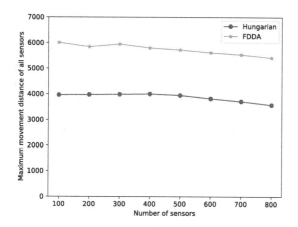

Fig. 6. Maximum movement distance vs number of sensors.

For the total movement distance of all sensors, although the Hungarian algorithm is the optimal solution that achieves the minimum total movement distance, the proposed algorithm also presents a sub-optimal results, as shown in Fig. 7, the optimal total movement distance of Hungarian algorithm is only 28% lower than that of the proposed algorithm.

The duration depicts the length of the time that the construction of the underwater strong k-barrier coverage continues. For Hungarian algorithm, the duration includes the computing time consumed by the central unit to assign a final position per sensor, and movement time consumed by all sensors to move from their initial positions to their final positions along a straight line. While for the proposed algorithm, the duration mainly includes the movement time consumed by all sensors to move from their initial positions to their final positions along a zigzag route. As shown in Fig. 8, the duration of Hungarian algorithm increases sharply as the number of sensors increases, while the increase of the duration of the proposed algorithm is relatively flat. This is because the optimal

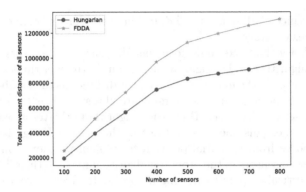

Fig. 7. Total movement distance vs number of sensors.

computational complexity of Hungarian algorithm is $O(n^3)$, while for the proposed algorithm, each sensor in three main activities only needs to communicate with at most $(n-1)$ sensors at each step, hence the total time complexity of the proposed algorithm is $O(n)$. It is worth noting that, the duration of Hungarian algorithm is a little less than that of the proposed algorithm when the number of sensors is less than 500, this is because, in this case, the central unit of the Hungarian algorithm only consumes relatively little time, while the maximum movement distance of the proposed algorithm is larger than that of Hungarian algorithm, which results in more time needed for the proposed algorithm to move the sensors to their desired positions. In general, the proposed algorithm outperforms Hungarian algorithm in terms of duration.

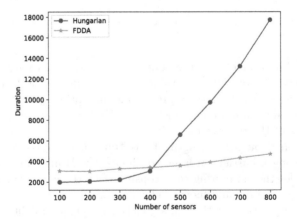

Fig. 8. Duration vs number of sensors.

6   Conclusion

In this paper, we devised a distributed algorithm for constructing maximum level underwater strong k-barrier coverage with available mobile sensors in 3D underwater environment. Considering that an underwater strong k-barrier coverage is constituted with k underwater strong 1-barrier coverage which is referred to as layer in this work, we first derive the optimal positions of the sensors in each layer, then we proposed a distributed algorithm for constructing maximum level underwater strong k-barrier coverage with available mobile sensors layer by layer from left to right in 3D underwater environment. By extensive simulations, we showed that the proposed algorithm outperforms the optimal centralized approach (i.e., Hungarian algorithm) in terms of duration, and achieves performance close to Hungarian algorithm with respect to two performance metrics, namely maximum movement distance of any sensor and total movement distance of all sensors.

References

1. Balister, P., Bollobás, B., Sarkar, A.: Barrier coverage. Random Struct. Algorithms **49**(3), 429–478 (2016). https://doi.org/10.1002/rsa.20656
2. Ban, D., Yang, W., Jiang, J., Wen, J., Dou, W.: Energy-efficient algorithms for k-barrier coverage in mobile sensor networks. Int. J. Comput. Commun. Control **5**(5), 616–624 (2010)
3. Barr, S.J., Wang, J., Liu, B.: An efficient method for constructing underwater sensor barriers. J. Commun. **6**(5), 370–383 (2011)
4. Chen, A., Lai, T., Xuan, D.: Measuring and guaranteeing quality of barrier-coverage in wireless sensor networks. In: Proceedings of the 9th ACM International Symposium on Mobile Ad Hoc Networking and Computing, MobiHoc 2008, 26–30 May 2008, Hong Kong, China, pp. 421–430 (2008). https://doi.org/10.1145/1374618.1374674
5. Dobrev, S., et al.: Weak coverage of a rectangular barrier. CoRR abs/1701.07294 (2017). http://arxiv.org/abs/1701.07294
6. Fan, H., Li, M., Sun, X., Wan, P.J., Zhao, Y.: Barrier coverage by sensors with adjustable ranges. ACM Trans. Sens. Netw. (TOSN) **11**(1), 14 (2014)
7. Gage, D.W.: Command control for many-robot systems. Technical report, DTIC Document (1992)
8. Ganesan, V., Chitre, M., Brekke, E.: Robust underwater obstacle detection and collision avoidance. Auton. Robots **40**(7), 1165–1185 (2016)
9. He, S., Gong, X., Zhang, J., Chen, J., Sun, Y.: Barrier coverage in wireless sensor networks: from lined-based to curve-based deployment. In: Proceedings of the IEEE INFOCOM 2013, 14–19 April 2013, Turin, Italy, pp. 470–474 (2013). https://doi.org/10.1109/INFCOM.2013.6566817
10. Kuhn, H.W.: The Hungarian method for the assignment problem. Naval Res. Logistics Q. **2**(1–2), 83–97 (1955)
11. Kumar, S., Lai, T., Arora, A.: Barrier coverage with wireless sensors. In: Proceedings of the 11th Annual International Conference on Mobile Computing and Networking, MOBICOM 2005, Cologne, Germany, 28 August–2 September 2005, pp. 284–298 (2005). https://doi.org/10.1145/1080829.1080859

12. Li, S., Shen, H.: Minimizing the maximum sensor movement for barrier coverage in the plane. In: 2015 IEEE Conference on Computer Communications, INFOCOM 2015, 26 April–1 May 2015, Kowloon, Hong Kong, pp. 244–252 (2015). https://doi.org/10.1109/INFOCOM.2015.7218388
13. Li, S., Shen, H.: Minimizing the maximum sensor movement for barrier coverage in the plane. In: 2015 IEEE Conference on Computer Communications (INFOCOM), pp. 244–252. IEEE (2015)
14. Liu, B., Dousse, O., Wang, J., Saipulla, A.: Strong barrier coverage of wireless sensor networks. In: Proceedings of the 9th ACM International Symposium on Mobile Ad Hoc Networking and Computing, MobiHoc 2008, 26–30 May 2008, Hong Kong, China, pp. 411–420 (2008). https://doi.org/10.1145/1374618.1374673
15. Rafferty, J.P., et al.: Oceans and Oceanography. Britannica Educational Publishing, Chicago (2010)
16. Saipulla, A., Liu, B., Xing, G., Fu, X., Wang, J.: Barrier coverage with sensors of limited mobility. In: Proceedings of the 11th ACM International Symposium on Mobile Ad Hoc Networking and Computing, MobiHoc 2010, 20–24 September 2010, Chicago, IL, USA, pp. 201–210 (2010). https://doi.org/10.1145/1860093.1860121
17. Silvestri, S., Goss, K.: Mobibar: an autonomous deployment algorithm for barrier coverage with mobile sensors. Ad Hoc Netw. **54**, 111–129 (2017)
18. Wang, Z., Chen, H., Cao, Q., Qi, H., Wang, Z., Wang, Q.: Achieving location error tolerant barrier coverage for wireless sensor networks. Comput. Netw. **112**, 314–328 (2017)
19. Wu, F., Gui, Y., Wang, Z., Gao, X., Chen, G.: A survey on barrier coverage with sensors. Frontiers Comput. Sci. **10**(6), 968–984 (2016)
20. Yang, G., Qiao, D.: Multi-round sensor deployment for guaranteed barrier coverage. In: 29th IEEE International Conference on Computer Communications, Joint Conference of the IEEE Computer and Communications Societies, INFOCOM 2010, 15–19 March 2010, San Diego, CA, USA, pp. 2462–2470 (2010). https://doi.org/10.1109/INFOCOM.2010.5462074

Constructing Underwater Weak k-Barrier Coverage

Weiqiang Shen[1], Chuanlin Zhang[1]([✉]), Jinglun Shi[2], and Ruiyan Han[2]

[1] College of Information Science and Technology, Jinan University,
Guangzhou 510632, China
weiqshen@gmail.com, tclzhang@jnu.edu.cn
[2] School of Electronic and Information Engineering,
South China University of Technology, Guangzhou 510641, China
shijl@scut.edu.cn, ruiyanh@163.com

Abstract. Most achievements on barrier coverage are based on an assumption that the sensors composing the barrier coverage are finally connected as a two-dimensional (2D) terrestrial wireless sensor network, where a barrier is a chain of sensors from one end of the deployment region to the other end with overlapping sensing zones of adjacent sensors. However, the 2D assumption cannot directly be applied in three-dimensional (3D) application scenarios, e.g., underwater wireless sensor networks, where sensors are finally distributed over 3D underwater environment. In this paper, we investigate weak k-barrier coverage problem in underwater wireless sensor networks. We first analyse how to construct 3D underwater weak k-barrier coverage with minimum sensors, then we propose a parallel movement manner, based on which an effective algorithm is proposed for constructing weak k-barrier coverage with minimum sensors while minimizing the total movement distance of all sensors in underwater wireless sensor networks. Extensive simulation results validate the correctness of our analysis, and show that the proposed algorithm outperforms the GreedyMatch algorithm.

Keywords: Underwater wireless sensor networks · Barrier coverage · Underwater weak k-barrier coverage · 3D sensor barrier

1 Introduction

Barrier coverage, which is a critical problem in wireless sensor networks (WSNs), is garnering more and more attention in recent years. Compared to area coverage, barrier coverage does not necessarily cover every point of the monitored region, but rather only needs to cover the monitored region border to detect intruders that cross the border [12]. Therefore, it is more cost-efficient for large-scale deployment of wireless sensors and has been widely employed in practical security applications such as international border surveillance, intrusion detection and critical infrastructure protection.

© ICST Institute for Computer Sciences, Social Informatics and Telecommunications Engineering 2019
Published by Springer Nature Switzerland AG 2019. All Rights Reserved
B. Li et al. (Eds.): IoTaaS 2018, LNICST 271, pp. 195–210, 2019.
https://doi.org/10.1007/978-3-030-14657-3_19

In existing literature, most of the works on barrier coverage assume that sensors are deployed in a 2D long thin belt region, where a barrier is a chain of sensors from one end of the region to the other end with overlapping sensing zones of adjacent sensors. Based on the 2D assumption, the sensors composing a sensor barrier are finally resided on a 2D plane. This assumption may be reasonable in a terrestrial wireless sensor network where the height of the network is usually negligible as compared to its length and width. However, the 2D assumption cannot cover all application scenarios, e.g., underwater wireless sensor networks (UWSNs), where sensors are finally distributed over 3D underwater environment.

In 3D underwater environment, a barrier is not a chain of sensors from one end of the region to the other end with overlapping sensing zones of adjacent sensors any more. Instead, a barrier in 3D UWSNs should be a set of sensors with overlapping sensing zones of adjacent sensors that covers an entire (curly) surface that cuts across the 3D space [1]. For the weak k-barrier coverage in 3D UWSNs, it can be seem from Fig. 3 that underwater weak k-barrier coverage guarantee the detection of intruders crossing deployment region along orthogonal paths, this is similar to the 2D case. Although underwater weak k-barrier coverage can not guarantee the detection of intruders crossing deployment region along any paths, in real life applications, an intruder is more likely to choose an orthogonal crossing path than other paths if he does not know the positions of the sensors [11]. This phenomenon may be related to the following two reasons. On one hand, the orthogonal crossing paths are the shortest crossing paths, through which an intruder can cross the interest region with the least amount of time. On the other hand, these non-orthogonal crossing paths have larger intrusion detection areas, which are likely to contain more sensors. In this sense, weak k-barrier coverage can meet the deployment requirements of most intrusion detection applications of 3D UWSNs.

In this paper, we focus on weak k-barrier coverage problems in 3D UWSNs and aim to construct underwater weak k-barrier coverage with minimum sensors while minimizing the total movement distance of the sensors. We first analyse how to construct 3D underwater weak k-barrier coverage with minimum sensors, then we propose a parallel movement manner, based on which an effective algorithm is presented for constructing weak k-barrier coverage with minimum sensors while minimizing the total movement distance of the sensors in underwater wireless sensor networks. Extensive simulation results validate the correctness of our analysis, and show that the proposed algorithm (Hungarian Method-based sensor assignment algorithm, HMB-SAA for short) outperforms the GreedyMatch algorithm.

The rest of the paper is organized as follows. Next section reviews the related work. In Sect. 3, we explain the network model and provide the problem statement. Next, in Sect. 4, we analyse how to construct weak k-barrier coverage with minimum sensors while minimizing the total movement distance of all sensors in UWSNs. Further, in Sect. 5, an effective algorithm is presented for constructing weak k-barrier coverage with minimum sensors while minimizing the total movement distance of all sensors. Section 6 evaluates the performance of the proposed algorithm through simulations, and finally, Sect. 7 concludes the paper.

2   Related Work

As a critical issue in WSNs, barrier coverage problem has been studied intensively in the past decades. The concept of barrier coverage was first appeared in [5] in the context of many-robot systems. In [10], Kumar et al. developed theoretical foundations for laying barriers of wireless sensors. They defined two types of barrier coverage including weak barrier coverage, which guarantees to detect intruders moving along orthogonal paths, and strong barrier coverage, which guarantees to detect intruders no matter what crossing paths they choose. Liu et al. [14] studied the strong barrier coverage of a randomly-deployed sensor network on a long irregular strip region. They showed that in a rectangular area of width ω and length ℓ with the relation $\omega = \Omega(\log \ell)$, if the sensor density reaches a certain value, then there exist, with high probability, multiple disjoint sensor barriers across the entire length of the area such that intruders cannot cross the area undetected; On the other hand, if $\omega = o(\log \ell)$, then with high probability there is a crossing path not covered by any sensor regardless of the sensor density. He et al. [6] presented a condition under which line-based deployment is suboptimal, and proposed a new deployment approach named curve-based deployment. Wang et al. [17] explored the effects of location errors on barrier coverage on a 2D plane by considering two scenarios (i.e. only stationary nodes have location errors, stationary and mobile nodes both have location errors), and proposed a fault-tolerant weighted barrier graph to model the barrier coverage formation problem. Dewitt and Shi [3] incorporated energy harvesting into the barrier coverage problem, and studied the lifetime issues of the k-barrier coverage problem for energy harvesting WSNs.

With the advances of technology, sensor mobility has been incorporated into sensor deployment framework [18], which offers more flexibility for designing more efficient sensor deployment strategies to solve coverage problem in WSNs. Li and Shen [13] studied the 2D MinMax barrier coverage problem of moving n sensors in a 2D plane to form a barrier coverage while minimizing the maximum sensor movement for the sake of balancing battery power consumption. Dobrev et al. [4] studied three optimization problems related to the movement of sensors to achieve weak barrier coverage, i.e., minimizing the number of sensors moved, minimizing the average distance moved by the sensors, and minimizing the maximum distance moved by the sensors. Silvestri and Goss [16] proposed an original algorithm called MobiBar, which has the capability of constructing k-barrier coverage in WSNs, self-reconfiguration and self-healing. Saipulla et al. [15] explored the fundamental limits of sensor mobility on barrier coverage, and presented a sensor mobility scheme that constructs the maximum number of sensor barriers with the minimum sensor moving distance. Li and Shen [12] studied the 2D MinMax problem of barrier coverage in which the barrier is a line segment in a 2D plane and the sensors are initially resided on this plane.

However, none of the above works considered weak k-barrier coverage problem in 3D UWSNs. The most related work to ours is by Barr et al. [1] who considered constructing underwater sensor barriers to thwart illegal intrusion of submarines, but they still did not consider weak k-barrier coverage problem in UWSNs. Therefore, we, in this work, study weak k-barrier coverage problem in UWSNs and hope to provide insights into further researches in 3D wireless sensor networks.

3   Network Model and Problem Statement

3.1   Network Model

We consider an underwater wireless sensor network consisting of a set of mobile sensors deployed in 3D underwater environment to detect intruders crossing the deployed region. For the sake of easy presentation and obtaining analytical results to provide insights, we model the underwater deployment region as a cuboid of size $l \times w \times h$, where l, w, and h denote the length, the width, and the height of the cuboid, respectively. We assume that sensors are homogeneous and have the same sensing radius r. Since we only consider coverage problem in this paper, we also assume that the sensor's communication range is reasonably large that guarantees the connectivity of the deployed UWSN. For simplicity, we assume an ideal 0/1 sphere sensing model that an object within (outside) a sensor's sensing sphere is detected by the sensor with probability one (zero). All sensors are able to identify their current locations by existing underwater localization algorithms [2, 7], let coordinates (x_i, y_i, z_i) denote the position of sensor s_i, and each sensor is able to reposition itself from its initial position to another specified position in 3D underwater environment.

In the initial configuration, sensors are uniformly and independently distributed in the cuboid. Without loss of generality, we assume that the illegal intruders move along the direction of cuboid length with the start point at the cuboid's left face and end point at the cuboid's right face, as shown in Fig. 1, O_1 and O_2 denote intruder1 and intruder2, respectively.

3.2   Problem Statement

In 2D WSNs, weak k-barrier coverage can guarantee that the intruders crossing the deployment region along orthogonal crossing paths are detected by at least k sensor(s), where the orthogonal crossing paths are straight lines perpendicular to the long side of the deployment region, as shown in Fig. 2.

Similarly, in 3D UWSNs, weak k-barrier coverage guarantees that the intruders crossing the 3D deployment region along orthogonal crossing paths are detected by at least k sensor(s). Since we assume that the intruders cross the monitored region along the orthogonal crossing paths with start point at the cuboid's left face and end point at the cuboid's right face, the orthogonal crossing paths are straight lines perpendicular to the cuboid's right face, as shown in Fig. 3.

According to the aforementioned assumptions, we give the formal definition of our problems as follows.

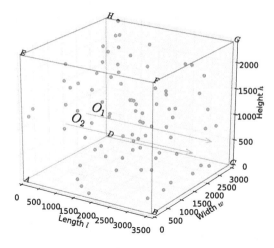

Fig. 1. The 3D underwater space is modeled as a cuboid, sensors are uniformly and independently distributed in the cuboid.

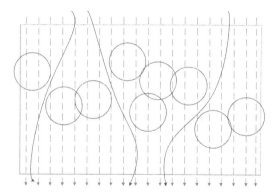

Fig. 2. An illustration of weak k-barrier coverage in 2D WSNs (in this figure, k = 1): the intruders crossing the deployment region along orthogonal crossing paths (dash lines) are detected by at least 1 sensor. However, uncovered paths (solid curves) may exist.

Problem 1. 3D MiniSum underwater weak k-barrier coverage construction problem (3D MiniSum underwater weak k-BC construction problem). Given an underwater cuboid region of size $l \times w \times h$, where l, w and h denote the length, the width and the height of the cuboid, respectively. What is the minimum number of sensors required to construct weak k-barrier coverage in the underwater cuboid region? And if the minimum sensors are distributed in this cuboid and compose a connected network, how to construct weak k-barrier coverage with these sensors while minimizing the total movement distance of all sensors?

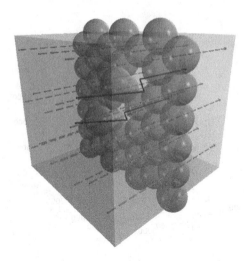

Fig. 3. An illustration of weak k-barrier coverage in 3D UWSNs (in this figure, k = 1): the intruders crossing the 3D deployment region along orthogonal crossing paths (dash lines) are detected by at least 1 sensor. However, uncovered paths (solid curves) may exist.

4   Problem Analysis

In this section, we analyze how to construct weak k-barrier coverage with minimum sensors while minimizing the total movement distance of all sensors in UWSNs. To achieve our goal, we first need to derive the minimum number of required sensors, and then consider how to construct underwater weak k-barrier coverage with the minimum sensors while minimizing the total movement distance of all sensors.

4.1   Minimum Number of Required Mobile Sensors

In 2D WSNs, a barrier is a chain of sensors from one end of the region to the other end with overlapping sensing zones of adjacent sensors, the minimum number of sensors for constructing one barrier is $\lceil \frac{l}{2 \times r} \rceil$, where l is the length of the deployment region. When considering k-barrier coverage problem, we need at least $k \times \lceil \frac{l}{2 \times r} \rceil$ sensors for constructing weak k-barrier coverage. In 3D UWSNs, however, a barrier is a set of sensors with overlapping sensing zones of adjacent sensors that covers an entire (curly) surface that cuts across the 3D space. Notice that if the cuboid's right face is completely k-covered after perpendicularly mapping sensors onto the cuboid's right face, then the deployed UWSN provides weak k-barrier coverage. Therefore, in our work, the minimum number of required sensors for constructing weak k-barrier coverage in a cuboid of size $l \times w \times h$ is equal to the minimum number of circles with radius r that completely k-cover a rectangle of size $w \times h$, where w and h are the width and the height of the cuboid, respectively. Hence, our work in this subsection is to answer the question 'What is the minimum number of required circles with radius r to completely k-covered a rectangle of size $w \times h$?".

For the sake of simplicity, we first derive the minimum number n of required circles with radius r that completely 1-cover a rectangle of size $w \times h$, and for the k-coverage problem, it is straightforward that the minimum number is equal to $k \times n$. Actually, in term of the optimal deployment problem in 2D plane, Kershner [8] has proved that the regular triangular tessellation is the optimal tessellation which results in a set of regular hexagons completely cover a 2D plane without any overlap. In the context of our work, we can set the circumradius of regular hexagon equal to the sensor's sensing radius, if we move the sensors to the center points of regular hexagons and make each center point occupied by at least one sensor, then the rectangle is completely 1-covered by these sensors, as shown in Fig. 4. Hence, the minimum number of required sensors for completely 1-covered a rectangle is equal to the number of regular hexagons that completely 1-covered this rectangle without overlap. We have Theorem 1 as follows:

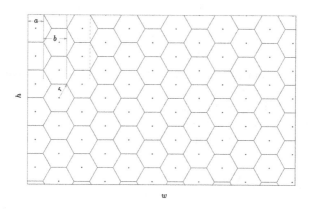

Fig. 4. A rectangle of size $w \times h$ completely covered with minimum regular hexagons.

Theorem 1. *The minimum number of required regular hexagons with circum-radius r that completely 1-cover a rectangle of size $w \times h$ is:*

$$f_s(w,h,r) = \lceil \frac{h}{r \times \sqrt{3}} \rceil \times \lceil \frac{\lceil \frac{2 \times (w-r)}{3 \times r} \rceil + 1}{2} \rceil + $$
$$(\lceil \frac{h - \frac{r\sqrt{3}}{2}}{r \times \sqrt{3}} \rceil + 1) \times \lfloor \frac{\lceil \frac{2 \times (w-r)}{3 \times r} \rceil + 1}{2} \rfloor. \tag{1}$$

Proof. As shown in Fig. 4, given a rectangle of size $w \times h$, in the length direction, we divide the rectangle into C columns, the first column width $a = r$, the 2th~ $(C-1)$th column width $b = \frac{3 \times r}{2}$, and the last column width is included in interval $(0, b]$, then the number of columns is:

$$f_c(w,r) = \lceil \frac{2 \times (w-r)}{3 \times r} \rceil + 1. \tag{2}$$

In the width direction, the number of rows of odd-number columns is:

$$f_o(h, r) = \lceil \frac{h}{r \times \sqrt{3}} \rceil. \tag{3}$$

The number of rows of even-number columns is:

$$f_e(h, r) = \lceil \frac{h - \frac{r \times \sqrt{3}}{2}}{r \times \sqrt{3}} \rceil + 1. \tag{4}$$

Combining Eqs. (2), (3) and (4), we obtain the minimum number of regular hexagons:

$$\begin{aligned} f_s(w, h, r) &= f_o(h, r) \times \lceil \frac{f_c(w, r)}{2} \rceil + f_e(h, r) \times \lfloor \frac{f_c(w, r)}{2} \rfloor \\ &= \lceil \frac{h}{r \times \sqrt{3}} \rceil \times \lceil \frac{\lceil \frac{2 \times (w-r)}{3 \times r} \rceil + 1}{2} \rceil + \\ &\quad (\lceil \frac{h - \frac{r \times \sqrt{3}}{2}}{r \times \sqrt{3}} \rceil + 1) \times \lfloor \frac{\lceil \frac{2 \times (w-r)}{3 \times r} \rceil + 1}{2} \rfloor. \end{aligned} \tag{5}$$

The above derivation obtains the minimum number $f_s(w, h, r)$ of required sensors that completely 1-covered a rectangle of size $w \times h$. Therefore, for the weak k-barrier coverage problem in our work, the minimum number of sensors required for weak k-barrier coverage is equal to $k \times f_s(w, h, r)$, where r is the sensor's sensing radius, w and h are the width and the height of the cuboid.

4.2   Construction of Weak k-Barrier Coverage

In the following, we consider a scenario where $k \times f_s(w, h, r)$ mobile sensors are distributed in 3D underwater environment and compose a connected network. As forementioned assumption, we model the 3D underwater environment as a cuboid of size $l \times w \times h$. Our goal is to construct underwater weak k-barrier coverage with these sensors while minimizing the total movement distance of all sensors. One naive solution is to move all these sensors to the cuboid's right (or left) face and make it k-covered. However, this solution may not be practical because it will waste enormous energy of the deployed sensors. According to the analysis in previous subsection, by adopting the optimal regular triangular tessellation which results in a set of regular hexagons completely cover a 2D plane without any overlap, we obtain the minimum number of required sensors, thus, if each regular hexagon center point is occupied by at least k virtual sensors (i.e., the sensor's projections), then the deployed UWSN provides weak k-barrier coverage. For clearness, we implement the optimal regular triangular tessellation on the cuboid's right and left face, and connect two hexagon center points, both of which have the same y, z-coordinate, with a line, as shown in Fig. 5. If each line contains at least k sensors, then the deployed UWSN provides weak k-barrier coverage. Therefore, in order to minimize the total movement distance of all sensors and thus minimizing the total energy consumption of the deployed UWSN, the optimal manner is to move the sensors to the lines in parallel, this energy-efficient movement manner is referred to as parallel movement.

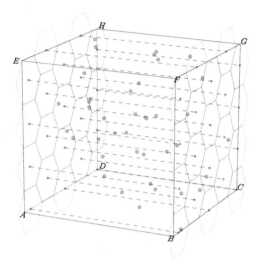

Fig. 5. Moving sensors to dash lines by parallel movement manner. If each line contains at least k sensors, then the deployed UWSN provides weak k-barrier coverage.

In the context of our work, the parallel movement means that each sensor's x-coordinate will not be changed in the moving process. Based on the parallel movement manner, we present a novel scheme with three steps to construct weak k-barrier coverage in UWSNs as follows.

1. Perpendicularly mapping sensors distributed in a cuboid onto the cuboid's right face.
2. In the cuboid's right face, we adopt the optimal deployment proved by Kershner's in [8] and compute the coordinates of hexagon center points $P'' = \{p_1'', p_2'', ..., p_m''\}$, as shown in Fig. 4.
3. We refer to the sensor's projections on the cuboid's right face as the virtual sensors. Then we devise an assignment algorithm to assign all virtual sensors to P'', and make each P_i'' contains and only contains k virtual sensors. Finally, for each sensor s_i distributed in the cuboid with coordinates (x_i, y_i, z_i), if its projection p_i' on the cuboid's right face finally is assigned to p_j'' whose coordinates are (x_j'', y_j'', z_j''), then we move sensor s_i from its initial position (x_i, y_i, z_i) to its final position (x_i, y_j'', z_j'').

Obviously, the core of the proposed scheme in above are (1) computation of the coordinates of hexagon center points $P'' = \{p_1'', p_2'', ..., p_m''\}$ in step 2; (2) and the sensor assignment algorithm in step 3.

In the following, we derive the coordinates of hexagon center points in the cuboid's right face. In order not to cause confusion, we consider all point's positions in 3D Cartesian coordinate system.

For each center point of regular hexagon in the cuboid's right face, its x-coordinate equals to the cuboid's length l. For y-coordinate and z-coordinate, we divide a rectangle of size $w \times h$ into several columns, the number of columns can be obtained via Eq. (2).

Notice that, the first column width $a = r$, the 2-th $\sim (f_c(w, r) - 1)$-th column width $b = \frac{3 \times r}{2}$, and the last column width $\in (0, b]$. It is straightforward to derive y-coordinate of each column via Eq. (6), where i denote the i-th column.

$$f_y(w, r, i) = \begin{cases} \frac{r}{2}, & i = 0 \\ \frac{r}{2} + \frac{i \times r}{2}, & \frac{r}{2} + \frac{(f_c(w,r)-1) \times r}{2} < w \\ w, & \frac{r}{2} + \frac{(f_c(w,r)-1) \times r}{2} \geq w. \end{cases} \tag{6}$$

For odd-number columns, let h denote the cuboid height, r denote the sensing radius, and j denote the j-th row, where $0 \leq j < \lceil \frac{h}{r \times \sqrt{3}} \rceil$, we have,

$$f_z(w, r, j) = \begin{cases} h - \frac{r \times \sqrt{3}}{2} - j \times r \times \sqrt{3}, & \text{odd-number columns} \\ h - j \times r \times \sqrt{3}, & \text{even-number columns.} \end{cases} \tag{7}$$

Consequently, by combining the x-coordinate l of the cuboid's right face, we obtain the coordinates $(l, f_y(w, r, i), f_z(w, r, j))$ of all center points of the regular hexagons on the cuboid's right face.

As described in step 3 in the above proposed scheme, after deriving the coordinates of hexagon center points P'', we need to devise an assignment algorithm to assign all virtual sensors (i.e., the projection points P' of sensors) to the final positions (i.e., the hexagon center points P''), and make each final position contains and only contains k virtual sensors while minimizing the total distance between virtual sensors and their corresponding final positions. Actually, this is a classical assignment problem, many approaches have been proposed in the existing works. In this work, to minimize the total movement distance of all sensors, we solve our assignment problem based on Hungarian Method [9]. The proposed algorithm will be given in Sect. 5.

5   Constructing Weak k-Barrier Coverage

In this section, we present how to construct underwater weak k-barrier coverage with minimum mobile sensors while minimizing the total movement distance of all sensors in UWSNs.

5.1   Algorithm Summary

According to the analysis in Sect. 4, mapping sensors distributed in 3D space to 2D plane could help us solve the underwater weak k-barrier construction problem efficiently, and after mapping sensors to the cuboid's right face, the final positions of all sensors are actually the center points of regular hexagons, as shown in Fig. 4. Thus, to construct weak k-barrier coverage with minimum mobile sensors while minimizing the total movement distance of all sensors, we devise an algorithm called HMB-SAA with three steps as follows.

1. Perpendicularly mapping sensors distributed in a cuboid onto the cuboid's right face. Let $P = \{p_1, p_2, ..., p_n\}$ denote the positions of all sensors distributed in the cuboid, (x_i, y_i, z_i) denote the coordinates of position p_i of sensor s_i. For each sensor s_i, the coordinates of its projection on the cuboid's right face are (l, y_i, z_i), where l is the length of the cuboid. Thus, after mapping all sensors to the cuboid's right face, we can easily get their positions $P' = \{p'_1, p'_2, ..., p'_n\}$ on the cuboid's right face by replacing the x-coordinate with cuboid length l. For example, if the coordinates of position p_i of sensor s_i are (x_i, y_i, z_i), then the coordinates of position p'_i, which is the projection position of s_i on the cuboid's right face, are (l, y_i, z_i).
2. In the cuboid's right face, we adopt the optimal deployment proved by Kershner's in [8] and compute the coordinates of hexagon center points $P'' = \{p''_1, p''_2, ..., p''_m\}$, where m is the minimum number of sensors for completely 1-cover the cuboid's right face, as shown in Fig. 4. Notice that in this work, we aim to construct weak k-barrier coverage with minimum sensors, hence, $k = \frac{n}{m}$.
3. In this step, we refer to point p'_i in P' as virtual sensor o'_i, and the points in P'' are the final positions of all virtual sensors, then we devise an assignment algorithm to assign all virtual sensors to the final positions, and make each final position contains and only contains k virtual sensors while minimizing the total distance between virtual sensors and their corresponding final positions. Finally, for each sensor s_i distributed in the cuboid with coordinates (x_i, y_i, z_i), if its projection p'_i on the cuboid's right face is finally assigned to p''_i whose coordinates are (x''_i, y''_i, z''_i), then we move sensor s_i from its initial position (x_i, y_i, z_i) to the final position (x_i, y''_i, z''_i) by parallel movement manner. After all sensors reach their final positions, the weak k-barrier coverage is constructed in UWSNs.

Obviously, step 1 is straightforward, and the coordinates of hexagon center points P'' in step 2 have been derived in Sect. 4.2. Therefore, we focus on how to assign all virtual sensors to the final positions in step 3 in next subsection.

5.2   Sensor Assignment

In this section, we aim to assign n sensors to m final positions and make each final position contains and only contains $k = \frac{n}{m}$ sensors while minimizing the total movement distance of all sensors. To achieve our goal, we present a Hungarian Method-based sensor assignment algorithm (HMB-SAA), as shown in Algorithm 1.

Algorithm 1. $HMB - SAA(P, w, h, r)$

Require:

Given a set of sensors $S = \{s_1, s_2, ..., s_n\}$ with initial positions $P = \{p_1, p_2, ..., p_n\}$ distributed in a cuboid of size $l \times w \times h$, where l, w and h denote the length, the width and the height of the cuboid, respectively. Let r denote the sensing radius.

Ensure:

Weak k-barrier coverage in UWSNs.

1: $P' \leftarrow map3dPointTo2d(P)$;

2: $n \leftarrow len(P')$;

3: $P'' \leftarrow the\ coordinates\ of\ hexagon\ center\ points$;

4: $m \leftarrow len(P'')$;

5: $k \leftarrow \frac{n}{m}$;

6: **for** $i = 0 \rightarrow m - 1$ **do**

7:     **for** $j = 0 \rightarrow k$ **do**

8:         $P_{ext} \leftarrow P''[i]$;

9:     **end for**

10: **end for**

11: $dm \leftarrow genDMatrix(P', P_{ext})$;

12: $L_{match} \leftarrow Hungarian(dm)$;

13: $n \leftarrow len(L_{match})$;

14: **for** $i = 0 \rightarrow n$ **do**

15:     $p'_t = L_{match}[i][0]$;

16:     $p''_t = L_{match}[i][1]$;

17:     In P, find sensor s whose projection is p'_t and then move s from its initial positions (x, y, z) to $(x, p''_t[1],\ p''_t[2])$;

18: **end for**

In Algorithm 1, we first perpendicularly map sensors distributed in a cuboid to the cuboid's right face in line 1, and then derive the coordinates of hexagon center points in line 3 (the deriving approach has been introduced in Sect. 4). Because the size m of P'' is not equal to the size n of P', that means Hungarian Method cannot be applied directly, we extend P'' to list P_{ext} of size n from line 6 to line 10, $P_{ext} = \{p''_1, p''_1, ...p''_2, p''_2, ..., p''_m, p''_m\}$, notice that there are k identical p''_i in list P_{ext}. In line 11, the function $genDMatrix(P', P_{ext})$ generates a distance matrix in which every value denotes the distance between two points in P' and P_{ext}, respectively. Line 10 calls Hungarian Method and returns the matching relationship L_{match} between sensors' projections and final positions, list L_{match} is similar to $\{((x'_0, y'_0, z'_0), (x''_0, y''_0, z''_0)), \ldots, ((x'_n, y'_n, z'_n), (x''_n, y''_n, z''_n))\}$. Finally, from line 14 to line 19, for each sensor s_i, whose 3D coordinates are (x_i, y_i, z_i), if its projection's corresponding final position in L_{match} are p''_i whose coordinates denoted by (x''_i, y''_i, z''_i), then we move it from its initial coordinates (x_i, y_i, z_i) to its final position (x_i, y''_j, z''_j) by parallel movement manner. After all sensors reach their final positions, the weak k-barrier coverage is constructed in UWSNs.

6   Performance Evaluation

In this section, we evaluate the performance of our proposed algorithm HMB-SAA by comparing with a greedy algorithm. We implemented the proposed approach in a Python-based simulator, each data point of our experiment results is an average value of the data collected by running the experiments 100 times.

6.1   Simulation Setup

According to aforementioned assumptions and the related context of our work, we setup the simulation environment as follows:

1. The underwater space where weak k-barrier to be constructed is modeled as a cuboid of length $l = 4000$ m, width $w = 3600$ m, and height $h = 3000$ m, respectively.
2. Initially, sensors are uniformly and independently distributed in the cuboid. All sensors have the same sensing range r, and they are able to relocate themselves from their initial positions to any specified positions in the cuboid.
3. Our goal is to construct weak k-barrier with minimum sensors, the minimum number of required sensors can be obtained via function $f_s(w, h, r)$ in Eq. (1). In the simulation experiments, we vary the sensing range r from 150 m to 300 m.

6.2   Simulation Results

As far as we know, we are the first to solve the 3D MiniSum weak-barrier coverage problem for the case that constructing weak k-barrier with minimum sensors. There is no prior work to be compared with directly. We choose a simple greedy algorithm, called GreedyMatch, in which each sensor move to the nearest final position by parallel movement manner.

First, the relationship between sensing radius and minimum number of required sensors is studied by considering two cases where $k = 2$ and 4, respectively. In each case, the sensing range r of sensor varies from 150 m to 300 m with a step 15 m. As shown in Fig. 6, when $k = 2$, the minimum number of required sensors varies from 424 to 116, and when $k = 4$, the minimum number of required sensors varies from 848 to 232. That is to say, the larger the sensing range, the less the minimum number of required sensors. Furthermore, the minimum number of required sensors when $k = 4$ is two times of that when $k = 2$, that is to say, the larger k, the more the minimum number of required sensors.

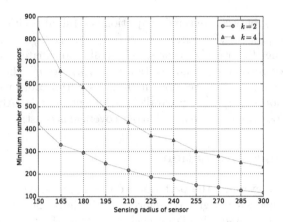

Fig. 6. Minimum number of required sensors versus sensing radius of sensor.

Then we study the impact of the sensing radius of sensor on the total movement distance of all sensors. In this scenario, we consider two cases, $k = 2, 4$, respectively. And in each case, the sensing radius r of sensor varies from 150 m to 300 m with a step 15 m. As Fig. 7 illustrates, as the sensing radius increases, the total movement distance tends to decrease. That is because we need less sensors when sensing radius increases. Furthermore, we also observe that as the parameter k increases, the total movement distance increases. That's because larger k means requirement of more sensors, thus the total movement distance increases. Overall, the total movement distance by our proposed algorithm HMB-SAA is always smaller than that by the GreedyMatch, which shows that our proposed algorithm outperforms the GreedyMatch.

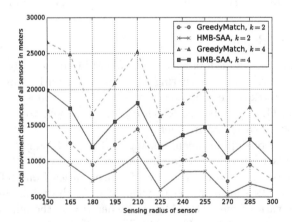

Fig. 7. Total movement distance of all sensors versus sensing radius of sensor.

7 Conclusion

In this paper, we investigated the underwater weak k-barrier coverage problem in 3D UWSNs. We first obtained the minimum number of sensors required for constructing underwater weak k-barrier coverage, and then analyzed how to construct 3D underwater weak k-barrier coverage with the minimum sensors. Furthermore, we propose a parallel movement manner, based on which an effective algorithm is presented for constructing underwater weak k-barrier coverage with minimum sensors while minimizing the total movement distance of all sensors in underwater wireless sensor networks. Extensive simulation results validate the correctness of our analysis, and show that the proposed algorithm HMB-SAA outperforms the GreedyMatch algorithm. To the best of our knowledge, this is the first result for the weak k-barrier coverage problem in UWSNs.

References

1. Barr, S.J., Wang, J., Liu, B.: An efficient method for constructing underwater sensor barriers. J. Commun. **6**(5), 370–383 (2011)
2. Das, A.P., Thampi, S.M.: Fault-resilient localization for underwater sensor networks. Ad Hoc Netw. **55**, 132–142 (2017)
3. DeWitt, J., Shi, H.: Barrier coverage in energy harvesting sensor networks. Ad Hoc Netw. **56**, 72–83 (2016)
4. Dobrev, S., et al.: Weak coverage of a rectangular barrier. CoRR abs/1701.07294 (2017). http://arxiv.org/abs/1701.07294
5. Gage, D.W.: Command control for many-robot systems. Technical report, DTIC Document (1992)
6. He, S., Gong, X., Zhang, J., Chen, J., Sun, Y.: Barrier coverage in wireless sensor networks: from lined-based to curve-based deployment. In: Proceedings of the IEEE INFOCOM 2013, 4–19 April 2013, Turin, Italy, pp. 470–474 (2013). https://doi.org/10.1109/INFCOM.2013.6566817
7. Isik, M.T., Akan, O.B.: A three dimensional localization algorithm for underwater acoustic sensor networks. IEEE Trans. Wirel. Commun. **8**(9) (2009)
8. Kershner, R.: The number of circles covering a set. Am. J. Math. **61**(3), 665–671 (1939)
9. Kuhn, H.W.: The Hungarian method for the assignment problem. Naval Res. Logistics Q. **2**(1–2), 83–97 (1955)
10. Kumar, S., Lai, T., Arora, A.: Barrier coverage with wireless sensors. In: Proceedings of the 11th Annual International Conference on Mobile Computing and Networking, MOBICOM 2005, 28 August–2 September 2005, Cologne, Germany, pp. 284–298 (2005). https://doi.org/10.1145/1080829.1080859
11. Li, L., Zhang, B., Shen, X., Zheng, J., Yao, Z.: A study on the weak barrier coverage problem in wireless sensor networks. Comput. Netw. **55**(3), 711–721 (2011)
12. Li, S., Shen, H.: Minimizing the maximum sensor movement for barrier coverage in the plane. In: 2015 IEEE Conference on Computer Communications, INFOCOM 2015, 26 April–1 May 2015, Kowloon, Hong Kong, pp. 244–252 (2015). https://doi.org/10.1109/INFOCOM.2015.7218388
13. Li, S., Shen, H.: Minimizing the maximum sensor movement for barrier coverage in the plane. In: 2015 IEEE Conference on Computer Communications (INFOCOM), pp. 244–252. IEEE (2015)

14. Liu, B., Dousse, O., Wang, J., Saipulla, A.: Strong barrier coverage of wireless sensor networks. In: Proceedings of the 9th ACM International Symposium on Mobile Ad Hoc Networking and Computing, MobiHoc 2008, 26–30 May 2008, Hong Kong, China, pp. 411–420 (2008). https://doi.org/10.1145/1374618.1374673
15. Saipulla, A., Liu, B., Xing, G., Fu, X., Wang, J.: Barrier coverage with sensors of limited mobility. In: Proceedings of the 11th ACM International Symposium on Mobile Ad Hoc Networking and Computing, MobiHoc 2010, 20–24 September 2010, Chicago, IL, USA, pp. 201–210 (2010). https://doi.org/10.1145/1860093.1860121
16. Silvestri, S., Goss, K.: Mobibar: an autonomous deployment algorithm for barrier coverage with mobile sensors. Ad Hoc Netw. **54**, 111–129 (2017)
17. Wang, Z., Chen, H., Cao, Q., Qi, H., Wang, Z., Wang, Q.: Achieving location error tolerant barrier coverage for wireless sensor networks. Comput. Netw. **112**, 314–328 (2017)
18. Yang, G., Qiao, D.: Multi-round sensor deployment for guaranteed barrier coverage. In: 29th IEEE International Conference on Computer Communications, Joint Conference of the IEEE Computer and Communications Societies, INFOCOM 2010, 15–19 March 2010, San Diego, CA, USA, pp. 2462–2470 (2010). https://doi.org/10.1109/INFOCOM.2010.5462074

Multi-objective Heuristic Multicast Routing Algorithm in NDN

Xuming An[1,2], Yu Zhang[1,2(✉)], Yanxiang Chen[1], and Yadong Wang[1]

[1] Beijing Institute of Technology, Beijing, China
{2120160725,yuzhang}@bit.edu.cn
[2] The Science and Technology on Information Transmission and Dissemination in
Communication Networks Laboratory, CETC-54, Shijiazhuang, China

Abstract. NDN naturally supports multicast better than the traditional Internet, and multicast plays an important role in NDN. Most researchs on multicast routing algorithms are focused on cost optimization without taking node cache into account. This paper constructs a mathematical model for joint optimization of delay and cost, which is more flexible in describing NDN than adding delay as a constraint to the model. Then, the heuristic multicast algorithm considering node cache for this model is proposed. Last, we analyze the delay performance of the algorithm by comparing it with the exact Algorithm and the classical STMPH algorithm.

Keywords: Multicast algorithm · Delay · Joint optimization · NDN

1 Introduction

Named Data Networking (NDN) [8]is a new network architecture. A user sends out an Interest packet whose name identifies the desired data. Routers forward the Interest packet based on their names. A Data packet with the matching name is returned to the requesting consumer. Because of node caching and explicitly naming data, NDN can realize multicast easier than TCP/IP.

There are several routing protocols for NDN: such as NLSR [3], CRoS [5], and OSPFN [4], but as best we know, there is no multicast algorithm in these protocols. The multicast routing problem can be interpreted as solving steiner tree, and there are many algorithms: the exact algorithm, such as STEA [7], TEA [7], DPA [7], LRA [7], the computation time of these algorithm increase exponentially as the network size increases; the approximate algorithm, authors of [1] propose a family of algorithms that achieves an approximation ratio of $i(i-1)k^{1/i}$ in time $O(n^i k^{2i})$ for any fixed $i > 1$, where k is the amount of destinations; the heuristic algorithm, such as KMB [7], MPH [7], ADH [7], KPP [6], BSMA [6], the

Supported by Science and Technology on Communication Networks Laboratory Foundation Project and Aerospace Field Pre-research Foundation Project (060501).

B. Li et al. (Eds.): IoTaaS 2018, LNICST 271, pp. 211–214, 2019.
https://doi.org/10.1007/978-3-030-14657-3_20

approximate algorithm and heuristic algorithm belong to the polynomial-time algorithm, and the calculated result is close to the optimal solution. Although KPP and other algorithms add delay constraints, this approach cann't describe NDN flexibly due to that the services in NDN have different QoS requirements. Therefore, this paper constructs a model for joint optimization of delay and cost to describe NDN more flexibly. Moreover, because there is no delay performance analysis for these algorithms, we analyze the delay performance of the algorithm.

2   Mathematical Model

In this section, we construct a mathematical model of joint optimization delay and cost in network:

$$\min \quad \lambda_1 * \sum_{arc(i,j)} c[i,j] * y[i,j] + \lambda_2 * D$$

$$\Lambda_+ = \{\lambda | \lambda \geq 0, \lambda_1 + \lambda_2 = 1\}$$

$$D = \sum_{arc(i,j)} x[i,j] * d[i,j]$$

$$\text{s.t.} \quad \sum_i x[i,j] - \sum_i x[j,i] = \begin{cases} K[s,i], & \text{if } j \neq s; \\ -\sum_t K[s,t], & \text{if } j = s. \end{cases} \quad (1)$$

$$BigM * y[i,j] \geq x[i,j]$$

$$y[i,j] \leq x[i,j] \quad (2)$$

$$y[i,j] * demand \leq cap[i,j] \quad (3)$$

$$x[i,j] \in \mathbb{N}$$

$$y[i,j] \in \{0,1\} \quad (4)$$

In this model, each link (i,j) has three parameters, namely bandwidth $b[i,j]$, cost $c[i,j]$ and delay $d[i,j]$. Moreover, the binary variable $K[s,t]$ indicates if the source s has data transmission to node t(1/0).

Formula 1 guarantees the flow conservation of the source node, the destination node, and the relay node, where the variable $x[i,j]$ represents the number of destination nodes that the content on link (i,j) flow to. Formula 3 ensures that the total bandwidth utilized on each link does not exceed its available bandwidth, where the variable *demand* denotes the requested bandwidth of the content. The formula 2 ensures that binary variable $y[i,j]$ controls the inclusion of link (i,j) in the solution. The values of λ_1 and λ_2 can be set according to the service type.

3   Multicast Algorithm

The steiner tree problem is NP-complete [7], the exact algorithm does not belong to the polynomial-time algorithm, it is impractical to solve route in NDN with the exact algorithm. Futhermore, the approximate algorithm and heuristic algorithm don't consider node cache. Thus, when designing multicast algorithm, we should consider node caching in NDN. The MOH (Multi-objective heuristic multicast routing algorithm) algorithm we proposed as follow:

Given a directed graph $G = (V, A)$, the source node S, dest set $M \subset V$, the requested bandwidth of the content: bw. Moreover, we define the link (i, j)'s distance: $\lambda_1 * c[i, j] + \lambda_2 * d[i, j]$. The multicast routing problem is that of finding a directed Steiner tree with S as the root $T = (V_t, A_t)$:

Step1: Start from the source node S, $k = 1$, taking S as T_1, then $T_k = T_1$, $V_k = V_1 = \{S\}$, M_k is the set of the destinations. Calculate the shortest path from the spanning tree T_k to all destinations in M_k. If the relay node has less than bw cache space, delete the relay node and look for the suboptimal path; if the available bandwidth of the link is less than bw, the link is deleted; Record the path and distance of the spanning tree T_k to each destination in M_k.

Step2: The nearest (the distance from the spanning tree to destiantion is the shortest) destination in M_k is selected, then the destination and all relay nodes that are in the shoretest path from the spanning tree T_k to the destination are added to T_k, after that, the destination is deleted from M_k.

Step3: The following procedure is performed for the new nodes in (2): Calculate the shortest distance from the new nodes to the remaining destiantions in M_k. If this distance is less than the remaining destiantion's distance from the spanning tree T_k, then this distance is taken as the distance from the spanning tree T_k to the destination, meanwhile the shortest path from the spanning tree T_k to the destiantion is also recorded.

Step4: Repeat **step2** and **step3** until M_k is empty. When the set is empty, all the destiantions are in the spanning tree T_k. T_k is the multicast tree that we wanted.

4   Performance Analysis

In order to effectively evaluate the delay performance of multicast algorithm, we simulate three algorithms: the MOH algorithm we proposed, the exact algorithm, the classical STMPH algorithm [2], by using a network models. In this models, the network size is fixed at 100 nodes and the node's degree are 10. Figure 1(a) shows that the average delay vs the multicasting group size. Figure 1(b) show the deviation of the two heuristic algorithms from the optimal value in the above three scenarios.

From above figures, we can see that when the scale of network is fixed, with the increase of group size, the two deviation become more obvious. After analyzing the two algorithms, we believe that the reason for this phenomenon should be: each time a terminal added, it is always added to the tree through the shortest path in the two algorithms. It is possible to miss those paths that are not the shortest path or whose cost are also the shortest but search order is later.

These paths are likely to implement reduce the overall delay of the entire multicast group. On the other hand, the STMPH's deviation is greater than MOH's. This is due to the fact that when you add a terminal, STMPH connects the new multicast terminal to the terminal or the source node in the multicast tree. This way only considers the shortest path between the terminals that already in the tree and the remaining multicast terminals. Thus, STMPH's search depth is far less than MOH's, which results in missing more paths. Moreover, because of the way of STMPH adding new terminal, it is likely to lead multiple multicast terminals to be in a chain structure on one path. This will result in an increase of overall transmission delay. Therefore, our proposed algorithm's delay performance better than the classcial STMPH algorithm.

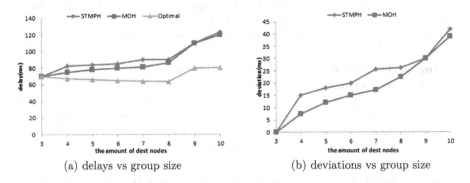

(a) delays vs group size   (b) deviations vs group size

Fig. 1. Delay performance when the node's degree is 10

References

1. Charikar, M.: Approximation algorithms for directed Steiner problems, pp. 192–200 (1998)
2. Li, H., Yu, J., Xie, W.: Locally searching minimum path cost heuristic. Acta Electron. Sinica **28**(5), 92–95 (2000)
3. Hoque, A.K.M., Amin, S.O., Alyyan, A., Zhang, B., Zhang, L., Wang, L.: NLSR: named-data link state routing protocol. In: ACM SIGCOMM Workshop on Information-Centric NETWORKING, pp. 15–20 (2013)
4. Wang, L., Hoque, M., Yi, C., Alyyan, A., Zhang, B.: OSPFN: an OSPF based routing protocol for named data networking (2012)
5. Torres, J.V., Ferraz, L.H.G., Duarte, O.C.M.B.: Controller-based routing scheme for named data network (2013)
6. Wang, B., Hou, J.C.: Multicast routing and its QoS extension: problems, algorithms, and protocols. IEEE Netw. **14**(1), 22–36 (2000)
7. Winter, P.: Steiner problem in networks. In: Conference on Computer Networks (1987)
8. Zhang, L., et al.: Named data networking (NDN) project, vol. 1892, no. 1, pp. 227–234 (2014)

Performance Evaluation of Dual Connectivity in Non-standalone 5G IoT Networks

Liangrong Zhao[1,2($\boxtimes$)], Zhangdui Zhong[1], Yong Niu[1], and Hong Zhou[2]

[1] State Key Laboratory of Rail Traffic Control and Safety,
Beijing Jiaotong University, Beijing, China
`liangrongzhao@bjtu.edu.cn`
[2] Nokia Bell Labs, Beijing, China

Abstract. Internet of Things (IoT) is considered as a major emerging technology with huge potential in social efficiency and civilian market. However the scarce available spectrum poses obstacles for the increasing amount of IoT devices. The coming 5G new radio (NR) of next generation is expected to exploit the spectrum of super high frequency by utilizing a heterogeneous network comprising the currently Long Term Evolution (LTE) systems and the 5G NR via dual connectivity (DC). This paper gives an overview of the new DC features introduced in Release 15 and an outline of the features and operation procedures in comparison with DC in LTE systems. We also tune the key parameters for different scenarios and it is demonstrated in system level simulations that the performance of DC for 5G NR deployment is significantly improved as compared to the DC in LTE.

Keywords: Dual connectivity · E-UTRA · New radio ·
Non-standalone · Edge User Equipment (UE)

1 Introduction

As the IoT is developing repidly, an increasing number of IoT devices are being connected to the Internet, which makes it difficult to meet the growing requirements for a higher user capacity. Aimed to achieve Ultra-Reliable Low Latency Communication use-case, non-standalone (NSA) NR is introduced in Release 15 as the first stage of 5G NR [1,2]. The initial specification work is concentrated on NSA NR mode, which is an interim deployment configuration using DC to achieve a smooth transition to 5G.

On top of Evolved Universal Terrestrial Radio Access (E-UTRA) DC in Release 12 [3], Multi-RAT Dual Connectivity (MR-DC) is standardized in order to achieve a hybrid networking consisting of E-UTRAN and NR nodes. The basic concept of DC is given in [4] and [5]. Depending on the type of core network,

Supported by Nokia Bell Labs, Beijing, China.

MR-DC can be divided into two categories. The MR-DC with Evolved Packet Core (EPC) is E-UTRA-NR DC (EN-DC), where the control plane (C-plane) is anchored on LTE while the user plane (U-plane) data is transmitted via master node (MN) and secondary node (SN) configured with LTE and 5G NR respectively to boost data-rates and reduce latency. The completion of MR-DC with next generation core is targeted for June 2018.

Based on different radio protocol architectures, 3GPP has standardized three types of EN-DC solutions: Option 3, Option 3a and Option 3x, which all have DC specified in [3] as baseline for interworking between NR and E-UTRA.

This article gives an overview of EN-DC features introduced in Release 15 in comparison with previous specifications on DC. The details of three solutions in EN-DC are outlined with radio protocol architecture, network interface and operation procedures. Performance results of system-level simulations carried out in configuration with unbalanced bandwidth at macro and micro layers are presented to show the benefits of EN-DC.

The rest of the paper is organized as follows: Sect. 2 introduces the protocol architectures and the network interface of solutions in EN-DC. Section 3 outlines the changes on operation procedures in EN-DC and Sect. 4 presents the simulation setup with a set of tuned parameters, as well as the performance results. Finally, Sect. 5 highlights the conclusions.

2 Radio Protocol Architecture and Functionality

2.1 User Plane

From the perspective of UE, the radio bearers for U-plane connection in MR-DC, fall into three categories, which are MCG bearer, SCG bearer, and Split bearer. The radio protocol architecture for three bearers from UEs prospective in EN-DC is of no difference with that in E-UTRA DC.

While these three bearers, from the perspective of network, can be terminated either in MN or SN, which means PDCP entity is located in the corresponding node. PDCP is not necessarily in the same node with RLC entity whose location determines whether the bearer is MCG bearer or SCG bearer. This is a huge difference between EN-DC and E-UTRA DC. The radio architecture for three bearers from a UE prospective in EN-DC is shown in Fig. 1.

2.2 Control Plane

A new RRC state RRC_INACTIVE is introduced in 5G NR mainly to minimize signaling and consumption on top of two current RRC states, RRC CONNECTED and RRC IDLE. Specifically, when an RRC connection is established, UE is in RRC_CONNECTED or RRC_INACTIVE, which is an intermediate state designed to avoid frequent switching between RRC_CONNECTED and RRC_IDLE. The concrete details are not completed yet and EN-DC currently does not support RRC_INACTIVE state [6].

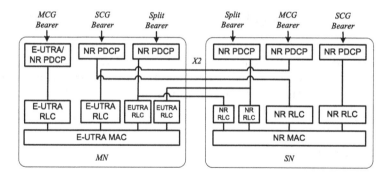

Fig. 1. Radio protocol architecture for MCG, SCG and split bearers from a network perspective

There is not much a difference between C-plane network interface in MR-DC and E-UTRA DC as the coordination of C-plane connection is still based on the non-ideal backhaul X2-C between the MN and SN.

3   Operation Procedural Aspect

The operation procedures defined in [3] are mostly applicable for Option 3/3a only with some minor changes on SN Addition and SN Modification as follows. Further necessary enhancements for Option 3x are given on top of that as the specifications for SN terminated split bearer are not given.

3.1   SN Addition

The major changes in SN Addition procedure for Option 3/3a are mostly because of the newly introduced split SRB and SRB3. When MN sends gNB addition request to gNB, MN will request SN to allocate radio resource for split SRB if MN wants to configure the bearer. MN always provide SN with all the necessary security information to establish SRB3 based on the decision of SN.

3.2   MN Initiated SN Modification

The MN, in this procedure, initiates SCG configuration changes to perform handover within the same MN while keeping the SN. If bearer type needs to be changed, it may result in adding the new bearer configuration and releasing the old one for the respective E-RAB in the procedure. In case of intra-MN handover, UE needs to apply the new configuration and apply synchronization to the MN after the MN initiates RRC Connection Reconfiguration, which makes random access necessary for UE before MN replies with RRCConnectionReconfigurationComplete.

3.3   SN Initiated SN Modification

It is different from the SN Modification in previous specifications that SN can initiate SN Modification without the involvement of MN. When SRB3 is established by SN, SN can directly send RRCConnectionReconfiguration message via SRB3 to UE without the need to send a request to the MN. It is also up to the SN whether to initiate Random Access.

4   Performance Evaluation

The performance of EN-DC is illustrated in the comparison with E-UTRA DC in the same scenarios and radio architectures, i.e. Option 3 in EN-DC is compared with 3C in E-UTRA DC and Option 3a with 1A. Considering there is no counterpart in E-UTRA DC for Option 3x with a SN terminated split bearer, we put Option 3x in comparison with Option 3a and 3C as they all have a split bearer.

4.1   Simulation Setup

The system-level simulation is mostly based on the HetNet scenarios in [7], where macro and small cells are deployed at different frequency layers. The network topology is a wrap-around model that consists of 7 three-sector MNs with 21 macrocells deployed at 2.6 GHz. Each macrocell has a condensed cluster of 4 randomly deployed small cells within a circular area with a 50 m radius, operating at 3.5 GHz and 28 GHz respectively in E-UTRA DC and EN-DC. The inter-site distance for macrocells is 500 m (ISD) [8].

Carrier bandwidth at macro layer in both E-UTRA DC and EN-DC scenarios is 20 MHz, while the bandwidth configured for small cells is 20 MHz and 100 MHz in E-UTRA DC and EN-DC respectively. Cell selection is based on the measurement results of the reference signal received quality (RSRQ) in A4 event which is triggered when the RSRQ of neighbouring cell is better than threshold [9,10].

For scenarios where a split bearer is configured, a request-and-forward flow control algorithm is applied aiming to match the data rate experienced in the SN [11]. In the simulation, we evaluate the performance by throughput of edge UE and medium UE, in the unit of Kb/s.

Each call has a fixed payload size of B = 5 Mb and will be terminated if its payload is successfully received by the UE, in which case the corresponding UE will be also removed from the simulation.

4.2   Analysis of Key Parameters

It is important to find a proper set of parameters to optimize the performance results of EN-DC, including Option 3/3a/3x. We first evaluate the performance of Option 3 and Option 3x in comparison with 3C over the traditional backhaul

Table 1. Flow control parameter settings for X2 latency

Parameters	Settings					
X2 latency (ms)	1	2	5	10	20	50
Target buffering time (ms)	10	10	15	20	30	50
Flow control periodicity (ms)	1 or 5					

with different X2 latency along with corresponding target buffering time and flow control periodicity shown in Table 1.

Flow control periodicity denotes how often the data request is sent and the number on LTE side is 5 ms while NR side is 1ms. In Option 3, for instance, data request is sent from SN, which is a NR node, so the flow control periodicity for Option 3, in this case, is 1ms. Target buffering time increases with a higher X2 latency to compensate for the fast variations of the user throughput in SN. It is found in [12] that the optimal setting of target buffering time depends on the X2 latency and flow control periodicity and an approximate expression for target buffering time is found to be

$$\theta \approx \min\{\frac{\Delta + 40}{3}, 20\} + 5\log_2(\frac{\rho}{5}) \tag{1}$$

where θ is target buffering time, Δ is X2 latency, and ρ is flow control periodicity.

It is observed in Fig. 2 that the bearable X2 latency for Option 3 and Option 3 x is within 5ms or the performance, especially the performance of Option 3, will be heavily compromised.

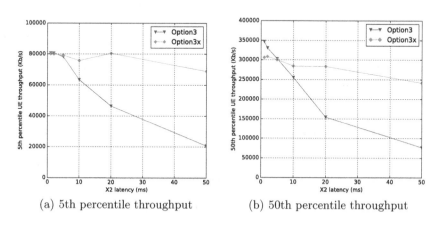

(a) 5th percentile throughput   (b) 50th percentile throughput

Fig. 2. 5th and 50th percentile throughputs with different X2 latency

We next analyze how to tune the setting of thresholds in −dB for RRC event A4, which is the minimum RSRQ a UE should reach when connected with SN to be a DC UE. A4 Threshold determines whether a UE is served by MN and SN simultaneously or only MN and consequently has a great influence on the performance of edge UEs. Optimal A4 threshold is supposed to get edge UEs a tolerable throughput with a balanced load between MN and SN and the A4 threshold for 1A and 3C is consequently assumed to be −14 dB, which results in over 90% of the UEs, with various offered loads, configured in DC and less than 10% only with MN. In order to find the most optimized A4 threshold for EN-DC, the performance results of three options in EN-DC are given with the offered load per macro area at 120 Mbps.

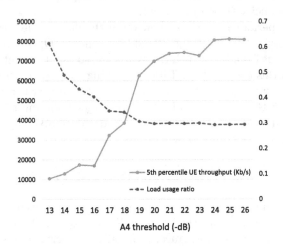

Fig. 3. 5th percentile throughput and load status with different A4 thresholds in Option 3

It is shown in Fig. 3 that the throughput of the edge UEs (lowest 5%) is improved as A4 threshold increases when the number is under 24 and there is not much a noticeable enhancement in throughput after the A4 threshold is over 24. Due to the high bandwidth at micro layer, migrating more UEs to SCG does not increase the load of SN as shown in the Fig. 3. Even if the A4 threshold is low enough for most of the UEs to be configured with DC, there are always a few UEs whose performance is too bad to be connected to SN. Consequently, they cause a huge load burden on MN as they can only be connected to MN. As a result, we consider the optimized A4 threshold for Option3 is −24 dB, with which the load of MN is around the lowest and the throughput is almost at the peak.

Similarly, the best optimized A4 threshold for Option 3a and Option 3x is −20 dB and −22 dB, respectively.

4.3   Performance Results

The benefits of using EN-DC are illustrated in comparison between E-UTRA DC and EN-DC. Figure 4 shows the 5th percentile and the mean UE throughput

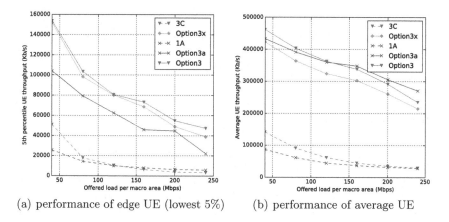

(a) performance of edge UE (lowest 5%)   (b) performance of average UE

Fig. 4. Performance comparison of edge and average UE between EN-DC and E-UTRA DC

with different offered traffic per macrocell area, which denotes the throughput experienced by at least 95 percent of the UEs and all on average, respectively.

Quite obviously, the UE throughput of EN-DC (including Option 3/3a/3x) in general is markedly higher than E-UTRA DC with an average gain of roughly 500% (including 1A and 3C) due to a larger bandwidth and higher transmission power at micro layer.

The throughput improvement on edge UEs (lowest 5%) is not as considerable as how it is on the average ones and the increasing offered load degrades the performance of edge UEs faster than it does mediocre UEs. It is observed from Fig. 4(a) that the maximum tolerable offered load per macrocell area increases enormously if there is a target of minimum throughput for at least 95% of the UEs to experience. The highest tolerable offered load for a throughput no less than 40 Mb/s increases from approximately 50 Mbps in 3C to about 210 Mbps in Option 3a and 240 Mbps in Option 3x, corresponding to a capacity gain of 320% and 380% respectively.

5   Conclusion

In this paper, we have summarized the key features of the EN-DC in NSA NR for IoT network deployment. A proper set of key parameters for the best performance in each option is analyzed and the UE throughput of EN-DC has been compared with E-UTRA DC to evaluate the potential of EN-DC. Carried out in configuration with unbalanced bandwidth at macro and micro layers, system-level simulation results show that EN-DC can provide a significant improvement in the average performance and a huge capacity gain for a target outage throughput.

Acknowledgment. This work is supported by Nokia Bell Labs, Beijing, China, the Fundamental Research Funds for the Central Universities Grant 2016RC056, the State Key Laboratory of Rail Traffic Control and Safety (Contract No. RCS2017ZT009), Beijing Jiaotong University and by the China Postdoctoral Science Foundation under Grant 2017M610040.

References

1. Rosa, C., et al.: Dual connectivity for LTE small cell evolution: functionality and performance aspects. IEEE Commun. Mag. **54**(6), 137–143 (2016)
2. 3GPP TS37.340: Evolved Universal Terrestrial Radio Access (E-UTRA) and NR; Multi-connectivity; Stage 2 (Release 15). v. 2.0.0 December 2017
3. 3GPP TS36.300: Evolved Universal Terrestrial Radio Access (E-UTRA) and Evolved Universal Terrestrial Radio Access Network (E-UTRAN); Overall Description; Stage 2 (Release 12). v. 12.4.0, December 2014
4. Zakrzewska, A., Lopez-Perez, D., Kucera, S., Claussen, H.: Dual connectivity in LTE HetNets with split control- and user-plane. In: IEEE Proceedings of Globecom Workshops, December 2013
5. Ishii, H., Kishiyama, Y., Takahashi, H.: A novel architecture for LTE-B :C-plane/U-plane split and phantom cell concept. In: IEEE Proceedings of Globecom Workshops, December 2012
6. 3GPP TS38.331: NR; Radio Resource Control (RRC) protocol specification (Release 15). v. 15.1.0, March 2018
7. Du, L., Zheng, N., Zhou, H., Liu, Y., Zhao, Z.: C/U split multi-connectivity in the next generation new radio system. In: 2017 IEEE 85th Vehicular Technology Conference (VTC Spring), pp. 1–5, June 2017
8. ITU-R: Guidelines for evaluation of radio interface technologies for IMT-advanced. (Report ITU-R M.2135-1), December 2009
9. 3GPP TS36.214: Evolved Universal Terrestrial Radio Access (E-UTRA); Physical layer; Measurements. v. 15.1.0, March 2018
10. Wang, H., Rosa, C., Pederson, K.I.: Dedicated carrier deployment in heterogeneous networks with inter-site carrier aggregation. In: IEEE Proceedings of WCNC, April 2015
11. Wang, H., Rosa, C., Pederson, K.I.: Inter-eNB flow control for heterogeneous networks with dual connectivity. In: Vehicular Technology Conference (VTC Spring), pp. 1–5, May 2015
12. Wang, H., Rosa, C., Pederson, K.I.: Dual connectivity for LTE-advanced heterogeneous networks. Wirel. Netw. **22**(4), 1315–1328 (2016)

Multiple Access and Communication Technologies for IoT

Slot Assign Algorithm with Threshold Based on Irregular Repetition Slotted ALOHA (IRSA)

Huaicui Zheng[1], Changle Li[1(✉)], Ni Tian[1], and Jun Cheng[2]

[1] State Key Laboratory of Integrated Services Networks,
Xidian University, Xi'an 710071, China
`clli@mail.xidian.edu.cn`
[2] Department of Intelligent Information Engineering and Science,
Doshisha University, Kyoto 6008586, Japan
`jcheng@mail.doshisha.ac.jp`

Abstract. As successive interference cancellation (SIC) technology has become a research hotspot recently, random multiple access protocols based on SIC such as contention resolution diversity slotted ALOHA (CRDSA) and irregular repetition slotted ALOHA (IRSA) are put forward one after the other. Although throughput performance for these protocols has been improved significantly. Yet the throughput will drop dramatically when the load increases gradually. In this paper, in order to get a higher throughput in the case of high load, we propose a slot assign algorithm with threshold based on IRSA to obtain a throughput T which is more than 1.

Keywords: Random multiple access protocol · SIC · CRDSA · IRSA · Slot assign algorithm with threshold

1 Introduction

Traditional random multiple access protocol ALOHA [1] is proposed in 1971, followed by slotted ALOHA [2]. However, there exists many drawbacks such as high probability collisions and low throughput, which may result in great communication delay and a waste of resource. In [3], diversity slotted ALOHA (DSA) is proposed. By allowing a user to transmit multiple copies of the same packet, DSA has a slight enhancement in throughput and decrease in delay when the load is moderate. As a multi-user detection technology, successive interference cancellation (SIC) is adopted as early as in the third generation of wireless mobile telecommunications technology (3G) and used widely in code division multiple access (CDMA). It could play disruptive effect in guaranteeing high-reliability and low-latency and satisfy the requirement of Ultra-Reliable Low Latency Communication (URLLC) in 5G [4–6].

[7] applies SIC to multiple access protocol based on ALOHA and then proposes the contention resolution diversity slotted ALOHA (CRDSA) protocol. In

B. Li et al. (Eds.): IoTaaS 2018, LNICST 271, pp. 225–231, 2019.
https://doi.org/10.1007/978-3-030-14657-3_22

CRDSA, a user is allowed to transmit two replicas on a MAC frame. And each header of the replica contains the position information of the another replica. As soon as one of the replicas is recovered correctly, then interference caused by the other replica can be eliminated. And this procedure can be continued to recover the remain packets on a frame. As a result, the throughput T (defined as the normalized throughput) in CRDSA can reach about 0.55. In [8], the author changes the number of a user to transmit replicas of a packet and then the irregular repetition slotted ALOHA (IRSA) is put forward. In IRSA, a variable repetition rate is employed and each user can choose the repeat times according to the probability distribution function (PDF). Compared to the CRDSA, there is a significant increase in the throughput performance for IRSA and throughout T can reach about 0.88. In order to achieve a higher throughput when the load increases, we propose a slot assign algorithm with threshold based on IRSA. An upper bound for the number of packets which can be decoded successfully in each slot will be set. In other words, as long as the number of packets in one slot is less than the given threshold, the packet can be recovered correctly.

The remainder of this paper is organized as follows. Section 2 will give a description of bipartite graph representation of the process of SIC. System model follows in Sect. 3. The simulation is conducted in Sect. 4 and then analysis is provided. At last, conclusions are shown in Sect. 5.

2 Bipartite Graph Representation of the Process of SIC

Considering the condition that a frame is composed of n slots, within which m users will transmit packets. There is no coordination among users when transmitting packets. Every user has one packet to transmit and l replicas of a generic packet can be transmitted within one frame. The slots occupation information is in the header of packets. Assuming the threshold of packets decoded successfully in each slot is s_{up}, which can be realized by power control [9]. When decoding in some slot, if the number of unrecovered packets in the slot is no more than the threshold s_{up}, all packets in the slot can be recovered successfully. At the same time, due to slots occupation information is included in the header of the recovered packets, these recovered packets in the slot can be removed from the other slots which they have been transmitted into.

It's convenient to use bipartite graph $G = \{P, S, E\}$ to represent the process of SIC, in which $P = \{p_1, p_2, p_3, \ldots, p_m\}$ and $S = \{s_1, s_2, \ldots, s_n\}$ denotes the set of slots and packets respectively. The set of E is $E = \{e_{ij}|i = 1, 2, \ldots, m; j = 1, 2, \ldots, n\}$, representing edges between slots and packets. If and only if a replica of i-th which is denoted by p_i is transmitted to j-th slot which is denoted by s_j, the p_i and s_j are connected with the edge e_{ij} in bipartite graph as shown in Fig. 1. It is obvious that l replicas of a packets are connected with l edges. The number of connection edges for slots and packets is referred to as their degree distribution in the bipartite graph. The definition of the slot and packet degree distribution is given by Eq. (1), in which $\Lambda(x)$ represents the slots degree distribution and $\Psi(x)$ represents packets degree distribution respectively. According to the bipartite graph in Fig. 1, the degree distribution of slots means the number of received packets and for packets the degree distribution means the number of replicas, i.e. the number of transmission for a packets [10]. Thus, the average transmission packets $average_{trans}$ for users can be derived by Eq. (2) and the average received packets in each slot $average_{rece}$ can be obtained by Eq. (3) within one frame. It is easy to prove that average number of transmission packets for one slot, which is also referred as to normalize offered load G, can be derived using the degree distribution of slots and packets by the Eq. (4).

$$\begin{cases} \Lambda(x) = \sum_l \Lambda_l x^l \\ \Psi(x) = \sum_l \Psi_l x^l \end{cases} \tag{1}$$

$$average_{trans} = \sum_l l\Lambda_l = \Lambda'(1) \tag{2}$$

$$average_{rece} = \sum_l l\Psi_l = \Psi'(1) \tag{3}$$

$$G = \frac{m}{n} = \frac{\Lambda'(1)}{\Psi'(1)} \tag{4}$$

If the packet is recovered in some slot, the connected edge between the packet and corresponding slot becomes dotted line and edges between the packet and the other slots also become dotted lines at the same time, which denotes the recovered packets are removed from the other slots. An example is given in Fig. 2 to better express the detailed process for SIC with an upper bound s_{up}. The example gives analysis for the condition that 6 packets will be transmitted within one frame divided into 4 same slots. Otherwise, the threshold s_{up} is set to 3. If one packet is recovered, we replace the edges with dotted lines, which are connected with the packet. The initial bipartite graph for the condition is shown in Fig. 1 and the SIC process starts by the Fig. 2(a), decoding in the first slot. In the first slot, packets p_2 and p_3 can be recovered on the account of the number of unrecovered packets being less than s_{up}. Once packets p_2 and p_3 are recovered in slot s_1, they will be removed from the other slots they occupy, i.e., they will be removed from the slots s_3 and s_4. That is to say the edges connected with the packets p_2 and p_3 become dotted lines in the bipartite graph as shown in Fig. 2(a). For the second slot s_2, though there is a collision, p_1 and p_4 still can be recovered due to the number of unrecovered packet is no more than 3. So the edges connected with p_1 and p_4 are replaced with dotted lines shown in Fig. 2(b). For the third slot, there is no collision because the packet p_2 is recovered in the first slot and has been removed from the slot s_3. Thus the packet p_5 can be recovered. The edges connected with packet p_5 become dotted lines in Fig. 2(c). For the last slot s_4, though the number of received packets is more than 3, the packets p_3, p_4 and p_5 is recovered in slot s_1, s_2 and s_3 respectively. Hence, the degree of the slot s_4 reduces to 1 and p_6 can be revealed. We can see all the packets can be recovered in the end shown in Fig. 2(d).

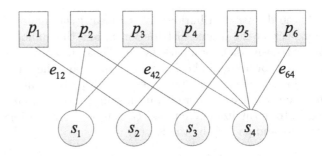

Fig. 1. The initial bipartite graph with 6 packets and 4 same packets

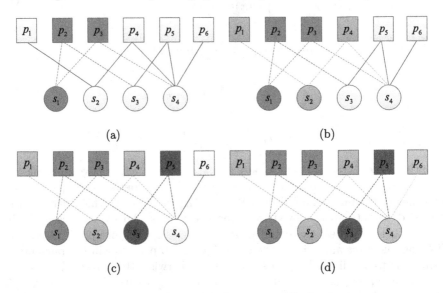

Fig. 2. The successive interference cancellation process

3   System Model

In the process of multi-user random access, each frame T_F for communication system is divided into n slots shown in Fig. 3, which is denoted by $s_j(j = 1, 2, \ldots, n)$. The length of each slot is defined as $T_n = T_F/n$ and m users transmit packets into the frame. Also there is no coordination among these users when they transmit their packets. That's to say when users transmit their packets, they are independent with each other. Each user starts to transmit their own packets at the beginning of each slot and the slot length is equal to the time of a packets transmission. Then we define system load $G = m/n$ (defined as the averaged number of packets in each slot) and throughout T (defined as probability of successful transmission packets).

In our proposed scheme, each user repeats their packets l times like IRSA shown in Fig. 3. And each header of replica carry the position information of

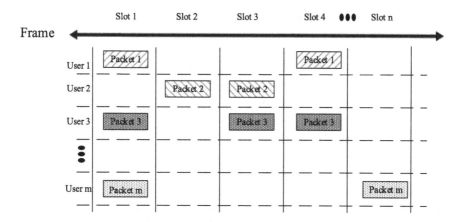

Fig. 3. The random access for IRSA and our proposed scheme

other replicas. If a replica is recovered, then corresponding replicas in other slots can be recovered. Then the received packets and upper bound of packets which can be recovered in each slot are denoted as $s_{j_{num}}(j = 1, 2, .., n)$ and s_{up} respectively. When decoding, the number of the packets in each slot will be counted. If $s_{j_{num}} \le s_{up}$, the packets in this slot can be recovered correctly according to the SIC. Otherwise, if the number is greater than the threshold and then these packets in current slot cannot be recovered directly. The packets which have been recovered in other slots should be found first and then those packets will be subtracted correspondingly. The number of packets in one slot may be counted again and if the number is still larger than the threshold, those packets in the slot will be thrown way. Then if the number is below the threshold, the process of SIC will be continued.

4   Simulation and Discussion

In this paper, simulations are performed in MATLAB. So as to observe the performance of different protocols, we choose $\Lambda(x) = 0.5x^2 + 0.28x^3 + 0.22x^8$ as the degree distribution in the simulation of IRSA and slot assign algorithm with threshold proposed in this paper. Then we assume that each frame shares the same length and the frame is divided into 2000 slots. More detailed simulation parameters are present in Table 1.

4.1   Results and Analysis of Throughout

Figure 4 are the simulation results for CRDSA and IRSA. From Fig. 4, we can see that when the load is about 0.6, the throughput can obtain a maximum of 0.55. Correspondingly, the throughput for IRSA can reach about 0.88 with the load is equal to 0.90. Compared to the throughput for CRDSA, the throughput for IRSA drops more dramatically than that for CRDSA after the throughput

Table 1. Simulation parameters

Simulation parameters	Configuration
The length of access frame	2000
The max iterations of sic	100
Rate of max repetition	8
Probability of ready packet	1%
Time of simulation time	100 ms
Bound of slot	3

reaches a peak. From relevant literature [8,11], when we adopt the distribution $\Lambda(x) = 0.5x^2 + 0.28x^3 + 0.22x^8$, if the load G is below the 0.938 (defined as the extreme value of G), the throughput T can almost reach the peak. However, this theoretical result can be obtained when the length of frame is infinite. Meanwhile, when the load G is larger than the extreme value, the reason why the throughput drops sharply is that packets in the slot can be recovered correctly and the SIC technology cannot work any more.

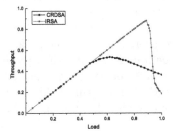

Fig. 4. Throughput for CRDSA and IRSA

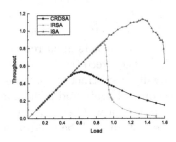

Fig. 5. Throughput for CRDSA, IRSA and ISA

Also, in the simulation above, the default of the upper bound (defined as packets which can be recovered correctly in each slot) is 1. Different from above, in Fig. 5, the upper threshold for the number of packets in one slot is set to 3. The process of recovering packets is slightly from that above and more detailed description has been shown in Sect. 3. From the result, we can also see the throughput for our proposed scheme known as ISA increases to 1 and even greater than 1. The peak of throughput occurs when the load is equal to 1.34. In this case, the throughput can reach about 1.13. Simulation results show that the improvement of throughput is obvious when the load is more than 1. In other words, by adding a decoding threshold to existing IRSA, performance of SIC can be improved significantly. What's more, our proposed can get a comparatively higher throughout compared to the CRDSA and IRSA.

5   Conclusion

In this paper, by analyzing throughput for existing random access protocol CRDSA and IRSA based on SIC, then we propose slot assign algorithm with threshold based on IRSA. Compared to known IRSA and CRDSA, we get a better throughput performance especially in the case of higher load and the value of throughput can be over 1. However, it also brings larger packet loss rate, which should be solved next. Also, we will carry out further simulation with actual communication scenario in network simulator OMNET++ to get a better performance evaluation such as delay and packet loss rate.

Funding Information. This work was supported in part by the National Natural Science Foundation of China under Grants U1801266, and 61571350, and in part by the Key Research and Development Program of Shaanxi under Contract 2017KW-004, Contract 2017ZDXM-GY-022, Contract 2018ZDXMGY-038, and Contract 2018ZDCXL-GY-04-02.

References

1. Abramson, N.: THE ALOHA SYSTEM - another alternative for computer communications. In: Proceedings of Fall Joint Computer Conference, pp. 281–285. ACM, New York (1970)
2. Davis, D.-H., Gronemeyer, S.-A.: Performance of slotted ALOHA random access with delay capture and randomized time of arrival. IEEE Trans. Commun. **28**(5), 703–710 (1980)
3. Gagan, L.-C., Stephen, S.-R.: Diversity ALOHA - a random access scheme for satellite communications. IEEE Trans. Commun. **31**(3), 450–457 (1983)
4. Liang, Y., Li, X., Zhang, J., Liu, Y.: A novel random access scheme based on successive interference cancellation for 5G networks. In: Wireless Communications and Networking Conference, pp. 1–6. IEEE, San Francisco (2017)
5. Stefanovic, C., Lazaro, F., Popovski, P.: Frameless ALOHA with reliability-latency guarantees. In: 2017 IEEE Global Communications Conference, pp. 1–6. IEEE, Singapore (2017)
6. Lzaro, F., Stefanović, Ć.: Finite-length analysis of frameless ALOHA with multiuser detection. IEEE Commun. Lett. **214**, 769–772 (2017)
7. Casini, E., De Gaudenzi, R., Del Rio Herrero, O.: Contention resolution diversity slotted ALOHA (CRDSA): an enhanced random access schemefor satellite access packet networks. IEEE Trans. Wirel. Commun. **6**(4), 1408–1419 (2007)
8. Liva, G.: Graph-based analysis and optimization of contention resolution diversity slotted ALOHA. IEEE Trans. Commun. **59**(2), 477–487 (2011)
9. Vem, A., Narayanan, K.-R., Cheng, J., Chamberland J.-F.: A user-independent serial interference cancellation based coding scheme for the unsourced random access Gaussian channel. In: 2017 IEEE Information Theory Workshop, pp. 6–10. IEEE, Kaohsiung (2017)
10. Narayanan, K.-R., Pfister, H.-D.: Iterative collision resolution for slotted ALOHA: an optimal uncoordinated transmission policy. In: 2012 7th International Symposium on Turbo Codes and Iterative Information Processing, pp. 136–139. IEEE, Gothenburg (2012)
11. Storn, R., Price, K.: Differential evolution - a simple and efficient heuristic for global optimization over continuous spaces. J. Glob. Optim. **11**(4), 341–359 (1997)

Joint Power and Splitting Factor Allocation Algorithms for Energy Harvesting Enabled Hybrid Cellular Networks

Jianjun Yang[✉], Zhiren Yao, Jie Hu, Longjiang Li, and Yuming Mao

School of Information and Communication Engineering,
University of Electronic Science and Technology of China,
Chengdu 611731, China
{jjyang,zhirenyao,hujie,longjiangli,
ymmao}@uestc.edu.cn

Abstract. In the hybrid cellular network with Simultaneous Wireless Information and Power Transfer (SWIPT), interference signal is a source of energy. In this paper, we develop a resource allocation scheme, which jointly optimizes transmit powers of base station (BS) and received power splitting ratios for energy harvesting and information processing at the users. Meeting the user's minimum throughput and energy harvesting rate, we perform with two different objectives to maximize the downlink information rate of small cell users and max-min their throughput. To solve the non-convex optimization problem, we propose to solve a series of geometric programming through the approach of successive convex approximation and devising iterative algorithms based on geometric programming. Numerical results are provided to demonstrate the effectiveness of proposed algorithm and its ability to improve network performance.

Keywords: Hybrid cellular networks · Resource allocation ·
Wireless information and power transfer

1 Introduction

With the rapid growth of mobile data traffic, especially the extensive use of the Internet of Things, wireless networks are able to take on massive data and huge access requirement. Hybrid cellular network, a kind of heterogeneous network structure formed through the overlapping and cooperation of different cellular networks, plays an important role in meeting the increasing demand of wireless coverage, amount of access, and high traffic volume [1], which is also the network topology adopted by 5G to meet access requirements.

Although hybrid cellular networks have a broad prospect for development and boosting the network capacity, but one of the direct challenges is how to maintain service quality requirements for users under strong interference from neighbor base station [1, 2]. If different networks are employed in a specific wireless channel, it will lead to inefficient use of the spectrum, while deployment in the same frequency will generate the co-layer and cross-layer interference. A lot of research works have been done to address that. In [3], the authors consider the interference from the macro cell station as white noise.

© ICST Institute for Computer Sciences, Social Informatics and Telecommunications Engineering 2019
Published by Springer Nature Switzerland AG 2019. All Rights Reserved
B. Li et al. (Eds.): IoTaaS 2018, LNICST 271, pp. 232–245, 2019.
https://doi.org/10.1007/978-3-030-14657-3_23

A semi-distributed interference management scheme based on joint clustering and resource allocation for femtocells is also proposed by authors of [4] to deal with management of both the cross-tier and the co-tier interference. In [5], authors proposed a joint resource allocation and admission control framework for an orthogonal OFDMA-based cellular network composed of a macro cell overlaid by small cells.

Meanwhile, in hybrid cellular network, while the distance between the mobile devices and the BS becomes shorter, the aggregate interference could be a beneficial energy source. That means mobile devices with limited battery can make use of the interference as energy by Energy harvesting (EH) technology. As one of EH technologies, Simultaneous Wireless Information and Power Transfer (SWIPT) is a current popular research topic. The device needs to be designed to decode the information while integrating energy collection circuits. Power splitting in time domain or in power domain is the most realistic current method to achieve cooperation between the two parts. In the [6], the author compares the efficiency of different energy-receiving methods and proposes a universal receiver operation, namely dynamic power allocation. With a similar topology of hybrid cellular network, it is also a good idea to extend the transmission distance of information and energy through the relay method [7]. Manny resource allocation algorithms have been put forward to maximize throughput performance, while ensuring a minimum energy collection rate [8].

In SWIPT networks, harvesting energy and mitigating inference are two contradicting requirements. Controlling base station power results in lower interference and greater throughput, but decreasing the harvested energy from the RF signal, vice versa. Therefore, it is necessary to analyze the performance of the hybrid cellular network which applies SWIPT technology. In [9], an optimal downlink power allocation with a fixed power spilt ratio is devised for the trade-off between information rate and energy harvesting rate in two-tier cellular networks. However, [9] ignores the noise from the signal processing circuit, then the optimization problem has changed into a conventional heterogeneous network resource allocation problem.

In this paper we also consider a hybrid cellular network with simultaneous infor mation and energy transfer, where there is a macro base station within a number of small cells and some connecting users of the specific base station. What we do first is to jointly optimize the downlink power allocation and power split ratio to maximize the sum of small cell station users' rate, while the minimum rate requirement is ensured for all users and the minimum harvested energy for small cell users should also be guaranteed. Since do not ignore the noise of the signal processing circuit, our optimization model approaches to the fact better. Taking account of fairness between small cells, we formulate a max-min fairness problem where the throughput of the most disadvantaged cell is maximized. The simulation result provides a trade-off relationship between information and energy transfer.

The main contributions can be summarized as follows.

- In hybrid cellular network with EH-enabled user's devices, we formulate two resource optimization problems, namely, sum-rate maximization and minimum-throughput maximization. Considering that noise can not be ignored in actual communication, our formula contains the user's demodulation noise, and the outcome is more reasonable.

- The objective is to jointly optimize the transmit powers at the BSs and the optimal power splitting ratio at user's devices. Our formulations target the multi-cell interference, while meeting users' requirements for energy harvesting and throughput at the same time. The formulated problems are not convex due to interference and many nonlinear multiplying terms of the optimization variables. Then, we jointly adopt successive convex approximation (SCA) and geometric programming (GP) method to obtain the solutions.
- Simulation results confirm that our joint optimization solutions significantly outperform those where the radio resource parameters are individually optimized.

The rest of this paper is organized as follows. Section 2 presents the system model and formulates the joint resource optimization problems. Section 3 proposes the GP-based SCA solutions for nonconvex resource allocation problems. In Sect. 4, numerical results confirm the advantages of our proposed algorithms. Finally, Sect. 5 concludes the paper.

2 System Model and Problem Formulations

Downlink transmission of a hybrid cellular Network is considered, which consist of one macro base station (MBS) at the center of the macro-cell overlaid by N small cells as shown in Fig. 1. Each small base station (SBS) has only one scheduled small cell user equipment (SUE) randomly distributed in the cell. We assume there is a microcell user equipment (MUE) randomly located in the macro-cell and all BSs use same frequency band.

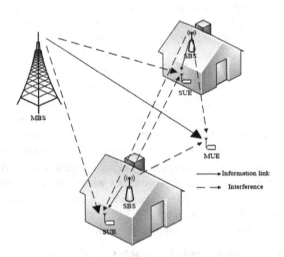

Fig. 1. Hybrid cellular network

The SUEs can harvesting energy from the microwave while receiving information by using power splitting. We consider a power split ratio α, the fraction of the total received power is used for information decoding and remaining fraction $1 - \alpha$ is used for energy harvesting. Energy harvesting capability particularly exists in SUEs for the short distance away from the SBS. Harvested energy can make a significant contribution to its uplink transmission. Uplink transmission and analysis is not considered in our paper.

Due to the capability of harvesting energy from the received microwave, there is a tradeoff between energy harvested and information received, not only in terms of power splitting ratio but also in terms of power allocation. For example, when the power of the macro base station becomes high, the SINR of the SUEs increases, which leads to the information rate decreasing. However, because of the increase of the total input power at this moment, the energy harvest rate improves. Needless to say, power splitting ratio α directly determines the energy allocation of the SUE. Therefore, we have implemented a joint optimal power allocation in two different cases.

Let h_i and h_M be the channel gains from i-th SBS and MBS to their scheduled users respectively. Similarly $h_{i,M}$ and $h_{M,i}$ be the channel gains from i-th SBS and MBS to the users in other cells respectively. If p_i is the transmit power of SBS and p_M is the transmit power of MBS then their signal-to-interference-plus-noise ratio (SINR) can be written as

$$SINR_M = \frac{p_M h_M}{\sum\limits_{i=1}^{N} p_i h_{i,M} + n_\omega + n_{sp}} \tag{1}$$

$$SINR_i = \frac{\alpha_i p_i h_i}{\alpha_i \left(p_M h_{M,i} + \sum\limits_{j \neq i}^{N} p_j h_{j,i} + n_\omega \right) + n_{sp}}, \forall i \tag{2}$$

n_ω and n_{sp} represent the Additive White Gaussian Noise (AWGN) and signal processing noise at the receiver. It should be noted that the signal processing noise remains unaffected by power splitting in SINR of SUEs. The information rate of MUE and SUE in each small cell are respectively given by

$$R_M = \log_2(1 + SINR_M) \tag{3}$$

$$R_i = \log_2(1 + SINR_i), \forall i \tag{4}$$

SUE harvests energy from the signals of all BS, therefore the energy harvesting rate of each SUE is given by (in Joules per second [Jps])

$$E_i = \eta(1 - \alpha_i)\left(\sum\limits_{j=1}^{N} p_j h_{j,i} + p_M h_{M,i} + n_\omega \right), \forall i \tag{5}$$

Where $\eta \in (0, 1)$ is the efficiency of energy conversion of the SUE. We see that the information rate and energy harvesting rate are greatly affected by cross-layer interference. Information rate of SUE decreases with the increase of interference, but the energy collection rate increases with the increase of interference.

In this paper, we aim to devise an optimal tradeoff of three parameters, the transmit power p_M of MBS, transmit power $P = [p_1, p_2, \ldots, p_N]$ at SBSs and power splitting factor α at SUEs to maximize the performance of the hybrid cellular network under consideration. Specifically, we will study the following problems which jointly optimize (p_m, P, α) for two different design objectives.

2.1 Problem (P1): Sum-Rate Maximization

We formulate maximize the sum information rate of all SUEs as follows.

$$
\begin{aligned}
&\max_{p_s, P, \alpha} \sum_{i=1}^{N} R_i \\
&subject\ to: \\
&C_1 : R_i \geq R_{\min(i)}, \forall i, \qquad C_2 : R_M \geq R_{\min(M)}, \\
&C_3 : E_i \geq E_{\min}, \forall i, \qquad C_4 : p_M \leq p_{\max(M)}, \\
&C_5 : p_i \leq p_{\max(i)}, \forall i, \qquad C_6 : p_i, p_M \geq 0, \forall i, \\
&C_7 : 0 \leq \alpha_i \leq 1, \forall i,
\end{aligned} \tag{6}
$$

Where C_1 and C_3 ensure that the throughput and energy harvesting rate of SUEs can satisfy their minimum rate constraint for each one in the worst case. C_2 is the minimum information rate constraint for MUE. C_4 and C_5 are maximum transmit power constraints of each MBS and SBS, respectively. C_6 is the non-negativity constraint on power variables. C_7 is the constraints for the power splitting factors for all SUEs.

2.2 Problem (P2): Max-Min Throughput Fairness

In Problem (P1), the network total rate is maximized regardless of the actual throughput achieved by the individual users. Max-min fairness can sufficiently improve the performance of users in the worst case and thus lead to a high level of fairness. Achieving the max-min fairness, we focus on solving the problem given in (7).

$$
\max_{p_M, P, \alpha} \ \min_{i \in N} R_i \\
s.t\ C_1 - C_7. \tag{7}
$$

Problems (P1) and (P2) are nonconvex in (p_m, P, α) because the throughput R_i in (4) is nonconvex in those variables. Then, SCA and GP method will be adopted to solve Problems (P1) and (P2) as described in Algorithm 1. The main idea of SCA is to approximate a non-convex problem into a series of solvable problems, to obtain the solution satisfying KKT conditions of the original problem [10]. GP can be used to solve a special-form non-convex problem, which can be reduced to a convex optimization problem through a logarithmic change of variables [11–13].

3 Power and Splitting Factor Allocation Scheme

3.1 SCA and GP-Based Solution for Problem (P1)

First, we express the objective function in (6) as

$$
\max_{p_M, P_i, \alpha} \sum_{i=1}^{N} R_i = \max_{p_M, P_i, \alpha} \log_2 \prod_{i=1}^{N} (1 + SINR_i)
$$

$$
\equiv \min_{p_M, P_i, \alpha} \prod_{i=1}^{N} \frac{1}{1 + SINR_i},
$$

(8)

Where

$$
\frac{1}{1 + SINR_i} = \frac{\alpha_i \left(p_M h_{M,i} + \sum\limits_{j \neq i}^{N} p_j h_{j,i} + n_\omega \right) + n_{sp}}{\alpha_i \left(p_M h_{M,i} + \sum\limits_{j=1}^{N} p_j h_{j,i} + n_\omega \right) + n_{sp}}.
$$

(9)

We get function (8) since $\log_2(\cdot)$ is monotonically increasing function. For the sake of brevity, let us define,

$$
u_i(\mathbf{x}) \triangleq \alpha_i \left(p_M h_{M,i} + \sum_{j \neq i}^{N} p_j h_{j,i} + n_\omega \right) + n_{sp},
$$

(10)

$$
v_i(\mathbf{x}) \triangleq \alpha_i \left(p_M h_{M,i} + \sum_{j=1}^{N} p_j h_{j,i} + n_\omega \right) + n_{sp}
$$

(11)

where $\mathbf{x} = [p_M, P, \alpha]^T \in \mathbb{R}_+^{2N+1}$, the objective function in (8) can be expressed as

$$
\min_{p_M, P, \alpha} \prod_{i=1}^{N} \frac{u_i(\mathbf{x})}{v_i(\mathbf{x})}.
$$

(12)

To transform Problem (P1) into a GP, we would like the objective function (12) to be a posynomial. We resort to SCA to approximate it into a series of problems in the following.

Based on Jensen's inequality, namely, the arithmetic mean is greater than or equal to the geometric mean for any set of positives [12], given the value of $\mathbf{x}^{[k-1]}$ at the kth iteration, we have

$$v_i(\mathbf{x}) \geq (\frac{v_i(\mathbf{x}^{[k-1]})\alpha_i p_M}{\alpha_i^{[k-1]}p_M^{[k-1]}})^{\frac{\alpha_i^{[k-1]}p_M^{[k-1]}h_{M,i}}{v_i(\mathbf{x}^{[k-1]})}} \times \prod_{j=1}^{N}(\frac{v_i(\mathbf{x}^{[k-1]})\alpha_i p_j}{\alpha_i^{[k-1]}p_j^{[k-1]}})^{\frac{\alpha_i^{[k-1]}p_j^{[k-1]}h_{j,i}}{v_i(\mathbf{x}^{[k-1]})}}$$
$$\times (\frac{v_i(\mathbf{x}^{[k-1]})\alpha_i}{\alpha_i^{[k-1]}})^{\frac{\alpha_i^{[k-1]}N_\omega}{v_i(\mathbf{x}^{[k-1]})}} \times v_i(\mathbf{x}^{[k-1]})^{\frac{N_{sp}}{v_i(\mathbf{x}^{[k-1]})}}. \tag{13}$$

For brevity, we define

$$\tilde{v}_i(\mathbf{x}) \triangleq (\frac{v_i(\mathbf{x}^{[k-1]})\alpha_i p_M}{\alpha_i^{[k-1]}p_M^{[k-1]}})^{\frac{\alpha_i^{[k-1]}p_M^{[k-1]}h_{M,i}}{v_i(\mathbf{x}^{[k-1]})}} \times \prod_{j=1}^{N}(\frac{v_i(\mathbf{x}^{[k-1]})\alpha_i p_j}{\alpha_i^{[k-1]}p_j^{[k-1]}})^{\frac{\alpha_i^{[k-1]}p_j^{[k-1]}h_{j,i}}{v_i(\mathbf{x}^{[k-1]})}}$$
$$\times (\frac{v_i(\mathbf{x}^{[k-1]})\alpha_i}{\alpha_i^{[k-1]}})^{\frac{\alpha_i^{[k-1]}N_\omega}{v_i(\mathbf{x}^{[k-1]})}} \times v_i(\mathbf{x}^{[k-1]})^{\frac{N_{sp}}{v_i(\mathbf{x}^{[k-1]})}}. \tag{14}$$

According to [13], using $\tilde{v}_i(\mathbf{x})$ to approximate $v_i(\mathbf{x})$ satisfies the conditions of SCA. Hence, the objective function $u_i(\mathbf{x})/v_i(\mathbf{x})$ in (12) is approximated by $u_i(\mathbf{x})/\tilde{v}_i(\mathbf{x})$. The latter formula is a posynomial because $\tilde{v}_i(\mathbf{x})$ is a monomial and the ratio of a posynomial to a monomial is a posynomial.

To make the problem fit into the GP standard form [13], the other constraints should also be transformed into GP standard type.

$$\tilde{C}_1 : \frac{\tilde{R}_{\min(i)}(\alpha_i(p_M h_{M,i} + \sum_{j \neq i}^{N} p_i h_{j,i} + n_\omega) + n_{sp})}{\alpha_i p_i h_i} \leq 1, \forall i,$$
$$\tilde{C}_2 : \frac{\tilde{R}_{\min(M)}(\sum_{i=1}^{N} p_i h_{i,M} + n_\omega + n_{sp})}{p_M h_M} \leq 1, \tag{15}$$
$$\tilde{C}_4 : \frac{1}{p_{\max(M)}}p_M \leq 1,$$
$$\tilde{C}_5 : \frac{1}{p_{\max(i)}}p_i \leq 1, \forall i \quad \tilde{C}_7 : \alpha_i \leq 1, \forall i,$$

where $\tilde{R}_{\min(i)} = 2^{R_{\min(i)}} - 1$ and $\tilde{R}_{\min(M)} = 2^{R_{\min(M)}} - 1$.

We will approximate constraint C_3 by a posynomial to fit into the GP framework. We lower bound posynomial $\eta(1 - \alpha_s)(\sum_{i=1}^{S} P_i h_{i,u_s} + P_m h_{m,u_s} + N_\omega)$. by a monomial. After replacing $1 - \alpha$ by an auxiliary variable t, we have,

$$w_i(\bar{\mathbf{x}}) \triangleq \eta t_i(\sum_{j=1}^{N} p_j h_{j,i} + p_M h_{M,i} + n_\omega), \tag{16}$$

$$w_i(\bar{x}) \geq \tilde{w}_i(\bar{x}) \triangleq \eta t_i \prod_{i=1}^{N} \left(\frac{w_i(\bar{x}^{[k-1]})p_j}{p_j^{[k-1]}}\right)^{\frac{p_j^{[k-1]}h_{j,i}}{w_i(\bar{x}^{[k-1]})}}$$

$$\times \left(\frac{w_i(\bar{x}^{[k-1]})p_M}{p_M^{[k-1]}}\right)^{\frac{p_M^{[K-1]}h_{M,i}}{w_i(\bar{x}^{[k-1]})}} \times w_i(\bar{x}^{[k-1]})^{\frac{N_\omega}{w_i(\bar{x}^{[k-1]})}}, \tag{17}$$

where $\bar{x} = [p_M, P, \alpha, t]^T \in \mathbb{R}_+^{3N+1}$. $w_i(\bar{x})$ can be used to get the relaxed constraint replacing C_3. The approximated subproblem can be formulated at the m-th iteration for Problem P1 as follows.

$$\min_{p_M, P, \alpha, t} \prod_{i=1}^{N} \frac{u_i(\bar{x})}{\tilde{v}_i(\bar{x})}$$
$$s.t \; \frac{E_{\min}}{\tilde{w}_i(\bar{x})} \leq 1, \forall i, \tag{18}$$
$$t_i \leq 1, \forall i, t_i + \alpha_i \leq 1, \forall i,$$
$$\tilde{C}_1, \tilde{C}_2, \tilde{C}_4, \tilde{C}_5, \tilde{C}_6, \tilde{C}_7.$$

We see that (18) is the form of a geometric program. Since the approximation from (12) to (18) satisfies the conditions of SCA, we can conclude that the optimal solution of (18) converges to the solution satisfying the KKT conditions of (12) [10].

3.2 SCA and GP-Based Solution for Problem (P2)

By introduce an auxiliary variable τ, (7) can be regarded as the problem of maximizing a common throughput,

$$\max_{p_M, P, \alpha} \tau$$
$$s.t \; R_i \geq \tau \geq 0, \quad \forall i \in N, \tag{19}$$
$$C_1 - C_7.$$

After carrying out simple algebraic manipulations, the first constraint of (19) can be rewritten as.

$$\frac{e^{\tau \ln 2}}{1 + SINR_i} \leq 1, \quad \forall i \in N; \quad and \quad \tau \geq 0, \tag{20}$$

By introducing the auxiliary variable t and $u_i(x)$, $v_i(x)$ and $\tilde{w}_i(\bar{x})$ defined in (10), (11) and (17), problem (P2) is approximate to the following GP problem with the similar approach in Sect. 3 – A.

$$\max_{p_M, P, \alpha, t} \tau$$
$$s.t \; \frac{u_i(\bar{x})e^{\tau \ln 2}}{\tilde{v}_i(\bar{x})} \leq 1, \forall i \in N,$$
$$\frac{E_{\min}}{\tilde{w}_i(\bar{x})} \leq 1, \forall i, \tag{21}$$
$$t_i \leq 1, \forall s,$$
$$t_i + \alpha_i \leq 1, \forall i,$$
$$\tilde{C}_1, \tilde{C}_2, \tilde{C}_4, \tilde{C}_5, \tilde{C}_6, \tilde{C}_7.$$

3.3   Based on SCA and GP Algorithm for Joint Resource Allocation

GP problems (18) and (21) are the approximations of the original Problems (P1) and (P2), respectively. In Algorithm 1, we propose an SCA algorithm in which a GP is optimally solved at each iteration.

Algorithm 1: Based on SCA and GP Scheme

1: Set tolerance ε, iteration number k= 1.

2: Choose a feasible point ($\overline{x}^{[0]} \triangleq (p_\mu^{[0]}, P^{[0]}, \alpha^{[0]}, t^{[0]})$).

3: Compute the value of $v_i(\overline{x}^{[0]}), \forall i \in N$, according to (11).

4: **repeat**

5: Using $v_i(\overline{x}^{[k-1]})$, form the approximate monomial $\tilde{v}_i(\overline{x})$ according to (14).

6: Solve (18) and (21) to obtain the kth iteration approximated solution ($\overline{x}^{[k]} \triangleq (p_\mu^{[k]}, P^{[k]}, \alpha^{[k]}, t^{[k]})$) solution using interior-point method.

7: Compute the value of $v_i(\overline{x}^{[k]}), \forall i \in N$, according to (11).

9: until Convergence of $\overline{x}$ or no further improvement in the objective value after a certain number of iterations.

---

4   Performance Evaluation

In this section, we present simulation results of the proposed power and split ratio allocation models in a hybrid cellular network. A hybrid cellular network consists of one MBS at the origin and four indoor SBSs on the coordinate axis. The distances of those SBSs from the MBS are 27 m, 80 m, 90 m, and 50 m. SUEs are all in the room. The MUE could move anywhere within the macro cell. For universal reasons, we design two scenarios. Scenario 1: the MUE is in the room and close to the first indoor SUE, e.g. coordinate value is (27, −5); Scenario 2: the MUE is away from four SBSs, e.g. coordinate value is (17, 25). The channel model comes from [14]. Detailed parameter value in the simulation is summarized in the table below. The default parameter means that this part of the parameter may be used as an argument in a later but is set by default if not specified. If microwave pass through the wall L_w is added to the channel gain (Table 1).

Table 1. Simulation parameter value

$p_{\max(M)}$	43 dBm	$p_{\max(i)}$	30 dBm	$R_{\min(i)}$	1 bps/Hz
L_w	10	η	1	Default $R_{\min(M)}$	1 bps/Hz
N_ω	−130 dBm	N_{sp}	−67 dBm	Default $E_{\min}$	−40 dBm
The antenna gain of MBS	18 dBi	The antenna gain of SBS	5 dBi		

Figures 2 and 3 show the numerical convergence results of GP-based algorithm 1 under scenarios 2 with different. In the simulation, each iteration is to solve a GP problem in the algorithm by CVX. It's clear that this algorithm can converge within 4 iterations and then achieve the sub-optimal throughput from Figs. 2 and 3. These prove the feasibility of the algorithm.

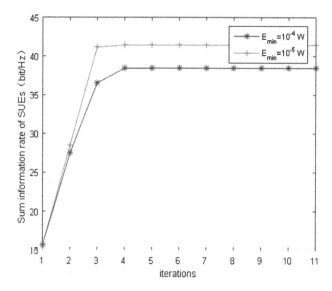

Fig. 2. Convergence of the algorithms in a problem (P1)

Next, we plot the sum of information against the minimum energy requirement and the minimum information requirement of MUEs, F_{min} and $R_{min(M)}$, respectively in Figs. 4 and 5. The sum rates of SUEs all decrease as the independent variables increases. When the minimum energy harvesting rate E_{min} is very low, more power can be assigned to information processing and the sum information rate of SUEs can be larger. When $R_{min(M)}$ increases, p_M has to increase to satisfied the minimum information demand. The larger p_M means that SUEs can harvest more energy and get more noise which may reduce the information rate. However, this part can be compensated by the power split ratio of the SUE. Another interesting phenomenon is that as the MUE is farther away from a small base station, the rate usually is higher. If the MUE is relatively far away from small cell station, that means the most influential small cell station's ability to cause interference to the MUE is reduced and transmit the power of micro station can be increased to achieve a higher rate. In contrast, curves with fixed $\alpha = 0.5$ were also drawn. Obviously, the information rates with fixed α are much lower than that of the optimal solution, which shows the necessity of our algorithm.

Finally we simulate max-min throughput fairness problem. The lowest information rate among SUEs in problem P1 is the comparison. Figure 6 precisely compare the numerical solution of problem P2 with the smallest rate of the SUE in problem P1

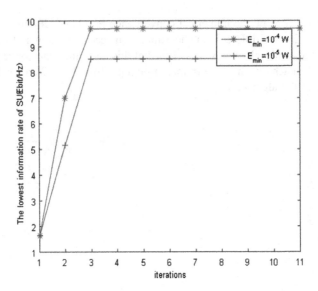

Fig. 3. Convergence of the algorithms in a problem (P2)

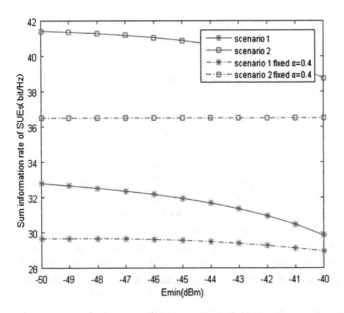

Fig. 4. Sum information rate vs $E_{\min}$

under different scenarios. Numerical results confirm the effectiveness of our proposed algorithm. The optimal solution in problem P2 is higher than the lowest rate of SUE in problem P1, which means that the fairness between users is guaranteed. It should be

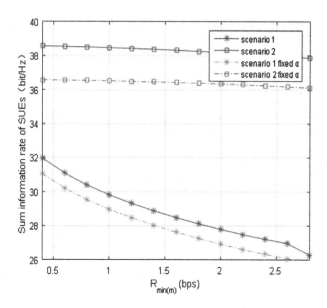

Fig. 5. Sum information rate vs $R_{\min(M)}$

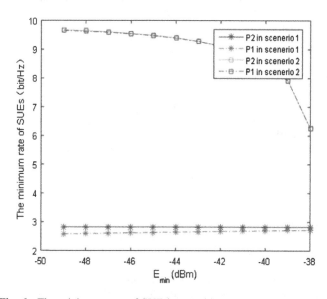

Fig. 6. The minimum rate of SUE in a problem (p1) and (P2) vs $E_{\min}$

noted that in scenario 2, the two curves almost coincide. Due to the long distance far from the small station plus the significant fading caused by walls, the channel gain from MUE to SBS is very low. Each SBS only has a small impact on the MUE, which is almost negligible in value. It means that the MUE and the SBSs are in an independent state at this moment. And we can solve the problem P1 to get the approximate solution to the problem P2 in this scenario.

5 Conclusion

In this paper, we have taken the problem for joint optimization of the downlink transmit power and power splitting ratio into consideration in the hybrid cellular network. The signal sent by the MBS will become cross-layer noise for other devices, affecting its information rate. In addition, it will also become the source of energy for wireless devices to harvest. Considering maximum sum-throughput and max-min throughput of SUEs, joint optimization of the downlink power and the power split ratio is performed. In the maximum sum-throughput problem, the outperformance for joint optimization is confirmed with the comparison of fixed α. On the other hand, a trade-off between information and energy transmission capabilities is observed, which means the achievable information rate demand is limited by the energy harvesting rate demand. Moreover, the effectiveness of fairness algorithm is also proved in the simulation.

Acknowledgement. This work is supported by the National Natural Science Foundation of China under Grant No. 61273235, the Fundamental Research Funds for the Central Universities of China (No. ZYGX2016J001), the Defense Advance Research Foundation of China under Grants 61401310105 and the Chongqing Research Program Basic Research Frontier Technology (No. cstc2017jcyjA1246).

References

1. Hossain, E., Rasti, M., Tabassum, H., Abdelnasser, A.: Evolution toward 5G multi-tier cellular wireless networks: an interference management perspective. IEEE Wirel. Commun. **21**(3), 118–127 (2014)
2. Idachaba, F.E.: 5G networks: open network architecture and densification strategies for beyond 1000x network capacity increase. In: 2016 Future Technologies Conference, FTC, San Francisco, CA, pp. 1265–1269 (2016)
3. Kim, J., Cho, D.: A joint power and subchannel allocation scheme maximizing system capacity in indoor dense mobile communication systems. IEEE Trans. Veh. Technol. **59**(9), 4340–4353 (2010)
4. Abdelnasser, A., Hossain, E., Kim, D.I.: Clustering and resource allocation for dense femtocells in a two-tier cellular OFDMA network. IEEE Trans. Wirel. Commun. **13**(3), 1628–1641 (2014)
5. Abdelnasser, A., Hossain, E.: Joint resource allocation and admission control in OFDMA-based multi-tier cellular networks. In: 2014 IEEE Global Communications Conference, Austin, TX, pp. 4689–4694 (2014)
6. Zhou, X., Zhang, R., Ho, C.K.: Wireless information and power transfer: architecture design and rate-energy tradeoff. In: 2012 IEEE Global Communications Conference, GLOBECOM, Anaheim, CA, pp. 3982–3987 (2012)
7. Lu, X., Wang, P., Niyato, D., Kim, D.I., Han, Z.: Wireless networks with RF energy harvesting: a contemporary survey. IEEE Commun. Surv. Tutorials **17**(2), 757–789 (2015)

8. Liu, L., Zhang, R., Chua, K.: Wireless information transfer with opportunistic energy harvesting. IEEE Trans. Wirel. Commun. **12**(1), 288–300 (2013)
9. Lohani, S., Hossain, E., Bhargava, V.K.: Downlink power allocation for wireless information and energy transfer in macrocell-small cell networks. In: 2016 IEEE Wireless Communications and Networking Conference, Doha, pp. 1–6 (2016)
10. Marks, B.R., Wright, G.P.: A general inner approximation algorithm for non-convex mathematical programs. Oper. Res. **26**(4), 681–683 (1978)
11. Boyd, S., Vandenberghe, L.: Convex Optimization. Cambridge University Press, Cambridge (2004)
12. Boyd, S., Kim, S.J., Vandenberghe, L., et al.: A tutorial on geometric programming. Optim. Eng. **8**(1), 67–127 (2007)
13. Papandriopoulos, J., Evans, J.S.: SCALE: a low-complexity distributed protocol for spectrum balancing in multiuser DSL networks. IEEE Trans. Inf. Theory **55**(8), 3711–3724 (2009)
14. 3GPP: Further advancements for E-UTRA physical layer aspects (Release9). 3rd Generation Partnership Project (3GPP), Technical report 36.814, March 2010

Multi-cell Cooperative Transmission for the Next Generation Millimeter-Wave WiFi Network

Biao Chen[1,2], Qi Yang[1], Bo Li[1], Mao Yang[1(✉)], and Zhongjiang Yan[1]

[1] School of Electronics and Information, Northwestern Polytechnical University,
Xian, China
cb2017@mail.nwpu.edu.cn, {libo.npu,yangmao,zhjyan}@nwpu.edu.cn
[2] Science and Technology on Communication Networks Laboratory,
Shijiazhuang 053200, China

Abstract. In this paper, we study the directional data transmission of multi-cells. In a high-density cell scenario, if nodes of multi-cell perform data transmission randomly in the Scheduling Period (SP) phase, there may be a large interference between the links of cells using the same frequency. In order to reduce the interference, improve the throughput of the network, and reduce the delay of the network and packet loss rate, we propose a multi-cell cooperative transmission scheme. In our proposed scheme, one cell is set as the primary cell, and one cell is the secondary cell, and a special new frame is proposed. When the nodes of the primary cell starts SP, the data sender of the cell sends the frame to synchronize between cells. The node of the other cell that received the frame performs data transmission at the appointed time point. Through our proposed scheme, the impact of other cells transmitting data on the acknowledgement (ACK) of the cell is greatly reduced. The simulation results show that the proposed scheme improves the anti-interference performance of the network including improving the network throughput and reducing the packet loss rate and packet transmission delay of the network compared with AP Clustering.

Keywords: Service period · Directional antenna · mmWave · Interference mitigation

1 Introduction

With the rapid growth of HD video, virtual reality and other ultra-high-speed services, traditional wireless LANs have become increasingly difficult to meet these needs. Since the 2.4 GHz and 5 GHz bands are used in the traditional WLAN, and the unlicensed spectrum resources in this band are less, the maximum information rate that can be provided is very limited. In recent years, the 60 GHz millimeter wave band WLAN with 5 GHz to 9 GHz unlicensed spectrum

B. Li et al. (Eds.): IoTaaS 2018, LNICST 271, pp. 246–254, 2019.
https://doi.org/10.1007/978-3-030-14657-3_24

resources has become a research hotspot in academia and industry. As mentioned above, the current WLAN development is difficult to meet these requirements. In the 60 GHz millimeter wave band, there are a large number of unlicensed spectrum resources, which can carry larger data information and can be used indoors, offices, etc. Therefore, it's a good solution to the current demand for Spectrum resource due to ultra-high-definition video and virtual reality. The wireless LAN protocol that more commonly used for the millimeter wave band is the 802.11ad standard.

In IEEE 802.11ad/ay, although the bandwidth of the millimeter wave and the transmission rate are very advantageous, the transmission loss is very serious due to the self-characteristic of the millimeter wave. Therefore, when the millimeter wave is used, the millimeter wave is oriented, and the directional beam can reduce the transmission loss in certain degree. This is why IEEE 802.11ad/ay specify that the MAC process is divided into two parts in a wireless LAN, we called Beacon Interval. The interior divided into two parts, Beacon Header Interval and Data Transmission Interval. The Beacon Header Interval (BHI) includes Beacon Transmission Interval (BTI), Association Beamforming Training (A-BFT), and Announcement Transmission Interval (ATI). Data Transmission Interval (DTI) is used for data transmission including Scheduling Period (SP) and Contented based access period (CBAP) [1–3]. This basic set of superframe structure is still used in 802.11ay (Fig. 1).

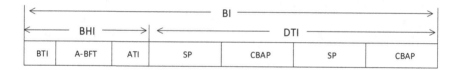

Fig. 1. A typical superframe structure.

In the existing IEEE 802.11ad standard and 802.11ay draft, if different cells are in the same SP phase, the senders don't randomly back off when transmitting data packets, which may result in interference from other cells in the same SP phase during transmission. It is not clearly stated in the standard whether there are Clear Channel Assessment (CCA) in the SP phase. If there is CCA, the links in different cells will be suppressed. If there is no CCA in SP, there will be great interference in the links in different cells [4,5].

The method for interference reduction of multi-cell in the existing protocol is mainly AP Clustering. AP Clustering interleaves the BTI or BHI phases of different cells to ensure the successful transmission of the beacon frame and the beam training is done correctly.

The work done in this paper: for the SP phase interference of multi-cell, an method is proposed based on the AP Clustering of the existing protocol standard. We propose a new frame called SPBeginBroad frame. This new frame can synchronize the transmission progress of different cells in SP phase. If the transmission progress can be synchronized, the interference to receiver can be reduced greatly. At the beginning of the SP, the initiator of the link directly broadcasts a new frame, which can reduce the interference in links of different cells. Specifically, we design a multi-cell cooperative transmission MAC protocol flow, and verify its performance through simulation. The simulation results show that the MAC protocol flow can improve throughput of the network and service quality of edge users.

The rest of this paper is organized as follows. In Sect. 2, we find the problems in the high-density cell scenario of the existing IEEE 802.11ad/ay protocol. In Sect. 3, a new algorithm is proposed to solve the problems. The Sect. 4 set the simulation configuration and gives the simulation results. Finally, Sect. 5 concludes this paper.

2 Motivation

In multiple cells, there are interferences between the cells using the same frequency band. We use AP Clustering principle to reduce the interference in the beam training phase and ensure that STAs can receive DMG Beacon frames. This paper uses distributed AP Clustering, which interlace the BHI period between multiple cells. While the nodes in a cell carries out information sending and beam training in BHI period, the nodes in other cells are in a special SP, Beacon SP, to keep silent, in order to ensure that the cell can successfully complete the information transmission and beam training [6,7].

In IEEE 802.11ad/ay, SP is introduced for scheduling transmission in the cell. When a AP assigns a certain SP to a certain STA for downlink transmission or uplink transmission, the SP of the cell can only be used for downlink transmission or uplink transmission with this STA. In SP phase nodes does not randomly back off to reduce interference from other cells. This leads to the interference of the same frequency channel between the cells specially when the multi cell are in same SP phase and the beam direction coincidence degree is high. In addition, if there are CCA in SP, when each cell uses the opposite direction of the beam, it will cause the STA to fail to send packet to complete link transmission process.

In order to solve the problems of the SP phase in the multi cell scenarios, firstly we set that the nodes in the SP do not carry out CCA. The data packet sender of the SP phase detects the specific packets from other cells. If the specific packets from other cells are received, the data transmission in this cell is synchronized with the other cell transmission according to the information in the specific packet. Thus, the interference in links which belong to different cells in the same SP is reduced, the concurrency of links in the high density scene of multi cells is improved, the anti-interference ability of multi cells is improved and the service quality of the edge users can be improved.

3 Method Description

3.1 Problem Description

AP Clustering guarantees the effectiveness of beam training and Beacon frame transmission. The concept of protection SP phase is proposed in 802.11ad protocol for multi-cell scenario. It means that when one cell is in the SP phase, other cells around it are not in the SP phase but in CBAP phase. If there is interference between two cells which are in SP phase and CBAP phase, respectively, the time required to access the channel of SP phase is short than CBAP phase, due to EDCF being used in CBAP phase and SP phase is scheduling [8]. It means that the data transmission in SP phase can be guaranteed more effectively than in CBAP phase [9,10].

STA11   AP1   AP2   STA21

Fig. 2. An example of multi-cell topology.

This paper proposes a new SP phase process based on the AP Clustering. For example, as shown in Fig. 2, the cell where the AP1 is located (the cell is the primary cell, and the information transmission of the cell is the highest priority) starts SP phase. Firstly, the STA in transmission state sends a broadcast frame named SPBeginBroad Frame to inform other cells. Other cells are secondary cells, and guarantee the information transmission of the self cell as much as possible. The SPBeginBroad contains the packet transmission start time and the size of the data packet and etc. After receiving the SPBeginBroad Frame, the STAs of other cells send the data packet synchronously with the cell where the AP1 is located, until the end of the SP phase belongs to itself. If the STAs of other cells do not receive the SPBeginBroad Frame, it means other cells do not interfere with the primary cell, so normal data transmission may be performed in these cells. In this paper, only the uplink is discussed.

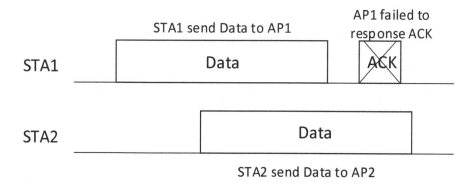

Fig. 3. The transmission failure in SP.

As shown in Fig. 3, in the SP phase of multi-cell, the transmission between cells is not controlled, and there is no CCA in the SP phase. The phenomenon shown in the figure will appear, and there may happen that one link is successfully transmitted, and the other links are always in a failed state. For example, when the STA11 sends a data packet, the data packet from the STA21 is received, and the data packet end time is after the ACK frame. Therefore, the STA11 fails to receive the ACK frame, and it is impossible to determine whether the Data is successfully received by the AP1, and then STA11 enters the retransmission state.

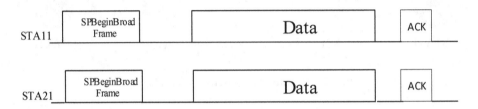

Fig. 4. The transmission progress in our method.

As shown in Fig. 4, we use the flowchart to describe the SP phase of multi cells. In combination with the topology of Fig. 3, after the cell of AP1 sending the SPBeginBroad Frame, STA11 starts the data transmission process after 2*SIFS, and STA21 receives this special frame. After receiving the frame, the STA 21 specifies the data transmission start time of the cell in which the AP1 is located, and starts the link data transmission process of the current cell after 2*SIFS; the data transmission process remains synchronized between the cells.

3.2   Frame Structure

Figure 5 is the SPBeginBroad frame structure. There are 4 parts of the frame: SPBegin Header, SPBegin MAC Header, Payload and FCS. We mainly need SP PacketSize, MCS ID and SP Duration to achieve synchronization of multiple cells.

SPBegin_Header			SPBegin_MAC_Header							
Data Size	TxVector	Data Premable	SP Packet size	MCS ID	SP Duration	DA	SA	TA	Payload	FCS

Fig. 5. SPBegin frame structure.

The cooperative transmission process ensures concurrency of different cells in the SP phase, effectively reduces interference and delay between cells, improves cell throughput.

4   Simulation and Discussion

4.1   Simulation Design and Implementation

We use the ns-3 simulation platform for the simulation, and ns-3 is one popular network simulation tool in the current. Based on the existing ns-3 simulation platform, the high frequency WiFi part is added for function development [11].

The simulation topology is set to a straight line. AP1 and STA11 on the left in Fig. 6 belong to the same cell, and AP2 and STA21 on the right belong to the same cell. AP1 is 0.5 m from STA11, AP1 is 1 m from AP2, and AP2 is 0.5 m from STA21. The simulation duration is 10s, the beacon interval is 100 ms, and the DTI phase is divided into 8 SPs, excluding Beacon SP. The simulation service adopts a uniform rate service. In this paper, the uniform rate service is used for performance discussion. The main simulation parameters are showed in Table 1.

STA11   AP1   AP2   STA21

Fig. 6. Simulation topology.

Table 1. Simulation parameters.

The number of AP	2	BI length	100 ms
The number of STA in one cell	1	Data MCS	DMGMCS10
Position of AP1	(1.5, 1, 0)	Management MCS	DMGMCS0
Position of AP2	(2.5, 1, 0)	Mobility model	Stationary
Position of STA11	(1, 1, 0)	Length of simulation time	10 s
Position of STA21	(3, 1, 0)	CBAP	no CBAP
Packet length	4096 bytes	SP	no RTS/CTS

There is no RTS/CTS in the process of transmitting data packets in the SP phase and we set that there is no CCA in SP. The MCS of data frame is DMGMCS10, and the MCS of management frame is DMGMCS0. The beacon interval of the two cells are staggered from each other, and Beacon SP is added to protect the BHI phase. As shown in Fig. 7, when AP1 is in the BHI phase, the cell where AP2 is located is in a silent state, does not interfere with the cell where AP1 is located. The SP is used in the DTI phase, and STAs send data to APs in SP.

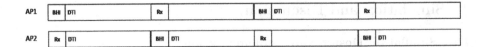

Fig. 7. Basic protocol framework.

In the SP phase, the STA broadcasts the SPBeaginBroad Frame before transmitting the data packet, and carries the information packet and the packet transmission start time of the SP. After broadcasting the SPBeginBroad Frame for 2*SIFS, the STA11 starts to send the Data packet. In ns-3, STA21 enters the scheduling timing after receiving the frame, and waits for 2*SIFS to perform synchronous transmission with the cell where AP1 is located. Since the specified data packets of the two cells are the same size, the two links can ensure a good synchronization effect in this SP phase.

4.2   Performance Evaluation

We analyzes the simulation results in this section.

In Fig. 8, we compare the throughput of the multi-cell cooperative transmission scheme and AP Clustering scheme. It can be seen from the figure that the throughput of the AP Clustering scheme is not much different from the multi-cell

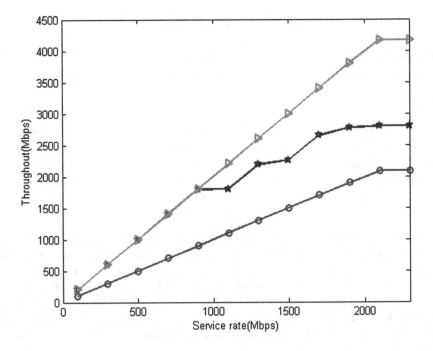

Fig. 8. Throughput of the network.

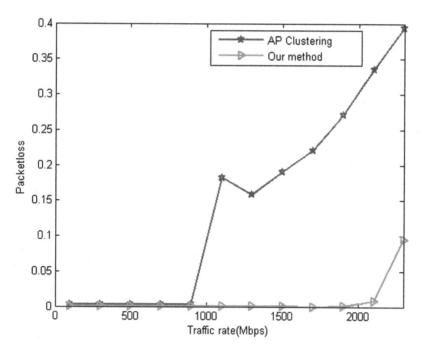

Fig. 9. Packet loss of the network.

cooperative transmission scheme when the traffic rate is low. As the service rate increases, the conflict between AP Clustering links increases, the trend of the throughput increases slowly, and the multi-cell cooperative transmission scheme shows better throughput gain. If the service is not a uniform rate service, the throughput performance gain is significant when the service is saturated.

In Fig. 9, we compare the packet loss of the multi-cell cooperative transmission to the AP Clustering schemes. It can be seen from the figure that when the service rate is low, the packet loss rate of the two schemes is not much different. As the service rate increases, the packet loss of the AP Clustering scheme increases significantly, and the packet loss rate of the multi-cell cooperative transmission scheme increases slowly.

5   Conclusions and Future Work

In this paper, based on AP Clustering protocol flow, a new multi-cell cooperative transmission MAC flow is proposed. Through principle analysis and ns-3 simulation verification, it shows that in the two-cell topology, when full traffic rate, the multi-cell cooperative transmission has a greater advantage than AP Clustering in throughput packet loss, delays and other aspects. This paper only discusses the uplink service in the two cells, and the multi-cell complex scenario is to be studied later.

Acknowledgement. This work was supported in part by the National Natural Science Foundations of CHINA (Grant No. 61771392, No. 61771390, No. 61501373, and No. 61271279), the National Science and Technology Major Project (Grant No. 2016ZX03001018-004, and No. 2015ZX03002006-004), the Fundamental Research Funds for the Central Universities (Grant No. 3102017ZY018), and the Science and Technology on Communication Networks Laboratory Open Projects (Grant No. KX172600027).

References

1. ISO/IEC/IEEE international standard for information technology-telecommunications and information exchange between systems-local and metropolitan area networks-specific requirements-part 11: wireless LAN medium access control (MAC) and physical layer (PHY) specifications amendment 3: enhancements for very high throughput in the 60 GHz band (adoption of IEEE Std 802.11ad-2012). ISO/IEC/IEEE 8802–11:2012/Amd.3:2014(E), pp. 1–634, March 2014
2. IEEE standard for information technology-telecommunications and information exchange between systems local and metropolitan area networks-specific requirements - part 11: wireless LAN medium access control (MAC) and physical layer (PHY) specifications. IEEE Std 802.11-2016 (Revision of IEEE Std 802.11-2012), pp. 1–3534, December 2016
3. IEEE draft standard for information technology-telecommunications and information exchange between systems local and metropolitan area networks-specific requirements part 11: wireless LAN medium access control (MAC) and physical layer (PHY) specifications-amendment: enhanced throughput for operation in license-exempt bands above 45 GHz. IEEE P802.11ay/D2.0, July 2018, pp. 1–673, January 2018
4. Nitsche, T., Cordeiro, C., Flores, A.B., Knightly, E.W., Perahia, E., Widmer, J.C.: IEEE 802.11ad: directional 60 GHz communication for multi-Gigabit-per-second Wi-Fi [invited paper]. IEEE Commun. Mag. **52**(12), 132–141 (2014)
5. Perahia, E., Gong, M.X.: Gigabit wireless LANs: an overview of IEEE 802.11 ac and 802.11 ad. ACM SIGMOBILE Mob. Comput. Commun. Rev. **15**(3), 23–33 (2011)
6. Hemanth, C., Venkatesh, T.G.: Performance analysis of contention-based access periods and service periods of 802.11 ad hybrid medium access control. IET Netw. **3**(3), 193–203 (2013)
7. Arora, K.K., Vyas, P., Rupani, A., Purohit, M.: Wi-Gig (IEEE 802.11ad): future, trends and era. In: 2017 2nd International Conference on Telecommunication and Networks (TEL-NET), pp. 1–4. IEEE (2017)
8. Rajan, M.N.U., Babu, A.V.: Theoretical maximum throughput of IEEE 802.11ad millimeter wave wireless LAN in the contention based access period: with two level aggregation. In: 2017 International Conference on Wireless Communications, Signal Processing and Networking (WiSPNET), pp. 2531–2536. IEEE (2017)
9. Chen, Q., Tang, J., Wong, D.T.C., Peng, X., Zhang, Y.: Directional cooperative MAC protocol design and performance analysis for IEEE 802.11ad WLANs. IEEE Trans. Veh. Technol. **62**(6), 2667–2677 (2013)
10. Saha, S.K., Koutsonikolas, D.: Towards multi-gigabit 60 GHz indoor WLANs. In: 2015 IEEE 23rd International Conference on Network Protocols (ICNP), pp. 470–472. IEEE (2015)
11. Riley, G.F., Henderson, T.R.: The ns-3 network simulator, pp. 15–34 (2010). http://www.nsnam.org/

Full Duplex Enabled Next Generation mmWave WiFi Network

Yue Li[1,2], Ping Zhao[1], Bo Li[1], Mao Yang[1(✉)], and Zhongjiang Yan[1]

[1] School of Electronics and Information, Northwestern Polytechnical University,
Xian, China
{libo.npu,yangmao,zhjyan}@nwpu.edu.cn
[2] Science and Technology on Communication Networks Laboratory,
Shijiazhuang 053200, China

Abstract. The full-duplex technology with the same frequency and the same time has developed rapidly in the wireless communication field. Compared with the traditional half-duplex technology, full-duplex technology can obtain the double spectral efficiency and throughput. However, in low-frequency asymmetric full-duplex communication, AP has full-duplex, and STA only has half duplex. And the interference between STAs results that we still cannot reach the theoretical multiple throughput. In this paper, we use the feature of directional propagation of mm wave to eliminate the interference between STAs in asymmetric full-duplex communication, and the full-duplex is applied to the WIFI high-frequency field for the first time. This paper introduces the BI frame of MAC architecture of IEEE 802.11 ay protocol in mmWave WIFI network, and designs the SP period of BI frame. So in SP period, the AP has full-duplex function and STAs still are half-duplex. We use Matlab to simulate, and get the theoretical multiple throughput.

Keywords: Full-duplex · IEEE 802.11ay · Triangle communication · Matlab

1 Introduction

Full-duplex (FD) communication is a new technology proposed in recent years, but it develops rapidly and is widely concerned in academia and industry. Traditional half-duplex (HD) cannot receive and send wireless signals at the same time and frequency, resulting in a waste of wireless spectrum resources [1]. And full-duplex can receive and send wireless signals at the same time and frequency, yet fundamentally avoid the waste of spectrum resources in half-duplex communication due to the orthogonality between the signal to send/receive. Theoretically, FD can exponentially improve the network capacity and spectrum efficiency. Full-duplex communication has great potential to improve spectrum resource utilization and user data throughput.

© ICST Institute for Computer Sciences, Social Informatics and Telecommunications Engineering 2019
Published by Springer Nature Switzerland AG 2019. All Rights Reserved
B. Li et al. (Eds.): IoTaaS 2018, LNICST 271, pp. 255–264, 2019.
https://doi.org/10.1007/978-3-030-14657-3_25

In wireless communication networks, the biggest technical challenge to the full-duplex is the interference. In fact, due to the nodes sending and receiving antennas are too close, sending signals will be leaked into receiving channel resulting in interference or even drowning the useful signal it wants to receive (Fig. 1) [2]. The full-duplex sender can know the relevant parameters of the signal it sends. In theory, it can eliminate the interference of the sending signal to the receiving signal, that is, self-interference elimination. In recent years, Self-interference canceling technology has also made great progress, which can gradually support full-duplex communication. Different levels of self-interference elimination technology have great influence on the capacity of full-duplex system, which is proportional to each other [3]. Next generation wireless network requires realize full-duplex both for AP and STA in order to improve the network throughput. But the cost of the STAs realization is too high, so we think it should be the AP is full-duplex mode, the STA still is half-duplex mode, which resulting in full-duplex mode is asymmetric full-duplex communication in the implementation [4]. In this model, as shown in Fig. 2, the AP sends and receives signals at the same time, and the STA only sends or receives signals. In low frequency wireless network communication, the AP and the STA often adopt omnidirectional communication. As a result, there is a serious problem in the asymmetric full-duplex communication: because of the low frequency omni-directional communication signals would radiate in all directions. When AP and STA2 are for uplink transmission, STA1 for downlink receiving will receive STA2s signal interference for uplink transmission. Therefore, in the asymmetric full-duplex communication, the full-duplex effect cannot reach the expected the twice network capacity, only has about 1.5 times in the current research [5,6].

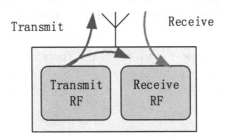

Fig. 1. Full duplex self-interference.

In the mmWave band, the beam is directional. In view of the problem for low frequency asymmetric full-duplex communication cannot gain twice the desired capacity, this paper puts forward that use of high frequency mmWave beam transmission characteristics to reduce the STA1s interference for STA2 to improve network capacity, expected to close to the theoretical spectrum gain of 2 times. The IEEE 802.11 ay protocol works in the ultra-high speed and low-interference 60 GHz frequency band, which means that in a cell, STA and AP can communicate with each other through directional beam realising point to

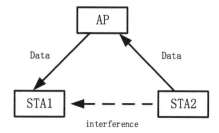

Fig. 2. Full duplex triangle communication self-interference.

point communication, with uplink and downlink functions [7]. Therefore, based on the IEEE 802.11 ay communication model, the design and simulation of high frequency full-duplex communication are carried out in this paper.

As far as we know, this paper will introduce the full-duplex into mmWave WiFi network for the first time, and provide the full-duplex communication MAC protocol support for the next generation mmWave WiFi network– the IEEE 802.11, and gain much better than the low frequency WiFi system. In low frequency, there must be different sectors for full-duplex. However, the intra-sectors and inter-sectors can both support full-duplex function in high frequency, especially the intra-sectors, which is of great significance.

The structure of this paper is as follows: Sect. 2 introduces the IEEE802.11ay high frequency directional communication MAC Layer; Sect. 3 introduces the implementation scheme of asymmetric full-duplex communication in high frequency directional communication. Section 4 presents the simulation results. Section 5 is the conclusion of this paper.

2   System Model

The IEEE 802.11 ay specification, based on the IEEE 802.11 standard, defines the physical layer PHY and the media access control layer MAC to provide wireless high-speed connectivity at the 60 GHz frequency band. Based on the IEEE802.11 ay communication model, the paper designs the MAC layer of high frequency.

The channel access process of IEEE 802.11 ay protocol is mainly carried out within Beacon Interval, and the sub-segment divided within Beacon Interval is called access period. Different access periods within beacon intervals have different access rules and are coordinated through schedules. The schedule is generated by STA as a PCP/AP, after which PCP/AP communicates with the device through Beacon and notification frames. The Non- PCP /AP device receives the time coordination information sent by PCP/AP and accesses the access channel according to the access mechanism of the corresponding stage within the coordinated time [8].

The channel time of IEEE 802.11 ay is in terms of hyperframe, called beacon interval, including four different access periods [9]: BTI: Beacon transmission interval, A-BFT: Association beamforming training, ATI: Announcement transmission interval and DTI:Data transfer interval, as shown in the figure. On the whole, it can be divided into BHI period and DTI period, beam training in BHI period and data transmission in DTI period. DTI is divided into competition based access period (CBAP) and Service period (SP).

During the beacon transmission period (BTI), the PCP/AP device sends one or more beacon frames containing location information, synchronization information, scheduling arrangements in this frame, etc. Non-PCP/AP devices synchronize with PCP/AP upon receipt of beacon frames and occupy channel as scheduled to communicate. During the BTI period, the beam will be trained to obtain the best transmission beam of PCP/AP (Fig. 3).

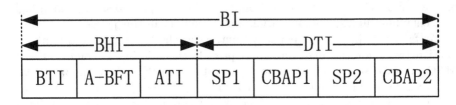

Fig. 3. Beacon interval frame structure.

After BTI, PCP/AP needs to wait for MBIFS time before starting the A-BFT process. The training process of beam forming in A-BFT includes the receiving sector sweep (RSS) and sector feedback sweep (SSW feedback). The best STA sending sector was obtained during the a-bft beam training.

ATI is usually done after A-BFT. During the ATI period, PCP/AP polls non-pcp /AP devices, which can send requests to obtain the allocation of time slot SP. After the A-BFT stage waiting MBIFS, PCP/AP sent Request frames to all STA, and before sending, it was necessary to check whether the CCA was idle. STA needs to reply after receiving the Request frame. If there is data to be transmitted, a Response frame is returned to PCP/AP after SIFS. If there is not data to be transmitted, an ACK frame is returned to PCP/AP after SIFS [10].

Beam refinement is between ATI and DTI, and the PCP/AP and non-pcp/AP devices in BRP stage interact many times to train their respective optimal receiving sectors.

Data exchanges between equipment during DTI period. Each beacon interval has one DTI. Non-pcp /AP devices communicate with others on CBAP they have competed and assigned SP in accordance with PCP/AP scheduling. PCP/AP presets a fixed start time and a fixed duration for each time slot, which is realized through the timing synchronization function. A time slot configuration can be a time slot SP where the ownership of the channel is granted to a single device. Or it could be a competing access time slot CBAP, where the right to use the channel is acquired through device competition.

3   mmWave Asymmetric Full-Duplex Scheme

Question: In general, at each SP stage, an AP only dispatches one STA to carry out uplink or downlink transmission with it. Not belonging to this SP, STA cannot communicate with AP within this SP time period, as Fig. 4.

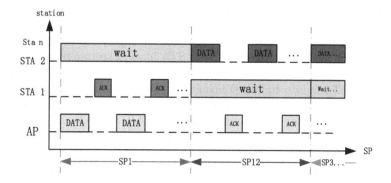

Fig. 4. Data transmisiion flow chart in SP period when half duplex.

Solution: In 802.11 ay, beam has directivity, so in SP stage, AP can consider scheduling multiple STA for uplink or downlink transmission. Respectively, the best sending or receiving sector for STA and AP must have different angles. In this case, the throughput for a SP stage will be multiplied. In this model, suppose AP has strong self-interference canceling ability.

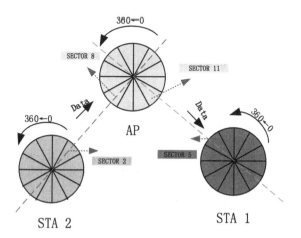

Fig. 5. Scenario 1.

Scenario 1: Inter-sector. The most common scenario is AP and the STA 1, STA 2 constituting asymmetric full-duplex communications, as Fig. 5: AP uses sector 8 communicating with the STA 1 with sector 2 for downlink transmission, at the same time, the AP uses sector 11 communicating with the STA 2 with 5 sectors for uplink transmission. STA 1 and STA 2 are visual for each other. In the low frequency they can interfere with each other, but in the high frequency mm wave, the beam directivity results that when STA 1 is in downlink receiving, interference of STA 2s upward transmission is nearly zero for the STA 1 direction.

Considering an extreme simple of scenario 1: the topological diagram location of the AP, STA 1, STA 2 as Fig. 6: AP and STA have 12 sectors respectively, and each sectors width is 30°. When AP and STA 1 are in downlink transmission, AP dispatch STA 2 into uplink transmission simultaneously. As shown in the figure, in the same SP stage, STA 1 with sector 1 transmits downlink with AP, while STA 2 with sector 7 transmits uplink and AP. The AP with the sector7 has downlink transmission with STA 1, at the same time AP with sector 2 has uplink transmission with the STA 1. Because of the two sectors of AP are facing, in the simulation the back of the interference of sectors are generally default as 0. Therefore, there is no interference between STA 1 and STA 2.

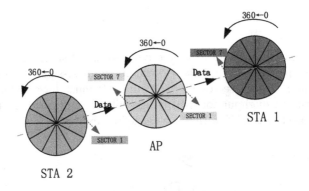

Fig. 6. Extreme simple of scenario 1.

Scenario 2: Intra-sector. In asymmetric full-duplex communication, when the two transmission paths between AP and STA 1 AP and STA 2 are very close to each other, it can be seen that AP-STA 1-STA 2 are on the same line. At this point, AP is equivalent to sending and receiving signals with the same beam, which improves efficiency on the basis of improving throughput. Although the sending and receiving sectors of STA 1 and STA 2 are in the same direction, the sending sector of STA 2 is on the back of STA 1, so there is little interference to STA 1.

Data Transmission Flow Chart. As shown in the Fig. 7, in SP1 period, the AP simultaneously schedules two STA for uplink and downlink transmission, no

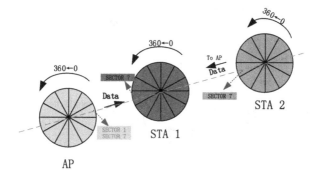

Fig. 7. Scenario 2.

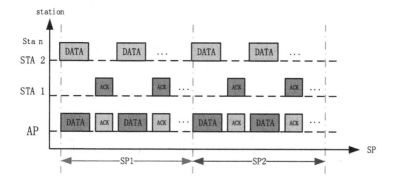

Fig. 8. Data transmission flow chart in SP period when full duplex.

matter in which scenario. The DATA for uplink or downlink both has the same transmission time and moment. The ACK time may not be the same. When STA 1 receives the downlink DATA packet from AP and AP receives the uplink DATA packet from STA 1, that is, there are two non-interference DATA packets transmitted at the same time in the SP1 stage, which improves the network capacity [11] (Fig. 8).

Integrated three scenarios, in asymmetric full-duplex communication, the interference between STA 1 and STA 2 becomes smaller because of the beam directivity characteristics of high frequency. The downlink receiving packets will not be affected, we can gain twice throughput of asymmetric full-duplex.

4   Simulation

The simulation scenario is STA intensive community. During the simulation, the size of DATA packets transmitted in the uplink or downlink is the same size and the ACK size is the same. The time interval between DATA and ACK is SIFS. In BI frame, BHI stage has completed beam training and has determined the best sending sector and the best receiving sector of STA and AP. The SP and CBAP

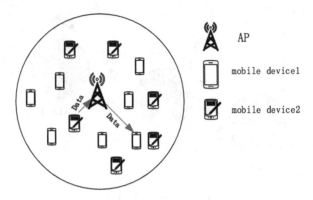

Fig. 9. AP schedule two STA simulation scene.

stages both transmite data through the best sector. In the DTI period, AP in SP stage dispatches two STA for full-duplex transmission, and CBAP keeps the original mode unchanged.

Simulation: Relationship Between SP Throughput and DateRate. The topology diagram of ap–sta 1–sta 2 is shown as Fig. 9. In a single-cell multi-sta intensive scenario, AP schedules one and two STA respectively. In SP period,

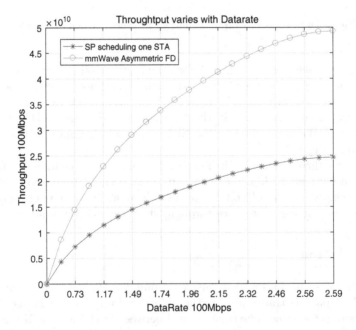

Fig. 10. Relationship between SP throughput and DateRate both on full or half duplex.

simulation time is 4s; packet is 8 Mbit; ACK is 304 bit, SIFS is 2.5/Ms, header is 416 bit. In SP period, throughput varies with the DateRate.

Analysis: From the simulation results, it can be seen that as the DataRate increases from 0, the throughput of the entire SP stage also increases under the two simulation conditions. When the DataRate reaches a fixed value 2.56, the increase trend of the throughput slows down and gradually increases to a value that remains basically unchanged, that is, the channel reaches saturation. The throughput of two STA scheduling in SP is always higher than that of one STA scheduled by AP, and is approximately twice as high as expected (Fig. 10).

5   Conclusions and Future Work

This paper mainly introduces the asymmetric full-duplex mode of high frequency mm wave WIFI network, and introduces full-duplex to high frequency mm wave WIFI network for the first time. According to the high frequency mm wave property and IEEE 802.11 ay MAC protocol, the data transmission of all asymmetric full-duplex communication in SP period is given, and we use MATLAB simulation tool to get the graph of the relationship between the throughput and data rate of all asymmetric full-duplex communication and semi-duplex communication in the high frequency WIFI network. The simulation results show that the asymmetric full-duplex of high frequency mmWave WIFI network solves the interference between low frequency asymmetric full-duplex STA very well, and gains the theoretical doubled throughput. This paper provides MAC protocol support for full-duplex communication for the next generation of WIFI IEEE 802.11ay, but the scheduling mode of participating in full-duplex when the AP schedus STAs in the SP stage needs to be improved and the efficiency needs to be improved.

Acknowledgement. This work was supported in part by the National Natural Science Foundations of CHINA (Grant No. 61771392, No. 61771390, No. 61501373, and No. 61271279), the National Science and Technology Major Project (Grant No. 2016ZX03001018-004, and No. 2015ZX03002006-004), the Fundamental Research Funds for the Central Universities (Grant No. 3102017ZY018), and the Science and Technology on Communication Networks Laboratory Open Projects (Grant No. KX172600027).

References

1. Chiao, B., Ma, M.: Analysis on full-duplex technology with the same frequency. Telecommun. Netw. Technol. (11), 29–32 (2013)
2. Liu, X.: Research on key technologies of full-duplex radio frequency wireless communication system. Beijing University of Posts and Telecommunications (2017)
3. Li, J.: Research and analysis on the effects of full-duplex system Zhonggan ratio on its performance. Harbin Institute of Technology (2014)

4. Wang, L., Xu, H.: Capacity analysis and comparison of traditional semi-duplex and pure full-duplex wireless communication systems. Mob. Commun. (18), 63–68 (2014)
5. Choi, J.I., Jain, M., Srinivasan, K., et al.: Achieving single channel full duplex wireless communication. In: International Conference on Mobile of Systems, pp. 301–312 (2011)
6. Duarte, M., Dick, C., Sabharwal, A.: Experiment-driven characterization of full-duplex wireless systems. IEEE Trans. Wirel. Commun. **12**, 4296–4307 (2012)
7. Zou, N.: IEEE802.11 AD standard and application. Inf. Technol. Stand. (3), 41–44 (2013)
8. Huang, X.L.: Construction of wireless network in laboratory based on IEEE802.11ad technology standard. J. Tonghua Normal Univ. (2017)
9. ISO/IEC/IEEE ISO/IEC/IEEE International Standard for Information technology-Telecommunications and information exchange between systems-Local and metropolitan area networks-Specific requirements-Part 11: Wireless LAN Medium Access Control (MAC) and Physical Layer (PHY) Specifications Amendment 3: Enhancements for Very High Throughput in the 60 GHz Band (adoption of IEEE Std 802.11ad-2012)
10. ISO/IEC/IEEE 8802–11:2012/Amd.3:2014(E) (2014)
11. Shi, L., Fapojuwo, A., Viberg, N., Hoople, W., Chan, N.: Methods for calculating bandwidth, delay, and packet loss metrics in multi-hop IEEE802.11 ad hoc networks. In: VTC Spring 2008 - IEEE Vehicular Technology Conference (2008)
12. Mohammadi, M., Chalise, B.K., Hakimi, A., Mobini, Z., Suraweera, H.A., Ding, Z.: Beamforming design and power allocation for full-duplex non-orthogonal multiple access cognitive relaying. IEEE Trans. Commun. **66**, 5952–5965 (2018)

ESR: Enhanced Spatial Reuse Mechanism for the Next Generation WLAN - IEEE 802.11ax

Yuan Yan, Bo Li, Mao Yang$^{(\boxtimes)}$, and Zhongjiang Yan

School of Electronics and Information, Northwestern Polytechnical University,
Xi'an, China
yanyuan2035@mail.nwpu.edu.cn, {libo.npu,yangmao,zhjyan}@nwpu.edu.cn

Abstract. Wireless local area network (WLAN) technology has become mature in the past few decades. However, with the development of social application forms, the situation of high-density deployment of cells has become more frequent in recent years, so the next-generation WLANs face the unprecedented challenge of quality of service (QoS) and quality of experience (QoE) of high-density cell users. In response to this challenge, The Overlapping Basic Service Set Power Detection (OBSS PD) mechanism and Spatial Reuse Parameter (SRP) mechanism are proposed in IEEE 802.11ax (Institute of Electrical and Electronics Engineers) to improve spatial multiplexing capability and certain throughput capabilities. But both of them possess inherent drawbacks. Based on this, this paper proposes an enhanced spatial reuse mechanism, named ESR (Enhanced Spatial Reuse), by limiting the value of the CCA threshold in the SRP mechanism to improve the quality of concurrent links, which further improves the average throughput. Compared to traditional SRP, simulation results show that ESR can increase throughput by 10% and reduce packet loss by 10%.

Keywords: WLAN · SRP · Clear Channel Assessment (CCA) · MAC protocol

1 Introduction

Due to the deep penetration of mobile Internet and the increasing enrichment of wireless network services, wireless communication and quality of service (QoS) requirements have increased dramatically in recent years [1]. Wireless LANs and cellular networks have been recognized as the most important carriers of today's network services due to their high speed, flexible deployment and low cost. According to statistics, from 2016 to 2021, the world's wireless data traffic will increase significantly, with an annual growth rate of 47%. According to the report, by 2021, the percentage of the world's wireless data traffic accounts for the total traffic will increase from 42% in 2015 to 49%. In order to cope with such rapid wireless service growth, the Institute of Electrical and Electronics

© ICST Institute for Computer Sciences, Social Informatics and Telecommunications Engineering 2019
Published by Springer Nature Switzerland AG 2019. All Rights Reserved
B. Li et al. (Eds.): IoTaaS 2018, LNICST 271, pp. 265–274, 2019.
https://doi.org/10.1007/978-3-030-14657-3_26

Engineers: IEEE 802.11 will soon release the next generation WLAN (Wireless local area network) standard amendment: IEEE 802.11ax [2].

It is precisely because of the huge increase in wireless demand that the business demand cannot meet the requirements in crowded places, such as companies, shopping malls and residential buildings. The wireless network scenarios that have emerged in recent years are collectively referred to as high-density deployment scenarios the focus of the next generation of WLAN protocol standard. High-density deployment of APs (Access Points) is bound to increase interference [3], and the quality of users' services is more difficult to guarantee.

In response to the above problems, the objective of improving spectrum reuse is proposed in the next generation protocol standard: IEEE 802.11ax. Thus, spatial reuse (SR) [4] is supposed to be a key technology of IEEE 802.11ax, which includes enhanced physical carrier sensing mechanism, enhanced virtual carrier sensing mechanism and power control technology. In the traditional scene, it is always only one transmission link will exist since both the physical carrier sensing and virtual carrier sensing are too conservative. However, allowing multiple links to be transmitted in concurrently with controlled interference will obviously further increase the transmission capacity. Therefore, IEEE 802.11ax [2] introduces two types of SR mechanisms: Overlapping Basic Service Set Power Detection (OBSS PD) mechanism and Spatial Reuse Parameter (SRP) mechanism.

OBSS PD mechanism. The protocol standard indicates that once the node receives the inter-BSS frame, it uses the OBSS PD value to listen to the channel state, which is slightly higher than traditional CCA (Clear Channel Assessment) threshold. If the channel energy is lower than OBSS PD, the node can ignore the virtual carrier sensing mechanism to start backoff and access the channel. The increase of the CCA threshold directly reduces the difficulty of link concurrency. OBSS PD also reduces interference between concurrent links by controlling the transmit power of the links [5]. However, since the current link cannot determine when the concurrent link terminates the transmission, the channel state will become confusing.

SRP mechanism. The protocol standard indicates that once the node receives the TF (Trigger Frame) from inter-BSS, if the SR is allowed, an uncontrollable CCA threshold is used to listen to the channel state for the data frame of the TF source cell. The SRP technology increases the concurrent opportunity by ignoring the interference of the current link to the concurrent link. Then, the current link performs certain control on the transmission power of the concurrent link to ensure that the concurrent link does not interfere with the primary link. However, due to the wide selection of the concurrent link, the problem of interference between concurrent links is serious, and the quality of service is greatly reduced.

For OBSS PD, the concurrent link is unknown and uncontrollable; for SRP, although the interference problem between the current and concurrent links is solved to some extent, the selection of links is really radical and the quality of the link cannot be effectively guaranteed because the physical carrier sensing is uncontrollable when concurrent links receive packets which from current link.

Since both OBSS PD and SRP mechanism possess inherent drawbacks as mentioned above, this paper will apply the idea of adjusting the CCA threshold based on SRP technology. When the node receives the TF, the CCA threshold is fixed from an uncontrollable value indicated in the IEEE 802.11ax to an appropriate value, so that the generation of the concurrent link is no longer the mode of receiving the TF sent by the current AP to continue to backoff and access the channel, but those nodes with low transmission possibilities in neighboring cells can participate in parallel transmission. Through the physical carrier sensing mechanism, the requirements for link concurrency are more strict, so as to further improve the quality of service.

The contributions of this paper can be summarized as follows:

– As far as we know, this is the first work to propose the idea of combining the SRP mechanism with a limited CCA threshold to achieve higher quality of service.
– The simulation results show that the scheme can guarantee a slight increase in throughput under the condition of multi-link parallel, and on the other hand, this mechanism supports SRP to select concurrent links with better quality to reduce packet loss rate.

The reminder of this paper is as follows: Sect. 2 illustrates the motivation and key idea of ESR. Section 3 depicts the simulation platform. Section 4 evaluate the performance. Finally, Sect. 5 concludes the paper.

2   SR Overview for IEEE 802.11ax

2.1   SRP Description

This article only discusses the uplink. The SRP mechanism is to send a trigger frame (TF) with control information before transmitting data, so that neighboring nodes can receive the trigger frame to continue to backoff and access the channel. The corresponding field of the HE-SIG-A in the TF carries a message indicating whether the current link transmission supports the SRP. If supported, the STAs that received the TF can continue to access channel to transmit data during the transmission time of the current link, as shown in Fig. 1.

While allowing the concurrent link, the TF received by STAs also obtain an SRP value, which is a power control information of the current link to the concurrent links, and the value is an adjustable parameter (the simulation of this paper is tentatively set to $-65\,\mathrm{dBm}$). $P = SRP - TFP_{RX}$, note that P is the transmit power of the concurrent link after the power control, TFP_{RX} is the received power of the TF. If $P \geq P_{\min}$ ($P_{\min}$ is the value of the minimum transmit power for the concurrent link, the value is tentatively set to $10\,\mathrm{dBm}$), the station is allowed to start the SRP backoff process, otherwise it will continue to hang. If the channel is successfully accessed, its actual transmit power is $min(P, P_{\max})$ ($P_{\max}$ is the maximum transmit power of the concurrent link, tentatively set to $15\,\mathrm{dBm}$). After power control, the concurrent link will only appear outside the

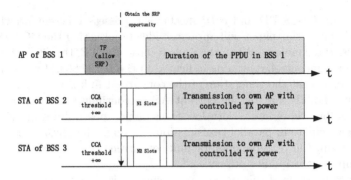

Fig. 1. SRP transmission mechanism.

area with the primary AP as the origin and d as the radius (d is the distance from the position where the dynamic transmission power equal to 10 dBm to the primary AP), as shown in Fig. 2, the STAs that can participate in the link are displayed in green, which ensures the quality of the primary link transmission.

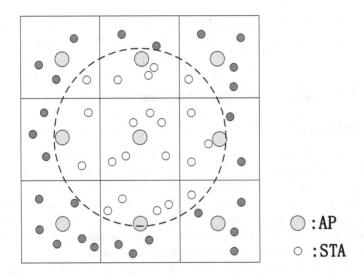

Fig. 2. SRP scene diagram. (Color figure online)

2.2   Description of SRP Problem

In the SRP model, the communication of current link is guaranteed through the power control. However, in the context of the concurrent link, although the dynamic transmission power causes the situation that STAs farther away from the AP, the transmission power is larger, but in a high-density deployment scenario, multiple users accessing the channel at the same time are more common, resulting in a large increase in interference between concurrent links. As shown in Fig. 3, STAs marked with red indicate those STAs with same backoff value. Because the cells are too close together, the simultaneous transmission of packets makes the data packet transmission probability failure, which is the main reason for the high packet loss rate of the SRP mechanism.

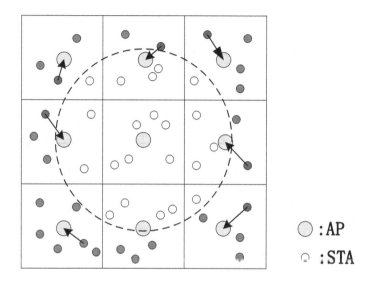

Fig. 3. SRP scene diagram. (Color figure online)

3   Introduction of ESR

3.1   Problem Solved

Considering that the value of the CCA threshold directly reflects the integrated transmission state of the neighboring cell, the more complex the application scenario, the setting of the CCA threshold is more important. If the scene in which the simultaneous transmission occurs is described as a problematic scenario, the following will give a solution to deal with the problematic scenario under an appropriate CCA_SRP.

As shown in Fig. 4, in the SRP mechanism, STA1 and STA2 can participate in parallel after receiving the TF from the current link. If STA1 and STA 2 get the same backoff value, these two transmissions are likely to fail because the two cells is too close. If the CCA threshold is set to SRP_CCA after the two STAs receive the TF, STA1 which is closer to current link will judge that the current state of the channel is busy because the physical carrier sense detects that the channel power accumulation exceeds the SRP_CCA threshold (-62 dBm). STA1 is unable to participate in this parallel transmission. However, STA2 is far away from the current link, the power accumulation of the channel does not exceed

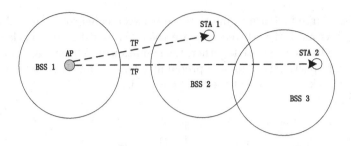

Fig. 4. Concurrent link interference scenario.

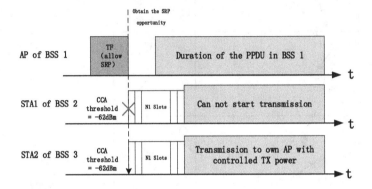

Fig. 5. ESR interference scene transmission mechanism.

the CCA_SRP threshold, and the state is idle. It can participate in this parallel transmission, thus avoiding the interference of multiple STAs simultaneously accessing the channel. As shown in Fig. 5.

Of course, the channel interference described in this scenario mainly comes from the current link. In a complex scene, multiple links affect the channel state, so the role of the CCA threshold is more significant.

3.2   Process

The design process of this paper is as follows:

Step 1. STAs that can transmit data under power control in other cells receive the TF from the current link.

Step 2. Set the CCA threshold from an uncontrollable value specified in traditional SRP to CCA_SRP.

Step 3. The STA detects the channel status. If it is busy, it sets the NAV and cannot send data. If it is idle, ignore the NAV and continue to backoff.

Step 4. If a new TF with the information that SR is not allowed during the backoff, the SR is immediately stopped and the CCA threshold is reset to the initial value until the arrival of the new TF frame allowing SR.

3.3   Frame Format

Because the CCA threshold set for the transmission of the primary link in the SRP mechanism is uncontrollable, that is, the current link cannot be detected effectively by physical carrier sensing in the concurrent cell, so the innovation of this paper is combined with the adjustment of the CCA threshold to improve the quality of concurrent links. This paper uses the new threshold CCA_SRP to represent the traditional CCA threshold when STAs receive the TF that allows SRP.

In this paper, four bit lengths are added to the Common Info field in the TF frame to indicate different SRP_CCA values. The frame format is shown in Fig. 6. The SRP_CCA value table is shown in Table 1.

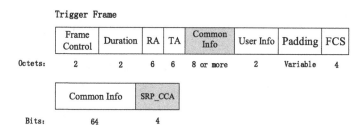

Fig. 6. Trigger frame format.

Table 1. SRP_CCA subfield encoding

SRP_CCA subfield value	Meaning
0	SRP_CCA_DISALLOW
1	−77 dBm
2	−72 dBm
3	−67 dBm
4	−62 dBm
5	−57 dBm
6	−52 dBm
7	−47 dBm
8–15	Reserved

4   Performance Evaluation

4.1   Simulation Design

In order to evaluate the performance of the above design and verify the expected effect, this paper is based on the NS-3 simulation platform [6].

The simulation configuration is designed according to the high-density deployment scenario, and the configuration is shown in Table 2 [7].

Table 2. ESR Scene configuration

Scene configuration	Value
Cell topology	4 * 8
Distance between cells	15 m
Number of users per cell	10
Traffic type	Uplink
Traffic rate	10 Mbps
Traffic packet size	1500Byte
Data MCS	VHTMCS7
Control frame MCS	VHTMCS0
Maximum transmission duration of A-MPDU	5.484 ms
Simulation time	15 s

4.2   Result and Analysis

This paper simulates the trend of system performance as SRP_CCA continues to increase. The average cell throughput is shown in Fig. 7. The average cell packet loss rate is shown in Fig. 8.

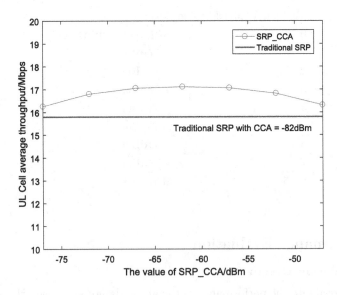

Fig. 7. Average throughput of the SRP_CCA mechanism.

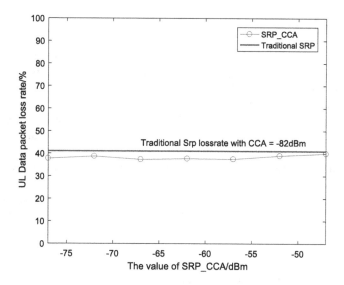

Fig. 8. Average packet loss rate of the SRP_CCA mechanism.

It can be seen from the results that as the SRP_CCA increases, the effect of the current link becomes less and less. It can be seen from Fig. 7 that when SRP_CCA = −62 dBm, the throughput reaches a maximum value. Then, with the SRP_CCA continues to increase, the throughput has a downward trend, and gradually tends to the traditional SRP baseline. According to the packet loss rate curve shown in Fig. 8, it can be seen that the packet loss rate is slightly lower than the traditional SRP mechanism after using the limited SRP_CCA, and as the SRP_CCA increases, the curve tends to the traditional SRP baseline.

In summary, according to the two curves, the increase of SRP_CCA controls the interference between concurrent links. Once the concurrent link responds to the interference of the current link, the problem scenario shown in Fig. 5 can be effectively solved., and as the SRP_CCA increases, the probability of concurrent links will gradually increase, and the problematic scene will also appear, slowly evolve into a traditional SRP model.

The transmission power of the current link is 20 dBm, and the maximum transmission power of the concurrent links is 15 dBm. After the power control, the node position that can participate in parallel are approximately in the area 30 meters away from the primary AP. The energy of the packets which are from the current link has been greatly attenuated [8,9] when these packets arrive in those areas and it is almost impossible to exceed the SRP_CCA value, so when SRP_CCA is higher than −42 dBm, it is almost identical to the traditional SRP model. This is also an important reason for the throughput and packet loss rate to go to the traditional SRP baseline.

5   Conclusion

Combined with the simulation results and analyses in the previous section, it can be considered that the change of the CCA threshold in the SRP mechanism can alleviate the serious problem of concurrent link interference. According to the above simulation results of different SRP_CCA values, at least the following conclusions can be summarized: (1) The value of the CCA threshold will directly affect the possibility of the node sending data. (2) Physical carrier sensing reflects the comprehensive judgment of different nodes on the transmission of neighboring cells. Especially in high-dense deployment scenarios, the effect is more obvious.

This paper only proposes the concept of flexible application of dynamic physical carrier sensing threshold to improve the quality of service in SRP. The author believes that the enhanced physical carrier sensing technology will be more widely used in the future wireless network environment.

Acknowledgement. This work was supported in part by the National Natural Science Foundations of CHINA (Grant No. 61771390, No. 61501373, No. 61771392, and No. 61271279), the National Science and Technology Major Project (Grant No. 2016ZX03001018-004, and No. 2015ZX03002006-004), and the Fundamental Research Funds for the Central Universities (Grant No. 3102017ZY018).

References

1. Drieberg, M., Zheng, F.C., Ahmad, R., et al.: An improved distributed dynamic channel assignment scheme for dense WLANs. In: International Conference on Information, Communications and Signal Processing, pp. 1–5. IEEE (2008)
2. Wireless LAN Medium Access Control (MAC) and Physical Layer (PHY) Specifications–Amendment 6: Enhancements for High Efficiency WLAN, IEEE Draft 802.11ax/D2.0, October 2017
3. Zhang, D., Mohanty, B., Sambhwani, S.D.: Scheduling based on effective target load with interference cancellation in a wireless communication system. US, US8676124 (2014)
4. Bellalta, B.: IEEE 802.11ax: high-efficiency WLANs. IEEE Wirel. Commun. **23**(1), 38–46 (2016)
5. Qu, Q., Li, B., Yang, M., Yan, Z., et al.: Survey and Performance Evaluation of the Upcoming Next Generation WLAN Standard - IEEE 802.11ax (2018)
6. Simulation and analysis of an integrated GPRS and WLAN network
7. Ha, D.V.: Network simulation with NS3 (2010)
8. Bae, D., Kim, J., Park, S., Song, O.: Design and implementation of IEEE 802.11i architecture for next generation WLAN. In: Feng, D., Lin, D., Yung, M. (eds.) CISC 2005. LNCS, vol. 3822, pp. 346–357. Springer, Heidelberg (2005). https://doi.org/10.1007/11599548_30
9. Chan, Z.H.: Investigation of next generation IEEE 802.11n wireless local area networks (WLAN) (2009)

PSR: Probability Based Spatial Reuse Mechanism for the Next Generation WLAN

Xuewei Cai, Bo Li, Mao Yang$^{(\boxtimes)}$, and Zhongjiang Yan

School of Electronics and Information, Northwestern Polytechnical University,
Xi'an, China
caixuewei@mail.nwpu.edu.cn, {libo.npu,yangmao,zhjyan}@nwpu.edu.cn

Abstract. Wireless local area network (WLAN) technology can be seen everywhere in people's lives, and technology itself is constantly developing, the coming new WLAN standard IEEE 802.11ax is excepted to increase the per-user throughput by 4 times than existing WLAN standards in residential, enterprise and other dense deployment scenarios. One of the key technologies of 802.11ax is spatial reuse (SR) technology, which improves area throughput by allowing concurrent links, and spatial reuse parameter (SRP)-based SR mechanism is introduced in 802.11ax. But in fact, the SRP mechanism lacks the ability to adapt to the scenario, and does not guarantee full gain from SR in any dense deployment, even exhibit a negative gain in some scenarios. This paper proposes a probability based SR (PSR) mechanism to achieve the full gain from SR with a probability value that reflects the scenario condition. Simulation results indicate that the PSR mechanism is a practical solution for SR in next generation WLAN (NGW).

Keywords: WLAN · IEEE 802.11 · Spatial reuse ·
Dense deployment · Next generation WLAN

1 Introduction

Wi-Fi is a common technology to access wireless local area network (WLAN) in people's daily life. In order to meet the communication needs of a large number of wireless network devices, outdoor and dense deployment scenarios [1] can also provide users with a good experience, the Wi-Fi community set up the next generation of WLAN IEEE 802.11ax working group in 2014 [2]. With the release of Draft 2.0 [3], the work of IEEE 802.11ax was nearing completion. The latest study demonstrates the per-user throughput of IEEE 802.11ax is up to 4.74 times of IEEE 802.11ac operating in the same band [4].

In WLAN, the adjacent links inevitably interfere with each other due to they are sharing the same wireless media, each link reduces the communication quality of other links, or even leads to failure; on the other hand, if the concurrent links can tolerate the interference caused by each other and transmit successfully,

© ICST Institute for Computer Sciences, Social Informatics and Telecommunications Engineering 2019
Published by Springer Nature Switzerland AG 2019. All Rights Reserved
B. Li et al. (Eds.): IoTaaS 2018, LNICST 271, pp. 275–283, 2019.
https://doi.org/10.1007/978-3-030-14657-3_27

then the gain can be obtained from the number of links to enhance the area throughput. In legacy IEEE 802.11, the carrier detection threshold is relatively conservative, on the one hand to ensure link quality, on the other hand, because the application scene is usually single cell with single access point (AP), and there is no gain in concurrent links; But in NGW, in order to pursue higher area throughput and frequency spectrum utilization under dense deployment, it is a very clear direction to obtain the gain from link numbers by controlling the interference between links through certain means. Therefore, a variety of spatial multiplexing (SR) techniques and two SR mechanisms are proposed in IEEE 802.11ax [3], and it has been proved by latest study that it can bring obvious gain in multi-BSS indoor scenarios [4]. The space reuse parameter(SRP)-based SR mechanism (SRP mechanism) is based on the combination of various technologies such as CCA enhancement, NAV enhancement, transmission power control and BSS Color mechanism, which effectively increases the number of concurrency links with controlled interference, and improves the area throughput and utilization of frequency spectrum.

Although the SRP mechanism controls the interference of the concurrent link to the current link by limiting the transmission power, it ignores that (1) the current link with default transmit power has greater interference with concurrent links; (2) collision problem caused by allowing STAs to access channel; (3) problem of hidden terminal [5] in multi-BSS scenarios. These problems will lead to too radical SR under the SRP mechanism and fail to achieve the desired results, or even destroy the transmission of the current link.

In order to solve the above problems, this paper proposes a probability based SR (PSR) mechanism, which adjusts the concurrent link access probability in SR according to the different scenarios, the number of concurrent links in the WLAN is maintained at a reasonable level so that the area throughput gain from SR is stable at a better level. The simulation results show that the PSR mechanism has stronger adaptability and is expected to give full play to the gain from SR in dense deployment scenarios. To the best of our knowledge, the probability based mechanism is first integrated into the SR mechanism in this study, and try to solve the problem that the existing SR mechanism cannot deal with the dense deployment scenarios with complicated channel conditions.

The reminder of this paper is as follows: Sect. 2 illustrates the motivation of this study. Section 3 details the key idea of PSR. Section 4 deploys several simulations and verifies the performance. Finally, Sect. 5 concludes the paper and plans the future work.

2   Motivation

2.1   Introduction to SRP-based SR Mechanism

In 802.11ax, uplink is triggered by trigger frame (TF) transmit by AP [6], assuming that all sites support SRP mechanism, then SRP mechanism flow is shown in Fig. 1. AP can set up whether to allow SR and control the interference from concurrent link power by configuring the SRP subfield in SIG-HE-A of TF. The SRP subfield containing 4 bits and encoded as shown in Table 1.

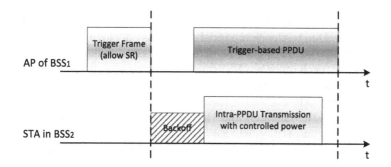

Fig. 1. SRP-based SR mechanism flow.

Table 1. SRP subfield encoding

Value	Meaning	Value	Meaning
0	SR disallow	8	SRP = −44 dBm
1	SRP = −80 dBm	9	SRP = −41 dBm
2	SRP = −74 dBm	10	SRP = −38 dBm
3	SRP = −68 dBm	11	SRP = −35 dBm
4	SRP = −62 dBm	12	SRP = −32 dBm
5	SRP = −56 dBm	13	SRP = −29 dBm
6	SRP = −50 dBm	14	SRP = −26 dBm
7	SRP = 470 dBm	15	Reserved

STAs in overlapping BSS (OBSS) can obtain the SRP after receiving the TF. If SR is allowed, the STA in OBSS can ignore CCA and NAV of current link and execute its backoff procedure until other concurrent link is detected, otherwise, it can start a transmission as the concurrent link if the backoff timer reaches 0. In addition, the following conditions need to be met from the concurrent link: (1) the concurrent link must end before the current link; (2) the transmission power must be controlled, that is, the transmission power of concurrent link is $P_{TX} \leq SRP - P_{RX}$, where P_{RX} is the receiving power of TF.

2.2 Problem Statement

(a) The interference of the current link to the concurrent link is brought by the sender of the current link, which is brought by the STA, but because the information that the concurrent link can obtain comes from the AP of the current link, the interference from the current link to the concurrent link is unpredictable, which will cause the concurrent link to be in a relatively fragile state.

(b) In the target indoor scenario defined by Task Group ax (TGax) for 802.11ax [7]. Due to the characteristics of dense deployment, there are often a large number of STA, such as enterprise scenario shown in Fig. 2. There are 32 offices, each office is a BSS, and 64 cubicles per office, each cubicle has more than one STA. The uplink transmission is triggered by TF in 802.11ax, so usually only AP will access the channel. But the SRP mechanism allows STA to access the channel, and the intensity of competition is difficult to predict, fierce competition will lead to the collision of the contention window and resulting in unexpected concurrent links if there are a large number of STA.

(c) In the multi-BSS scenario, the classical problem of hidden terminal may also lead to unexpected concurrent links, thus making the interference of current link higher than expected.

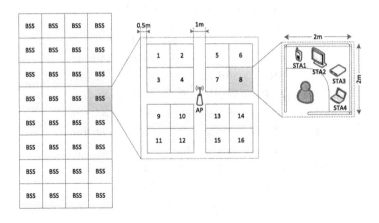

Fig. 2. Enterprise scenario.

In a word, the SRP mechanism has limited the concurrent link, but in fact, its ability to adapt to the scenario is weak, and there are serious drawbacks in the control scheme of the concurrent link, and the gain of SR cannot be fully obtained in any dense deployment scene.

3   Key Idea

3.1   Overview

In order to solve the problems of the SRP mechanism in the Sect. 2, this paper proposes a probability based SR mechanism. Within the PSR mechanism, each

AP maintains a probability of SR p, and it need to be carried in each TF, the STAs that receive the TF perform SR competition by probability p, it means there are some STAs that are not allowed to obtain the opportunity of SR at random. In general, the probability p reflects the intensity of SR competition, that is, the quantity information of STA in OBSS. The information can be generated by statistics of the transmission success rate of the BSS by AP, and can also be interacted through inter-AP communication. In addition, the transmission power of concurrent link still need to be controlled.

3.2   Mechanism

This section illustrates the workflow of the PSR mechanism through examples, and points out its source of gain. Figure 3 shows the common situations and problems encountered by SRP mechanism in the dense deployment: AP of BSS_1 send a SR allowed TF, and STA_1 in OBSS obtain this opportunity of SR because its backoff timer reaches 0; other STAs in OBSS (e.g. STA_2) pause their backoff process due to CCA mechanism; but in dense deployment scenario with a large number of STA, there is a greater possibility that contention window will collide, if there is an unexpected concurrent link just like the transmission from STA_3, it will cause uncontrollable interference to the other two links and reduce the benefit.

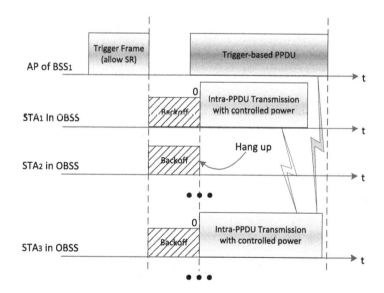

Fig. 3. The workflow of SRP mechanism under dense deployment.

In the same dense deployment scenario, the workflow of the PSR mechanism is shown in Fig. 4. The SR probability is p carried in the TF, and p roughly reflects the number of surrounding STA in OBSS. The STAs that receive the TF perform SR competition by probability p, it means the collision probability of contention window is directly related to p, so the SR probability p take an appropriate value can make the gain from SR fit the expectation.

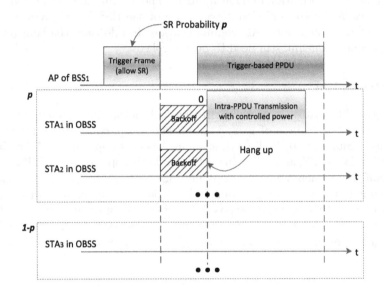

Fig. 4. The workflow of PSR mechanism under dense deployment.

4   Simulation

4.1   Simulation Environment

To test and compare the gain from SR with the SR probability p take different values by simulation, we use the NS-3 simulator [8]. The configuration of the simulation parameters is shown in Table 2. Link adaptation for STAs is allowed. The SR probability pp is fixed, the PSR mechanism will be degenerated to the SRP mechanism when p is taken as 1, and SR disallowed when p is taken as 1.

4.2   Performance Evaluation

Figure 5 illustrates that the gain from SR changed as the number of STAs in each BSS increased. When the number of STA in each BSS is small, especially 5–10, the gain from SR is obvious, up to more than 80%, but it is difficult to judge the optimal value of p; however, as the number of STA in each BSS increases, the gain from SR decreases, and the rate of decline is positively related to p; the SRP mechanism ($p = 1.0$) exhibit a negative gain with 25–30 STAs in each BSS, but considerable gain can still be achieved with PSR mechanism and set p equal to 0.2.

Table 2. The configuration of the simulation parameters

Parameter	Value
Service type	CBR (constant bit rate)
Traffic model	Poisson Model
Traffic rate	10 Mbps
Packet size	1472 octets
Trigger frame size	36 octets
Default data frame MCS	VHT MCS 7
Control frame MCS	VHT MCS 0
AP transmit power	20 dBm
Default STA transmit power	15 dBm
Lowest STA transmit power	5 dBm
SR probability	0, 0.2, 0.4, 0.6, 0.8, 1

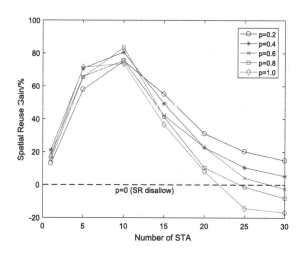

Fig. 5. Spatial reuse gain performance of PSR at different number of STA in each BSS.

Figure 6 illustrates that the collision probability changed as the number of STAs in each BSS increased. When SR is disallowed, only AP can access the wireless channel, so the collision probability is independent of the number of STA in each BSS and is maintained at a lower level; the collision probability increases with the increase of the number of STA in each BSS when SR is allowed. In general, the greater the p, the higher the collision probability.

From the simulation results above, it is observed that SR probability p has a significant impact on the gain from SR. The gain from SR varies with the changes in scenario, and the SRP mechanism is sometimes difficult to adapt to the changes, which can lead to the decrease of the gain or even exhibit a negative

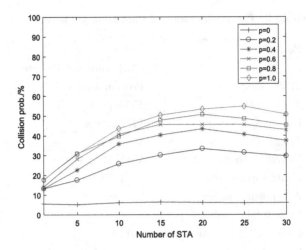

Fig. 6. Collision probability performance of PSR at different number of STA in each BSS.

gain. In this case, the use of the PSR mechanism with an appropriate value of p can effectively solve the problem.

5   Conclusions and Future Work

This paper proposes a PSR mechanism to increase the gain from SR under indoor scenario, and solve the problem that the existing SRP mechanism cannot dynamically adapt to the scenario changes, the simulation results show the feasibility of the scheme. In the future, we plan to make dynamic optimization strategies for SR probability in PSR mechanism based on the key idea described in Subsect. 2.1; we also plan to study the theoretical model of PSR mechanism and give a more perfect solution to obtain the fullest gain from SR.

Acknowledgement. This work was supported in part by the National Natural Science Foundations of CHINA (Grant No. 61501373, No. 61771390, No. 61771392, and No. 61271279), the National Science and Technology Major Project (Grant No. 2016ZX03001018-004, and No. 2015ZX03002006-004), and the Fundamental Research Funds for the Central Universities (Grant No. 3102017ZY018).

References

1. IEEE 802.11, Proposed 802.11ax Functional Requirements, IEEE 802.11-14/0567r7, July 2014
2. IEEE 802.11 "Status of Project IEEE 802.11ax". http://www.ieee802.org/11/Reports/tgax_update.htm
3. Part 11: Wireless LAN Medium Access Control (MAC) and Physical Layer (PHY) Specifications: Amendment 6: Enhancements for high efficiency in frequency bands between 1 GHz and 6 GHz, IEEE Standard 802.11ax draft 3.0 (2018)

4. Qu, Q., Li, B., Yang, M., Yan, Z., et al.: Survey and Performance Evaluation of the Upcoming Next Generation WLAN Standard - IEEE 802.11ax (2018)
5. Buehrer, R.: Code Division Multiple Access (CDMA), pp. 1–192. Morgan & Claypool, San Rafael (2006)
6. Deng, D.-J., Lien, S.-Y., Lee, J., Chen, K.-C.: On quality-of-service provisioning in IEEE 802.11ax WLANs. IEEE Access 4, 6086–6104 (2016)
7. Merlin, S., Barriac, G., Sampath, H., et al.: TGax simulation scenarios, doc. IEEE 802.11-14/0980r16 (2015)
8. The NS-3 Network Simulator. http://www.nsnam.org/

Backoff HARQ for Contention-Based Transmission in 5G Uplink

Xun Li[1,2], Zhangdui Zhong[1], Yong Niu[1(✉)], and Hong Zhou[2]

[1] State Key Laboratory of Rail Traffic Control and Safety,
Beijing Jiaotong University, Beijing, China
{16120101,zhdzhong}@bjtu.edu.cn, niuy11@163.com
[2] Nokia Bell Labs, Beijing, China
hong.1.zhou@nokia-sbell.com

Abstract. With extraordinary growth in the Internet of Things (IoT), the amount of data exchanged between IoT devices is growing at an unprecedented scale, which is an important requirement for Fifth generation (5G) networks. Current Long Term Evolution (LTE) system is not able to efficiently support massive connectivity, especially on the uplink (UL). The contention based transmission enables to transmit data packet without waiting for dynamic and explicit scheduling grant from base station, and is more efficient for small packet transmission in terms of lower signaling overhead, lower latency and lower energy consumption. However, this may lead to high collision rates and a large amount of retransmissions. New low complexity recovery mechanism of non-successful transmissions is needed. So in this paper, a new Hybrid Automatic Repeat reQuest (HARQ) mode, called Backoff HARQ, is proposed for contention-based (CB) transmission in 5G uplink. For Backoff HARQ mode, round trip time (RTT) is set to a longer random time. We can get more HARQ processes and put off the retransmission information for the longer random backoff time as we set. We evaluate the proposed backoff HARQ mode using system simulations and show that the new mode provides superior system performance.

Keywords: Internet of Things (IoT) · Fifth generation (5G) ·
Contention-based access · HARQ · Back off

1 Introduction

The Internet of Things (IoT) is an intelligent infrastructure of uniquely identifiable heterogeneous computing devices capable of communicating with each other, services, and people through the Internet without human interaction. Fortunately, the fifth generation (5G) network called New Radio (NR) by 3GPP will need to efficiently support a range of new services, which is featured with small packets but massive connections and low latency, high reliability [1,2]. The 4G network known as Long Term Evolution (LTE), which adopts strict scheduling

© ICST Institute for Computer Sciences, Social Informatics and Telecommunications Engineering 2019
Published by Springer Nature Switzerland AG 2019. All Rights Reserved
B. Li et al. (Eds.): IoTaaS 2018, LNICST 271, pp. 284–291, 2019.
https://doi.org/10.1007/978-3-030-14657-3_28

and control procedure, is designed to provide high data rate services for relatively small number of users. Current LTE uplink (UL) will not be sufficient to support massive connections and achieve higher spectrum efficiency as the significant signaling overhead of the scheduling. In additional, the UL scheduler's request and grant procedure introduce extra latency for the transmission.

The contention-based (CB) transmission is a promising solution for small data packets transmission in 5G UL with advantages of low signaling overhead, low latency, and support of massive connectivity [3–6]. But, the CB transmission efficiency will be limited by the packet collision rate, which is increased in proportion to the number of simultaneously transmitting packets and in inverse proportion to the transmission opportunities provided by the CB transmission channel. To further improve the transmission efficiency, nonorthogonal multiple access schemes with advanced receiver have been proposed to be employed with 5G UL CB transmission. A Non-Orthogonal Coded Access (NOCA) scheme, called ZC-NOCA, is used for CB transmission in 5G uplink. In the scheme, data bit is spread over OFDM symbols by using the non-orthogonal codes, which is generated by Zadoff-Chu (ZC) sequences [7]. The simulations in the paper is under contention-based NOCA scenario.

There are already some works on HARQ. A HARQ protocol, called Flexible HARQ, is proposed in [8], which allows flexible bandwidth allocation of retransmission data and new transmission data based on imperfect channel state information (CSI) and soft decoding information sent along with ACK/NACK feedback. [9] proposed an Adaptive HARQ (A-HARQ) scheme, where RTX are done on better quality sub-bands, with resources dynamically allocated based on Channel Quality Indicator (CQI) reports. [10] propose a novel HARQ scheme for 5G enhanced mobile broadband (eMBB) systems that also support ultra-reliable and low latency communications (URLLC). The proposed scheme performs retransmission of the lost part of a data packet when URLLC traffic, e.g., accident alert signal on the road, is transmitted immediately upon its creation by puncturing a part of the current data packet to avoid delay.

Now there are already two modes, asynchronous and synchronous HARQ mode, in LTE system. In this paper, Backoff HARQ mode is proposed for CB transmission in 5G uplink. For this mode, RTT is set to a longer random time. We can get more HARQ processes and put off the retransmission information for the longer random back-off time as we set. For stop and wait HARQ, in a HARQ process, whether the next transmission is to transmit new data or to retransmit old data is decided at a RTT after a transmission has occurred. So through a longer RTT, the interval between the same HARQ processes is increased for a random time. In general, it is important to minimize the HARQ buffer memory requirement at the UE. Similar objectives have been used in to optimize HARQ buffer memory management for a single carrier. And we evaluate the proposed backoff HARQ mode using system simulations to prove its superior system performance.

The rest of the paper is organized as follows, Sect. 2 gives the motivation of Backoff HARQ mode. Overview of Backoff HARQ mode is presented in Sect. 3.

The link level simulation of Backoff HARQ mode are presented in Sect. 4. Finally, we conclude the paper in Sect. 5.

2   Motivation of Backoff HARQ Mode

One proven way of increasing the reliability of transmission is by using HARQ, which can be sufficient for most applications, provided the given deadline is sufficiently long. However, it is possible for the system to reach the limit of 4 HARQ RTXs if the channel conditions are poor enough. HARQ is utilised as a Stop and Wait (SAW) process in LTE.

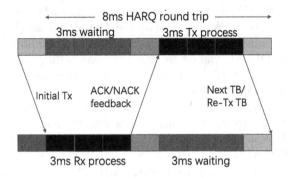

Fig. 1. The typical 8 ms LTE HARQ process.

A typical 8 ms LTE HARQ process follows as shown in Fig. 1. It involves the decoding of the received TB, Cyclic Redundancy Check (CRC), and the encoding of Acknowledgement (ACK)/Negative ACK (NACK) feedback for decoding success or failure respectively within 3 ms. The ACK/NACK feedback is then transmitted to the Tx. The Tx decodes the received ACK/NACK, constructs the next TB or RTX TB for decode success or failure respectively, and then encodes the next TB or RTX TB within 3 ms. The Tx then sends the next TB or RTX TB to the Rx to complete one LTE HARQ round trip. One problem with using a single HARQ SAW process, is that it waits for feedback for the majority of the 8 ms round trip. This is why LTE uses 8 parallel HARQ processes in the uplink (synchronous) or up to 8 in the downlink (asynchronous). However, for the contention-based transmission in 5G uplink, massive connectivity is required and they compete for the same time-frequency resources. The current LTE HARQ method can't cope well with the retransmissions of large amounts of data. So, a new method is needed.

3   Backoff HARQ Mode

HARQ uses stop-and-wait protocol to send data. In the stop-and-wait protocol, the sender sent a TB, it stopped waiting for confirmation. The receiver will acknowledge (ACK) or negate (NACK) the TB using 1-bit information. However, after each transmission, the sender stops and waits for acknowledgment, resulting in low throughput. So LTE uses multiple parallel stop-and-wait processes: when one HARQ process is waiting for acknowledgment, the sender can use another HARQ process to continue sending data. These HARQ processes jointly form a HARQ entity, which incorporates a stop-and-wait protocol while allowing continuous transmission of data. We already know that in order to make full use of 8 ms round trip time, LTE uses 8 parallel HARQ processes.

In order to deal with a large number of retransmission requests, RTT is set as a longer random time in backoff HARQ mode. For stop and wait HARQ, in a HARQ process, whether the next transmission is to transmit new data or to retransmit old data is decided at a RTT after a transmission has occurred. So through a longer RTT, the interval between the same HARQ processes is increased for a random time. And each HARQ entity has an independent backoff time.

For example in Fig. 2, if RTT is set to 10 TTIs, which means 2 TTIs longer than standard LTE RTT, the HARQ process used at TTI 1 is available at TTI 11. And we get two more HARQ processes and put off the retransmission information for two TTIs. Next, we will evaluate the proposed backoff HARQ mode using system simulations and see whether a better system performance is got.

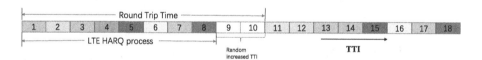

Fig. 2. Longer round trip time.

Table 1. Table captions should be placed above the tables.

Meaning	Symbol	Simulation parameter
Carrier frequency	**W**	4 GHz
Bandwidth	**B**	10 GHz
SubCarrier BW	**K**	15 kHz
Receiver in UL	**M**	LMMSE SIC
Modulation	**S**	QPSK
TTI duration	**T**	1 ms

4   Simulation Results

In this section, we analysis the performance of proposed Backoff HARQ mode using a detailed system simulator that used contention-based NOCA for 5G uplink transmission. In our simulations, UEs experience distance-dependent pathloss, shadow fading, and fast fading as per commonly-used 3GPP simulation assumptions. Each HARQ entity can experience independent round trip times for 8 to 12 TTIs. And we have tabulated the simulation parameters used in our simulation in Table 1.

4.1   Frame Error Rates (FER)

As shown in Figs. 3 and 4, in the NOCA scenario, when the number of receive antenna is one or two, and the max number of retransmissions is 6, the FER of asynchronous and synchronous HARQ mode are the same. Because when the arrival rate is high, the empty process does not exist, so there is no difference between the asynchronous and synchronous HARQ. And whatever the number of receive antennas and the packet arrive rate (PAR) of UEs are, the FER of Backoff HARQ mode is always lower than other modes. When the PAR is 200, the biggest difference among different HARQ modes is about 2.68%.

In Fig. 5, we observe FER of all the modes as we vary the max number of retransmissions when the PAR of UE is 100 and there is one receive antenna. As we can see, Backoff HARQ mode also shows its better performance. We find that the larger the max number of retransmissions is, the larger the gap is, as expected. And the biggest difference among different HARQ modes is 1.44%.

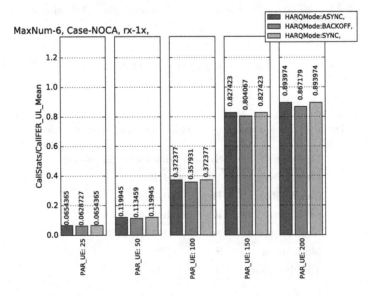

Fig. 3. FER of different HARQ mode under different PAR of UE when rx-1x.

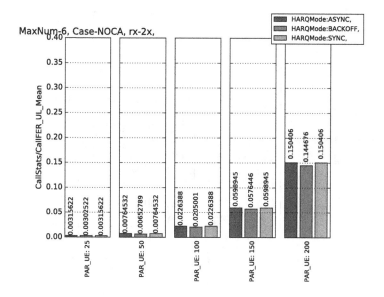

Fig. 4. FER of different HARQ mode under different PAR of UE when rx-2x.

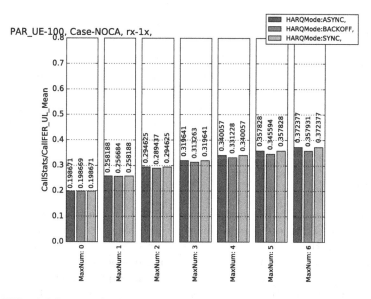

Fig. 5. FER of different HARQ mode under different max number of retransmissions.

As for when the number of receive antennas is two, the difference of all the HARQ modes is small, the next we will focus on the simulation results of one receive antenna. And for the similar reason, we will only observe asynchronous and Backoff HARQ mode.

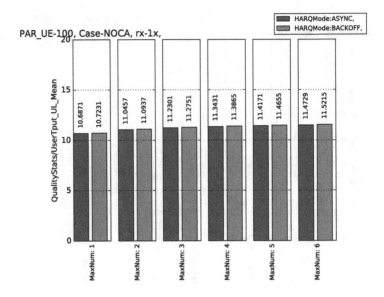

Fig. 6. Throughput of different HARQ mode.

4.2 Throughput

Next, our proposed Backoff HARQ mode is compared to the asynchronous HARQ mode in terms of user throughput of mean value. As we can see in the Fig. 6, The results are very close, but we can also find that the throughput of Backoff HARQ mode is always larger. And the bigger the max number of retransmissions is, the larger the throughput is. The biggest difference between the average of user throughput is 48.6 kbps.

5   Conclusion

In this paper, Backoff HARQ mode is proposed for CB transmission in 5G uplink. In contention-based access, because multiple users share the same time-frequency resource, non-orthogonal collisions of arriving signals at the receiver may be caused by the users. And collisions will be more serious when a large number of users access. In the stop-and-wait protocol, multiple parallel stop-and-wait processes are used. We try to solve the problem by increasing the RTT by a random back-off time (0 to 4 TTIs), for which we called Backoff HARQ mode. The link level simulation of Backoff HARQ mode has been implemented. The simulation results show the FER and throughput of our proposed Backoff HARQ are proved to be better than synchronous and asynchronous HARQ mode. The pursuit of better performance is the future work.

Acknowledgment. This study was supported by the National Natural Science Foundation of China Grants 61725101 and 61801016; and by the China Postdoctoral Science Foundation under Grant 2017M610040 and 2018T110041; and by National key research and development program under Grant 2016YFE0200900; and by the Beijing Natural Fund under Grant L172020; and by Major projects of Beijing Municipal Science and Technology Commission under Grant No. Z181100003218010.

References

1. 3GPP TR 38.913: Study on Scenarios and Requirements for Next Generation Access Technologies, v. 0.3.0, March 2016
2. Zhang, Z., et al.: A novel traffic generation framework for LTE network evolution study. In: IEEE 25th Annual International Symposium on Personal, Indoor, and Mobile Radio Communication (PIMRC), 2–5 September 2014, pp. 1315–1320 (2014)
3. 3GPP TSG-RAN WG1, R1–165022: Uplink contention-based access in 5G New Radio, Nokia, Alcatel-Lucent Shanghai Bell
4. Zhou, K., Nikaein, N., Knopp, R., Bonnet, C.: Contention based access for machine-type communications over LTE. In: 2012 IEEE 75th Vehicular Technology Conference (VTC Spring), Yokohama, Japan, 6–9 May 2012, pp. 1–5 (2012)
5. Singhy, B., Tirkkoneny, O., Lix, Z., Uusitalox, M.A.: Contention-based access for ultra-reliable low latency uplink transmissions. IEEE Wirel. Commun. Lett. **7**(2), 182–185 (2018)
6. Lin, D., Charbit, G., Fu, I.-K.: Uplink contention based multiple access for 5G cellular IoT. In: 2015 IEEE 82nd Vehicular Technology Conference (VTC Fall), Boston, MA, USA, 6–9 September 2015, pp. 1–5 (2015)
7. Sesia, S., Toufik, I., Baker, M. (eds.): LTE: The UMTS Long Term Evolution. Wiley, Hoboken (2009)
8. Zhuang, H., Sethuraman, V.: Flexible HARQ: a high throughput hybrid-ARQ protocol. In: 2017 IEEE International Conference on Communications (ICC), Paris, France, 21–25 May 2017, pp. 1–6 (2017)
9. Cabrera, E., Fang, G., Vesilo, R.: Adaptive hybrid ARQ (A-HARQ) for ultra-reliable communication in 5G. In: 2017 IEEE 85th Vehicular Technology Conference (VTC Spring), Sydney, NSW, Australia, 4–7 June 2017, pp. 1–6 (2017)
10. Yeo, J., Park, S., Oh, J., Kim, Y., Lee, J.: Partial retransmission scheme for HARQ enhancement in 5G wireless communications. In: 2017 IEEE Globecom Workshops (GC Wkshps), Singapore, 4–8 December 2017, pp. 1–5 (2017)

A Flexible Iterative Log-MPA Detector for Uplink SCMA Systems

Xiaojing Shi[1,2], Pinyi Ren[1,2(✉)], and Dongyang Xu[1,2]

[1] School of Electronic and Information Engineering, Xi'an Jiaotong University, Xi'an 710049, China
shixiaojing713@gmail.com, pyren@mail.xjtu.edu.cn, xudongyang@stu.xjtu.edu.cn
[2] National Simulation Education Center for Communications and Information Systems, Xi'an, China

Abstract. The development of the communications industry is changing with each passing day. At present, 5G research work is in full swing. In the SCMA (Sparse Code Multiple Access) system, this paper proposes a flexible iterative logarithmic domain message passing algorithm(FI-Log-MPA) aiming at the traditional logarithmic domain message passing algorithm to magnify the number of iterations in the codeword decision process, which results in a waste of running resources. The algorithm determines whether to perform subsequent iterations by adding a link to determine the convergence rate of the codeword during the update process of the variable node. If the codeword convergence rate is lower than a preset threshold, the iteration is stopped, otherwise the iteration is continued. Based on the principle of the optimization algorithm, we built a SCMA and NOMA (Non-Orthogonal Multiple Access) simulation platform to verify that the algorithm can effectively reduce the number of iterations. Although there is a loss in bit error rate and throughput performance, it remains at the same order of magnitude. In general, the complexity is reduced, and the decoding performance is optimized.

Keywords: 5G · Message passing algorithm ·
Sparse code division multiple access ·
Power domain non-orthogonal multiple access

1 Introduction

The innovation of each generation of mobile communication systems is accompanied by the evolution of communication access technologies. However, whether it is time or frequency division multiple access, or code division multiple access,

---

The research work reported in this paper was supported by the National Natural Science Foundation of China (NSFC) under Grant No. 61431011. Key Research and Development Program of Shannxi Province under Grant 2017ZDXM-GY-012.

B. Li et al. (Eds.): IoTaaS 2018, LNICST 271, pp. 292–302, 2019.
https://doi.org/10.1007/978-3-030-14657-3_29

it belongs to orthogonal multiple access technology, which is accessed through orthogonal resources [1]. However, orthogonal multiple access technology has some drawbacks. Not only the peak rate is far from being able to meet the requirements, but also the number of access users is still severely limited, and overload cannot be achieved. In order to achieve the high speed, high coverage and low latency requirements of 5G, it is urgent to introduce a new multiple access technology.

It is worth noting that the potential wireless access technology for 5G has flourished. This article mainly introduces two new wireless access technologies, SCMA and NOMA. In the SCMA system, the MPA (Message Passing Algorithm) is a key issue. In response to this problem, the industry recognizes two major challenges:

1. Number of iterations. In the current MPA algorithm, the number of iterations is generally a fixed value. That is, regardless of whether the decoded sequence is sufficiently convergent, a certain number of iterations must be completed, which will result in a waste of a large amount of running resources. Therefore, it is urgent to propose a new scheme to achieve an iterative number of effective times.
2. Iterative process. Since each iteration needs to complete the update of the variable node and the function node, the calculation process is too cumbersome and even contains some repeated calculations, so it needs to be simplified.

1.1 Related Works

In 2014, NTT DoCoMo, Japan's largest communications operator, proposed Non-Orthogonal Multiple Access (NOMA) technology, which significantly improves spectrum utilization. It implements power domain or code domain multiplexing at the transmitting end, and the receiving end uses Serial Interference Cancellation (SIC) demodulation. A number of documents have made in-depth analysis of the applicability of the most relevant LTE (Long Term Evolution) features, indicating that NOMA can be easily combined with SU-MIMO (Single User MIMO) and MU-MIMO (Multi-user MIMO). Effective benefits of high throughput. Further results can be found in the literature [2] and [3]. In summary, NOMA has good applicability and superiority for the Internet of Things. In 2015, at the Mobile World Congress (MWC), Huawei released the Sparse Code Division Multiple Access (SCMA) technology, which is one of the new air interface technologies for 5G. By adding the code domain dimension at the transmitting end, the receiver demodulates with a specific algorithm [4]. In the same year, Huawei and NTT DoCoMo cooperated to test the new 5G air interface technology. The results show that SCMA technology has excellent throughput performance. For the decoding algorithm MPA used by its receiving end, many articles have been studied in depth. Literature [5] proposes a SCMA low complexity decoding algorithm based on variable node serial strategy that can be accepted and transmitted simultaneously. Literature [6] also proposes a grouping threshold MPA algorithm based on serial update. Literature [7] proposed a SCMA multi-user detection system based on weighted message transfer.

Literature [8] proposed a SCMA multi-user detection algorithm based on the serial-to-parallel combination strategy. Literature [9] combines the literature [8] and the log-domain message passing algorithm, which not only has fast convergence speed, low algorithm complexity, but also can approach the optimal detector. The above is some improvement of the MPA algorithm. In general, both NOMA and SCMA are non-orthogonal technologies. Although the interference between users becomes larger, it can achieve overload and significantly improve the throughput of the system [10].

Motivation and Contribution. Since the solution of the above documents only considers the simplification of the iterative process, the optimization problem of the iteration number is not solved. However, in the process of codeword decision, the traditional logarithmic domain message passing algorithm uses a fixed number of fixed iterations, which will result in waste of running resources. Therefore, for the number of iterations, we propose a flexible iterative logarithmic domain messaging algorithm (Flexible Iterative-Log-MPA, FI-Log-MPA). Specifically, our contribution is to control the number of effective iterations while using the log-domain messaging algorithm to simplify the computational process. The optimization scheme is verified on the established SCMA platform, which shows that the scheme can achieve performance optimization.

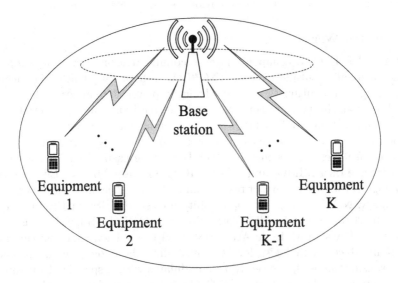

Fig. 1. Uplink SCMA system.

2   System Model

The Fig. 1 below shows the uplink SCMA system model. It is divided into three aspects: sender, receiver and uplink and downlink [11]. First we introduce

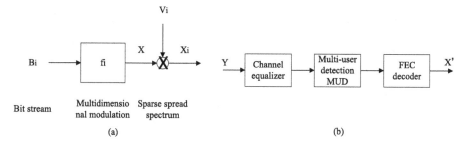

Fig. 2. (a) Basic model of the sender; (b) Receiver basic model.

the model of the sender. Its working principle generally includes the following three steps: error correction coding, multi-dimensional modulation and sparse spread spectrum [12]. The process of generating the SCMA codeword is shown in Fig. 2(a). The expression of the i-th bit multiplexed user SCMA codeword of the i-th layer is:

$$X_i = V_i f_i(B_i) \tag{1}$$

wherein, V_i represents a mapping matrix for spreading, and f_i represents a modulation function for designing a multi-dimensional constellation, and B_i represents information processed by error correction encoding. Next, we introduce the model of the receiver. At the receiving end, three steps of information processing are required: first, the received signal is passed through a channel equalizer, then multi-user detection (MUD) is performed, and finally, forward error correction (FEC) decoding is performed. The basic model of the receiving end is shown in Fig. 2(b). Finally, the system model of the uplink and downlink is introduced. As shown in Fig. 3(a) and (b). The channel matrix H_i and the codeword X_i are independent of each other. J is the number of users, k is the antenna, and N is the noise. When there are multiple antennas at the receiving end, the signal at the receiving end is:

$$Y^k = diag(H_i) \sum_{i=1}^{J} X_i^k + N^k \tag{2}$$

3   Proposed Detector

The number of iterations for the decoding algorithm in the SCMA system is fixed. We propose a flexible iterative log domain message passing algorithm (FI-Log-MPA). The scheme determines whether the decoding result after the iteration can achieve a sufficiently low error rate by adding a link determining the probability convergence rate of the codeword in each iteration process.

3.1   FI-Log-MPA

Considering that the current research on MPA algorithms mainly focuses on the optimization in each iteration process, the decision of each codeword needs to be

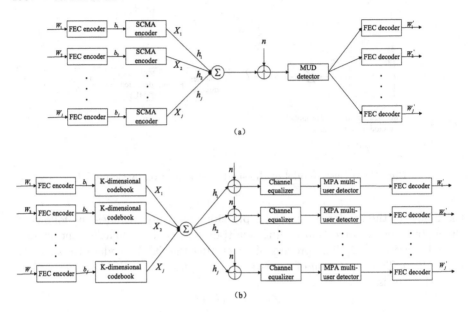

Fig. 3. (a) Basic model of the SCMA system uplink; (b) Basic model of the downlink of the SCMA system.

repeated a fixed number of iterations. In order to simplify the iterative process, a log-domain messaging algorithm Log-MPA capable of converting multiplication into addition is selected. For the number of iterations, since the codeword convergence of different time slots is different, the number of iterations required is different. Iterative redundancy is caused by a fixed number of iterations. In the paper [13], a message passing algorithm (Delete Redundant Iterations-MPA, DRI-MPA) that can remove redundant iterations is proposed for this problem. On this basis, we achieve similar results by optimizing the message passing algorithm in the log domain. The principle of the algorithm is to determine whether to perform subsequent iterative processes by supervising the convergence rate of each codeword probability, so as to ensure that the complexity can be reduced under the premise of achieving specific performance. The convergence rate of the codeword probability is defined as the ratio of the difference between the current iteration probability and the previous iteration probability to the previous iteration probability. The specific formula is as follows:

$$w^t(k) = |\frac{Q^t_{m,k} - Q^{t-1}_{m,k}}{Q^{t-1}_{m,k}}| \tag{3}$$

wherein w^t_k represents the codeword probability convergence rate of the k-th user in the t-th iteration process. m represents the number of codewords in each codebook. Q represents the probability of iteration, the specific formula is as follows:

$$Q^0_{m,k} = 1/M \tag{4}$$

$$Q_{m,k}^t = \log(1/M) + Igv_{m,k}^{n_1} + Igv_{m,k}^{n_2} \tag{5}$$

where M represents the number of codewords in each codebook, and $Igv_{m,k}^{n_1}$ and $Igv_{m,k}^{n_2}$ represent the function node update values corresponding to the user's two non-zero value sequence numbers, respectively. The function node update formula will be described in detail below. The algorithm compares the convergence rate of the codeword with the threshold set in advance when each iteration process is completed. If

$$w^t(k) \geq h_d \tag{6}$$

we continue the next iteration. h_d represents the threshold.

3.2 FI-Log-MPA

The model of the flexible iterative log domain message passing algorithm is shown in Fig. 4.

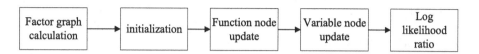

Fig. 4. FI-Log-MPA decoding flow chart.

In the first step, the factor graph calculation is completed according to the given codebook. The resulting factor matrix is as follows:

$$F = \begin{bmatrix} 0 & 1 & 1 & 0 & 1 & 0 \\ 1 & 0 & 1 & 0 & 0 & 1 \\ 0 & 1 & 0 & 1 & 0 & 1 \\ 1 & 0 & 0 & 1 & 1 & 0 \end{bmatrix} \tag{7}$$

According to the factor matrix, the mapping matrix corresponding to each layer user j is further solved. The solution process is: taking the j-th column of the factor matrix, and placing the corresponding values on the four subcarriers into the diagonal position of one diagonal matrix in turn, and the remaining elements are 0. After that, the all-zero column in the diagonal matrix is deleted, and the mapping matrix corresponding to user j is obtained:

$$V_1 = \begin{bmatrix} 0 & 0 \\ 1 & 0 \\ 0 & 0 \\ 0 & 1 \end{bmatrix}, V_2 = \begin{bmatrix} 1 & 0 \\ 0 & 0 \\ 0 & 1 \\ 0 & 0 \end{bmatrix}, V_3 = \begin{bmatrix} 1 & 0 \\ 0 & 1 \\ 0 & 0 \\ 0 & 0 \end{bmatrix}, V_4 = \begin{bmatrix} 0 & 0 \\ 0 & 0 \\ 1 & 0 \\ 0 & 1 \end{bmatrix}, V_5 = \begin{bmatrix} 1 & 0 \\ 0 & 0 \\ 0 & 0 \\ 0 & 1 \end{bmatrix}, V_6 = \begin{bmatrix} 0 & 0 \\ 1 & 0 \\ 0 & 1 \\ 0 & 0 \end{bmatrix} \tag{8}$$

The second step is to initialize the variable node, the function node, the iteration probability Q and the number of iterations. The third step is to complete the function node update during an iteration. The fourth step is to complete the variable node update in an iterative process, that is, to update the non-zero location information of all users. In this process, the calculated codeword convergence rate is compared with the threshold set in the main program. Finally, the maximum log likelihood ratio is calculated according to the Q value, that is, the iteration probability, and the bit error rate is obtained. The error rate calculation formula is as follows:

$$BER = \frac{N_{\text{err}}}{N_{bits}} \tag{9}$$

Among them, N_{err} represents the total number of errors, and N_{bits} represents the total number of bits. The simulation of the designed new decoding algorithm is based on the SCMA platform. The specific steps are as follows: **Step 1:** Select the appropriate codebook. The codebook selected here is a matrix. The first dimension of the codebook represents the number of orthogonal resources, the second dimension represents the number of codewords in each codebook, and the third dimension represents the number of users. **Step 2:** Complete the parameter initialization settings, such as the maximum number of bits, the maximum number of errors, and so on. **Step 3:** The rationality of the parameters, here is divided into two parts. One is to determine whether the number of bits is smaller than the maximum number of bits under a certain SNR; the second is to determine whether the number of errors is less than the maximum number of errors under the SNR. If they are all satisfied, follow the process. **Step 4:** The transmission sequence is randomly generated. **Step 5:** The sequence obtained in the previous step is propagated through the SCMA joint coding and fading channel to obtain the fading SCMA signal. **Step 6:** The signal is added with a Gaussian white noise corresponding to a specific signal to noise ratio. **Step 7:** The receiver uses a flexible iterative log-domain messaging algorithm FI-Log-MPA to decode the SCMA signal and output a Maximum Log Likelihood Ratio (LLR). Its calculation formula is as follows:

$$LLR_{j,k} = \log(\sum\nolimits_{b_{k,i}=0} Q^t_{m,k}/\sum\nolimits_{b_{k,i}=0} Q^t_{m,k}) \tag{10}$$

where $LLR_{j,k}$ represents the maximum log likelihood ratio of the j-th bit of the kth user, and $b_{k,i} = 0$ represents the codeword of the k-th user bit being zero. **Step 8:** The decoded sequence is XORed with the original sequence, and the error rate is calculated and obtained. Its throughput calculation formula is:

$$Throughput_{SCMA} = (1 - BER_{SCMA}) \cdot N_{bits} \cdot K_{SCMA} \tag{11}$$

where BER_{SCMA} represents the average error rate of users of the sparse code division multiple access system, and N_{bits} represents the number of binary bits carried by each user codeword. In the system, it is set to 4, and K_{SCMA} represents the number of accessible users of the system, which is set to 6.

3.3 Comparison Between FI-Log-MPA and Log-MPA

The log-domain messaging algorithm Log-MPA is an optimization scheme proposed by the traditional messaging algorithm MPA. The specific principle is that when the function node is updated, a method of evaluating the logarithm is adopted in the update process of different codewords in each codebook. The formula can be expressed as follows:

$$Igv = \sum_{i=1}^{M \times M} \log(e^{xx_i - \max(xx_i)}) + \max(xx_i) \tag{12}$$

xx_i represents the results obtained in the pre-link of the function update. Compared with the Log-MPA algorithm, FI-Log-MPA differs mainly in variable node update. Because the variable node is updated, the convergence rate of the codeword in each iteration is also calculated. Each iteration requires increasing the number of multiplications and comparison algorithms.

4   Simulation Results

The simulation parameter settings are shown in Table 1. Based on the SCMA system, this paper completes the performance comparison between two decoding algorithms Log-MPA and FI-Log-MPA. Figure 5(a) and (b) show the six user error rates for the two decoding algorithms, respectively (Table 2).

Table 1. Parameter initial value setting table.

Parameter	Set value
Number of orthogonal resources	4
User number	6
Number of codewords in each codebook	4
Number of frames	10,000
Eb/N0	020
Maximum number of bits	10,000,000
Maximum number of errors	100
Maximum number of iterations	10
Threshold	0

Table 2. Average bit error rate (20 dB).

Log-MPA	FI-Log-MPA
0.5×10^{-4}	0.8×10^{-4}

In Fig. 5(a), the error rates of the six users approach each other. This is because the graphs drawn are semi-logarithmic graphs. In fact, their true values

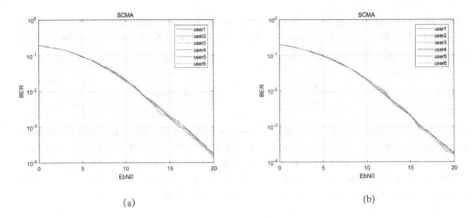

Fig. 5. (a) Bit error rate of Log-MPA; (b) Bit error rate of FI-Log-MPA.

differ greatly, but they are in the same order of magnitude, so the curves drawn are closer. Moreover, under the algorithm, as the signal-to-noise ratio increases, the bit error rate reaches an order of magnitude. This has great advantages in the application of actual systems. In Fig. 5(b), the bit error rate trend of the six users is roughly the same as that of Fig. 5(a), and it has also reached the order of magnitude, but slightly increased. However, the performance is somewhat lower than the original. In addition, in the process of improving the signal-to-noise ratio, the difference of bit error rate between the six users is larger than that of the unoptimized algorithm, but the effect is not significant.

Figure 6(a) and (b) show the average user error rate and throughput for the two algorithms, respectively. The average bit error rate and user throughput performance degradation percentage of the two algorithms are shown in Table 3.

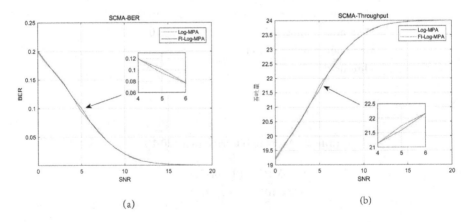

Fig. 6. (a) User average error rate of Log-MPA and FI-Log-MPA; (b) User throughput of Log-MPA and FI-Log-MPA.

Table 3. Percentage of average bit error rate and user throughput performance degradation (5 dB).

Average bit error rate	User throughput
8.69%	1.61%

Observing Fig. 6, we can find that the average error rate of the two algorithms is similar, but the error rate of the optimization algorithm is higher than the original algorithm in a certain interval, that is, the performance of the error rate after optimization is deteriorated. In addition, from Eq. (11), the throughput is inversely proportional to the bit error rate, so the throughput is slightly lower than the throughput of the original algorithm in some intervals. That is to say, the throughput performance is slightly worse after optimization, but the difference is small. In summary, after using the optimization algorithm, although the bit error rate and throughput performance have a certain loss, the difference is small and still can reach the same order of magnitude. Most importantly, after comparison, the complexity was found to be about 1/3 lower than Log-MPA. Therefore, the algorithm can be considered to achieve a balance between Bit Error Ratio (BER) and complexity.

5   Conclusion

In this paper, we propose a flexible iterative logarithmic domain message passing algorithm for the decoding algorithm MPA of SCMA system, and complete the construction and performance analysis of SCMA and NOMA platform, and verify the optimization algorithm. The results show that the complexity is reduced by about 1/3 compared to Log-MPA, and the performance is optimized. In the future, we can consider designing a more reasonable codeword probability convergence rate to achieve lower complexity.

References

1. Osseiran, A., Monserrat, J.F., Marsch, P.: 5G Mobile Wireless Communication Technology. People's Posts and Telecommunications Press, Beijing (2017)
2. Ali, S., Li, A., Hossain, E., Kim, D.I.: Non-orthogonal multiple access (NOMA) for downlink multiuser MIMO systems: user clustering, beamforming, and power allocation. IEEE Access **5**, 565–577 (2016)
3. Lin, H., Gharba, M., Siohan, P.: Impact of time and frequency offsets on the FBMC/OQAM modulation scheme. Signal Process. **102**, 151–162 (2013)
4. Mu, H., Ma, Z., Alhaji, M., Fan, P., Chen, D.: A fixed low complexity message pass algorithm detector for up-link SCMA system. IEEE Wirel. Commun. Lett. **4**(6), 585–588 (2015)
5. Ji, M., Zheng, J.: SCMA low complexity decoding based on serial node strategy of variable nodes. Telecommun. Sci. **9**, 50–57 (2017)

6. Wang, Y., Hao, X., Wang, L.: A grouping threshold MPA algorithm based on serial update. J. Hainan Univ.: Nat. Sci. Ed. (2018)
7. Zheng, J., Li, Y., Tian, M.: SCMA multiuser detection algorithm based on weighted message transfer. Comput. Eng. Des. **39**(4) (2018)
8. Wang, R.-N., Chang, J., Yu, J.: SCMA multi-user detection based on serial-to-parallel combination strategy. Telecommun. Technol. **58**(3), 245–250 (2018)
9. Zhang, X., Ge, W., Wu, X.: Multi-user detection message passing algorithm based on SCMA system. Comput. Eng. **44**(5) (2018)
10. Yang, Z.: Research on non-orthogonal multiple access and related technologies of 5G mobile network (2016)
11. Yao, Q.: System-level performance simulation analysis of non-orthogonal multiple access technology in 5G mobile communication network (2017)
12. Zeng, X., Liu, N., Sun, X.: Spread Spectrum Communication and Its Multiple Access Technology. Xi'an University of Electronic Science and Technology Press, Xi'an (2011)
13. Hao, S., Fan, B., Zhang, X.: SCMA multiuser detection algorithm based on deletion redundant iteration. Guangdong Communication Technology (2018)

Symmetry and Asymmetry Features for Human Detection

Xinchuan Fu$^{(\boxtimes)}$ and Shihai Shao

University of Electronic Science and Technology of China, Chengdu 611731, China
xinchuan.fu@foxmail.com, ssh@uestc.edu.cn

Abstract. Edge is a very important type of feature for human detection and Histogram of Oriented Gradient (HOG) is the most popular method to encode edge information since proposed. Because HOG detects edges based on intensity gradients, it is not invariant with respect to image illumination. In this paper, we propose three new types of features based on local phase: Local Phase based symmetry (LPS), Local Phase based Asymmetry (LPA), and Histogram of Oriented Asymmetry (HOA) for human detection. The LPA and HOA are similar with gradient magnitude and HOG features, but from different perspective. The key idea is the intensity around an edge point in an image is always asymmetry. Thus we can detect edges by measuring the asymmetry of the local structure at every point in the image. This is achieved by analyzing the phase of its constituent frequency components. This asymmetry measurement is invariant with respect to image contrast. After the asymmetry is computed, this value could be distributed to different orientation bins according to gradient orientation. We also measure symmetry around each point which yields LPS. This is useful to detect torso and limbs. These local phase induced features are combined with the classical Aggregated Channel Features (ACF) and are fed into the boosted decision tree (BDT) framework. Experiment shows that the proposed features are complementary to the ACF features and will increase the detection accuracy.

Keywords: Human detection · Symmetry · Asymmetry

1 Introduction

Human detection is an active research topic in recent years. The application includes Advanced Driver Assistance Systems (ADASs), visual surveillance and human-robot interaction, etc. Recently, deep learning methods have achieved the state-of-the-art accuracy [7,16], but these methods rely on high-end GPU device because of the high computation cost. On the other hand, hand-craft features together with boosted decision tree (BDT) methods are also competitive for its light-weight CPU implementation [3,10,19]. Among all the handcraft features,

Supported by the National Natural Science Foundation of China 61771107.

B. Li et al. (Eds.): IoTaaS 2018, LNICST 271, pp. 303–311, 2019.
https://doi.org/10.1007/978-3-030-14657-3_30

Histogram of Oriented Gradient (HOG) [4] is the most popular one for human detection.

The basic idea of HOG is to compute gradients in every point in the image. Then distribute the gradients magnitude to several orientation bins and form histogram in different cells. Currently, the HOG features are usually used together with gradient magnitude and CIE-LUV color. The resulting features are called Aggregated Channel Features (ACF) [5]. Many newly devised features are based on ACF, like LDCF [12], InformedHaar [17], Checkerboard [18] and MRFC [2], etc.

Though very effective for human and general object detection, the HOG features have an intrinsic limitation rooted in the gradient computation. In the implementation of HOG, the gradients are computed by the intensity difference, which is sensitive to local contrast. The computed edge of a person in the image will have different strength because of different illumination conditions, or just because the person wear clothes of different color. This is contrary to the goal of invariant representation in feature design. The existence of an edge should determined by the image structure, not the intensity difference. Though HOG compensates for the contrast variance by local normalization, this operation is coarse. For example, the strength of an edge with low contrast will be suppressed by an edge with high contrast in the local region.

Is there an alternative way to capture edge information? The answer is positive. By inspecting the edges in an image, we could see that the edge points usually locate about the midpoint of a transition ramp and the intensity around this point exhibit an approximate asymmetry. Based on this insight, we could detect edges by measuring the local asymmetry. Peter Kovesi [11] proposed a method measuring the symmetry and asymmetry based on local phase. For a point to be an symmetry point, all frequency components must have phase 0 or π. Conversely, for a point to be an asymmetry point, all frequency components must have phase $\pi/2$ or $3\pi/2$. Thus we could measure the extent of asymmetry by phase deviation from $\pi/2$ or $3\pi/2$. This type of features are called Local Phase based Asymmetry (LPA). We could also measure the extent of symmetry by phase deviation from 0 or π. This type of features are called Local Phase based Symmetry (LPS). The reason we compute LPS is that symmetry is a useful cue for human detection [1,3]. Following the idea of HOG, we distribute the estimated asymmetry level to different orientation bins. The resulting features are called Histogram of Oriented Asymmetry (HOA). The symmetry and asymmetry measurement only considers the structure of the local signal hence is invariant to local contrast. We add all these symmetry and asymmetry induced feature channels to ACF channels and put them into BDT to train a human classifier. The experiment result shows that the combined channels have superior performance than ACF.

2   Method

In this section, we first introduce the symmetry and symmetry measure for one dimensional signal, then we extend this method to color image, finally we introduce the HOA and show how these features are used for human detection.

2.1   Symmetry and Asymmetry Measure for One Dimensional Signal

As we stated in Sect. 1, to measure the symmetry and asymmetry around a point of a signal, we could compute the local phase deviation of its constituent frequency components of the signal. Suppose $I(x)$ is a one dimensional signal and $\phi_n(x)$ is the phase of the nth frequency component at point x. According to [11], the symmetry $S(x)$ and asymmetry $R(x)$ could be estimated as

$$
\begin{aligned}
S(x) &= \sum_n \frac{A_n(x)\lfloor|cos(\phi_n(x))| - |sin(\phi_n(x))|\rfloor}{A_n(x) + \varepsilon} \\
&= \sum_n \frac{\lfloor|e_n(x)| - |o_n(x)|\rfloor}{A_n(x) + \varepsilon}
\end{aligned}
\tag{1}
$$

$$
\begin{aligned}
R(x) &= \sum_n \frac{A_n(x)\lfloor|sin(\phi_n(x))| - |cos(\phi_n(x))|\rfloor}{A_n(x) + \varepsilon} \\
&= \sum_n \frac{\lfloor|o_n(x)| - |e_n(x)|\rfloor}{A_n(x) + \varepsilon},
\end{aligned}
\tag{2}
$$

where $\lfloor\cdot\rfloor = max(\cdot, 0)$, $A_n(x) = \sqrt{e_n(x)^2 + o_n(x)^2}$. ε is a small value to avoid zero denominator. In our experiment, ε is set to 0.001. $e_n(x)$ and $o_n(x)$ is the even and odd parts of the bandpass filtered signal, which are computed as

$$
e_n(x) = I(x) * g_n^e(x) \tag{3}
$$
$$
o_n(x) = I(x) * g_n^o(x), \tag{4}
$$

where $g_n^e(x)$ is a bandpass filter and $g_n^o(x)$ is its Hilbert transform. Log-Gabor filter [9] is usually chosen as the bandpass filter. The advantage of log-Gabor filter is that it has zero DC component for arbitrary large bandwidth. Note that in the original implementation [11] there is another term to compensate for the noise influence. But in our paper, the features extracted in this stage will be fed into the BDT and the decision tree will automatically choose a threshold for a feature. Thus the noise compensation is not necessary here.

2.2   Symmetry and Asymmetry for Color Images

To detect edges in an image which may exhibit various orientations, we can perform the previous operations with bandpass filters of different orientations and combine the result. But the more efficient way is to use the monogenic signal [8]. The monogenic signal is a generalization of analytic signal to 2D case with the Hilbert transform in the analytic signal replaced by Riesz transform. For 2D signal, the symmetry and asymmetry estimation is similar with Eqs. 1 and 2, except now there are two odd parts

$$o1_n(\mathbf{x}) = I(\mathbf{x}) * g_n^{o1}(\mathbf{x}) \tag{5}$$

$$o2_n(\mathbf{x}) = I(\mathbf{x}) * g_n^{o2}(\mathbf{x}), \tag{6}$$

where the frequency representation of $g_n^{o1}(\mathbf{x})$ and $g_n^{o2}(\mathbf{x})$ is given by the Riesz transform of $G_n(\boldsymbol{\omega})$

$$G_n^R(\boldsymbol{\omega}) = \frac{i\boldsymbol{\omega}}{|\boldsymbol{\omega}|} G_n(\boldsymbol{\omega}) = [G_n^{o1}(\boldsymbol{\omega}), G_n^{o2}(\boldsymbol{\omega})]. \tag{7}$$

The $o_n(x)$ in Eqs. 1 and 2 is then replaced by

$$o_n(\mathbf{x}) = \sqrt{o1_n(\mathbf{x})^2 + o2_n(\mathbf{x})^2} \tag{8}$$

For color image, there are three channels. We first convert the image from RGB space to LUV space, and then compute $S(\mathbf{x})$ and $R(\mathbf{x})$ for three channels separately. Finally we take the maximum value of the three computed channels as the final feature.

2.3   Histogram of Orientated Asymmetry

For human detection, it is beneficial not only detect edge, but also know the orientation of the edge. In the HOG implementation, the gradient magnitude is distributed to several predefined orientation bins, which encodes orientations of edges. In the same way, we could also distribute the asymmetry value to different orientations to indicate which axis the local structure is asymmetry to and the resulting features are HOA. As we stated in Sect. 1, the asymmetry could be taken as a similarity measure to an edge structure. Thus the distributed asymmetry value will represent the edge orientations. The orientation in each point could be estimated by the two odd filtered signals

$$ori(\mathbf{x}) = atan2(o1_n(\mathbf{x}), -o2_n(\mathbf{x})). \tag{9}$$

However, because in our experiment we will use the ACF features together with our features, the orientation has already been computed by the gradient based method. To save computation, we just use this gradient based orientation to compute HOA. The number of orientation is set to six, the same with that in ACF features. Together with the symmetry and asymmetry feature channels, we have $1+1+6=8$ new feature channels. As for ACF features, these channels are further aggregated with a shrinkage factor. In our experiment, we set the shrinkage factor to four.

3 Experiments

3.1 Experiments for One Dimensional Signal

We first show the symmetry and asymmetry features for a one dimensional signal from real data. We convert a human image (see Fig. 2(a)) from RGB color space to LUV color space and show the U channel in Fig. 1(a). A row from this channel is selected as our 1D signal (denoted by the red dash line). The first row of Fig. 1(b) is the selected 1D signal. Compared with the left image, we could see that the visual edge points are about the middle point of the transition ramp. The second row and the third row of Fig. 1(b) show the computed LPS and LPA. The peaks of the LPS correspond to the peaks or troughs of the original signal. The peaks of the LPA correspond to the transition ramp of the original signal, which correspond the edge of the image. Note that the value of the LPA range from 0 to 1 and a high value does not correspond to a high intensity difference. This is what we want because we need a descriptor to capture the edge structure which is invariant of the local contrast.

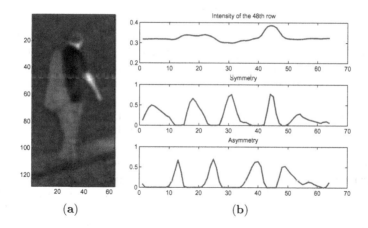

(a)   (b)

Fig. 1. Illustration of LPS and LPA for one dimensional signal. (a) The U channel of an human image. (b) From the top to bottom: the selected 1D signal, the LPS and the LPA. (Color figure online)

3.2 Experiment for a Color Image

In this section, we show the symmetry and asymmetry feature maps for a color image. The original color image is show in Fig. 2(a). The LPS feature map is show in Fig. 2(b). From the figure, we can see that the LPS feature map have high response at head, torso and limbs. These features are discriminative for human detection. Figure 2(c) shows the LPA feature map, which effectively capture the edge information of the image. Again, we find the asymmetry response is irrelevant with the local contrast, only correlated with the local structure.

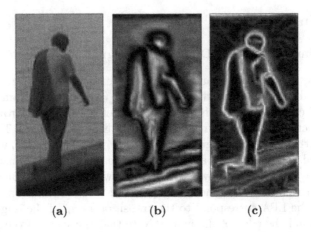

(a)   **(b)**   **(c)**

Fig. 2. Illustration of the LPS and LPA feature maps for a color image. (a) The original color image. (b) The LPS feature map. (c) The LPA feature map.

3.3 Evaluation on INRIA Dataset

In this section, we evaluate our proposed method on the INRIA person dataset. Miss Rate (MR) vs. False-Positive-Per-Image (FPPI) curve and log average miss rate are used as evaluation metric [6]. The log average miss rate is computed by averaging miss rate at 9 FPPI points that are evenly spaced in the log-space ranging from 10^{-2} to 10^{0}.

We incorporate the proposed features into the BDT framework to train a human detector. The model size is set as 128×64. Five frequency components are used to compute symmetry and asymmetry features whose center frequencies are $2\pi/\lambda_k$, where $\lambda_k = 10 * 1.5^{k-1}, k = 1, 2, ..., 5$. The training process includes three hard negative mining stages. The final classifier consists of 4096 level-3 decision trees. To show the effectiveness of our new features, two models are trained. The first model only use the 10-channel ACF features to train. For the second model, the first 2048 weak classifiers still use the ACF features to train. For the latter 2048 weak classifiers we replace the seven gradient based channels with our local phase based channels (but keep the color channels). Both models are evaluated on the test set, thus we can see how the detection accuracy evolves with the weak classifier number.

The test result is shown in Fig. 3. As shown in the figure, with increasing number of decision trees, the log average miss rate (the lower the better) gradually decreases (though with fluctuation). For both models, the first 2048 trees use the same features, hence the log average miss rate is always the same. However, as the gradient based channels are replaced by our local phase based channels for the later 2048 trees, the second model achieves higher accuracy than the first one. After 4096 decision trees the ACF features only achieve 13.25% log average miss rate, while our method achieves 11.39%. This improvement shows the effectiveness of our new features.

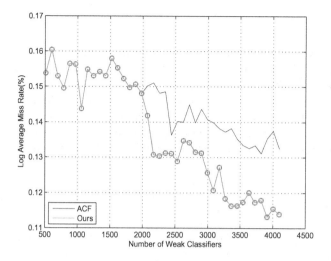

Fig. 3. Weak classifier number versus the log average miss rate on the INRIA test dataset.

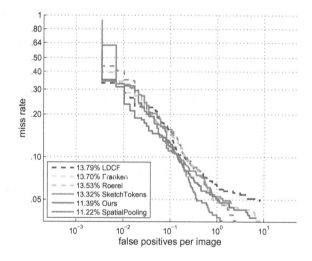

Fig. 4. Comparison with state-of-the-art non-CNN methods on the INRIA test dataset.

Next, we compare our detector with other non deep learning detectors using MR vs. FPPI curves in Fig. 4. For a fixed FPPI point, we prefer a lower MR, thus the curves close to the left-bottom corner are better. Log average miss rate is given at the legend to rank different method. From the figure, we can see that our detector outperforms other detectors, except a slightly worse than the SpatialPooling [14]. Note that we only use 18 feature channels in total, while SpatialPooling has used 259 feature channels. Except for ACF features, it also used other feature types like covariance [15] and LBP [13].

4 Conclusion

In this paper, we propose three types of local phase based features, LPS, LPA, and HOA. The LPA and HOA are alternatives for gradient magnitude and HOG. Compared with gradient magnitude and HOG, the advantage of LPA and HOA is its invariance with respect to local contrast. LPS is used for capture the symmetry structure of a human. Experiment for one dimensional signal and color image shows the visual reasonability of this method. Experiment on INRIA person data set shows that these features will boost the performance of the ACF features.

References

1. Cao, J., Pang, Y., Li, X.: Pedestrian detection inspired by appearance constancy and shape symmetry. IEEE Trans. Image Process. **25**(12), 5538–5551 (2016). https://doi.org/10.1109/TIP.2016.2609807
2. Costea, A.D., Nedevschi, S.: Semantic channels for fast pedestrian detection. In: 2016 IEEE CVPR, pp. 2360–2368 (2016). https://doi.org/10.1109/CVPR.2016.259
3. Costea, A.D., Varga, R., Nedevschi, S.: Fast boosting based detection using scale invariant multimodal multiresolution filtered features. In: 2017 IEEE Conference on Computer Vision and Pattern Recognition, CVPR 2017, 21–26 July 2017, Honolulu, HI, USA, pp. 993–1002 (2017). https://doi.org/10.1109/CVPR.2017.112
4. Dalal, N., Triggs, B.: Histograms of oriented gradients for human detection. In: 2005 IEEE Computer Society Conference on Computer Vision and Pattern Recognition, CVPR 2005, 20–26 June 2005, San Diego, CA, USA, pp. 886–893 (2005). https://doi.org/10.1109/CVPR.2005.177
5. Dollár, P., Appel, R., Belongie, S.J., Perona, P.: Fast feature pyramids for object detection. IEEE Trans. Pattern Anal. Mach. Intell. **36**(8), 1532–1545 (2014). https://doi.org/10.1109/TPAMI.2014.2300479
6. Dollár, P., Wojek, C., Schiele, B., Perona, P.: Pedestrian detection: an evaluation of the state of the art. IEEE Trans. Pattern Anal. Mach. Intell. **34**(4), 743–761 (2012). https://doi.org/10.1109/TPAMI.2011.155
7. Du, X., El-Khamy, M., Lee, J., Davis, L.S.: Fused DNN: a deep neural network fusion approach to fast and robust pedestrian detection. In: 2017 IEEE Winter Conference on Applications of Computer Vision, WACV 2017, 24–31 March, Santa Rosa, CA, USA, pp. 953–961 (2017). https://doi.org/10.1109/WACV.2017.111
8. Felsberg, M., Sommer, G.: The monogenic signal. IEEE Trans. Sig. Process. **49**(12), 3136–3144 (2001). https://doi.org/10.1109/78.969520
9. Field, D.J.: Relations between the statistics of natural images and the response properties of cortical cells. J. Opt. Soc. Am. A-Opt. Image Sci. Vis. **4**(12), 2379–2394 (1987)
10. Kim, H.K., Kim, D.: Robust pedestrian detection under deformation using simple boosted features. Image Vis. Comput. **61**, 1–11 (2017). https://doi.org/10.1016/j.imavis.2017.02.007
11. Kovesi, P.: Symmetry and asymmetry from local phase. In: Tenth Australian Joint Conference on Artificial Intelligence, pp. 2–4 (1997)

12. Nam, W., Dollár, P., Han, J.H.: Local decorrelation for improved pedestrian detection. In: Advances in Neural Information Processing Systems 27: Annual Conference on Neural Information Processing Systems 2014, 8–13 December 2014, Montreal, Quebec, Canada, pp. 424–432 (2014). http://papers.nips.cc/paper/5419-local-decorrelation-for-improved-pedestrian-detection
13. Ojala, T., Pietikäinen, M., Mäenpää, T.: Multiresolution gray-scale and rotation invariant texture classification with local binary patterns. IEEE Trans. Pattern Anal. Mach. Intell. **24**(7), 971–987 (2002). https://doi.org/10.1109/TPAMI.2002.1017623
14. Paisitkriangkrai, S., Shen, C., van den Hengel, A.: Strengthening the effectiveness of pedestrian detection with spatially pooled features. In: Fleet, D., Pajdla, T., Schiele, B., Tuytelaars, T. (eds.) ECCV 2014. LNCS, vol. 8692, pp. 546–561. Springer, Cham (2014). https://doi.org/10.1007/978-3-319-10593-2_36
15. Tuzel, O., Porikli, F., Meer, P.: Pedestrian detection via classification on Riemannian manifolds. IEEE Trans. Pattern Anal. Mach. Intell. **30**(10), 1713–1727 (2008). https://doi.org/10.1109/TPAMI.2008.75
16. Zhang, L., Lin, L., Liang, X., He, K.: Is faster R-CNN doing well for pedestrian detection? In: Leibe, B., Matas, J., Sebe, N., Welling, M. (eds.) ECCV 2016. LNCS, vol. 9906, pp. 443–457. Springer, Cham (2016). https://doi.org/10.1007/978-3-319-46475-6_28
17. Zhang, S., Bauckhage, C., Cremers, A.B.: Informed haar-like features improve pedestrian detection. In: 2014 IEEE Conference on Computer Vision and Pattern Recognition, CVPR 2014, 23–28 June 2014, Columbus, OH, USA, pp. 947–954 (2014). https://doi.org/10.1109/CVPR.2014.126
18. Zhang, S., Benenson, R., Schiele, B.: Filtered channel features for pedestrian detection. In: IEEE Conference on Computer Vision and Pattern Recognition, CVPR 2015, 7–12 June 2015, Boston, MA, USA, pp. 1751–1760 (2015). https://doi.org/10.1109/CVPR.2015.7298784
19. Zhao, Y., Yuan, Z., Chen, D., Lyu, J., Liu, T.: Fast pedestrian detection via random projection features with shape prior. In: 2017 IEEE Winter Conference on Applications of Computer Vision, WACV 2017, 24–31 March 2017, Santa Rosa, CA, USA, pp. 962–970 (2017). https://doi.org/10.1109/WACV.2017.112

Workshop on Edge Computing for 5G/IoT

Space Group Targets Detecting and Resolving Algorithm via Ultra-low Sidelobe Filtering

Hongmeng Chen[1($\boxtimes$)], Yaobing Lu[1], Jing Liu[1], Hanwei Sun[1],
Jiahao Lin[1], Xiaoli Yi[1], Heqiang Mu[1], and Zeyu Wang[2]

[1] Beijing Institute of Radio Measurement, Beijing 100854, China
chenhongmeng123@163.com, luyaobing65@163.com
[2] China Academy of Electronics and Information Technology,
Beijing 100041, China
beidou13579@163.com

Abstract. Detecting and resolving the space group targets in the main beam of radar is an urgent requirement for the air-defense and anti-missile radar system. Ground-based radar, as an important instrument for space surveillance, can be used to detect and track the space targets like grouped aircrafts, warheads and the decoys of the missiles. However, it is difficult to detect and resolve the dense targets due to the limit of the radar resolving power. To solve this problem, a space group targets detecting and resolving algorithm based on ultra-low sidelobe filtering is proposed. By exploiting the convex optimization into the pulse-Doppler radar, the problem of ultra-low sidelobe is converted into the problem of optimization. The key of this algorithm is to minimize the peak to sidelobe level (PSL) of the range sidelobes with a constraint of signal to noise ratio (SNR) loss. Then the ultra-low sidelobe filtering results are used to detect and resolve the space group targets in the main beam. Numerical and experimental results demonstrate the effectiveness of the proposed algorithm.

Keywords: Group targets · Ultra-low sidelobe · Convex optimization

1 Introduction

In modern war, formation flying is usually an important means of sudden attacks. In this case, the targets always appear in the form of groups (i.e., group targets) [1–4]. When the group targets are illuminated with the same radar beam, the group targets fall into the same beam of the radar. Different from single target, the returns of the group targets superpose with each other. Moreover, conventional pulse compression usually produced high range sidelobes, which may makes it difficult to detect all the group targets. Especially, the maiblobe of the weak one may be masked by the sidelobes [1, 4–6] of the strong one among the group targets, and miss alarm will occur. The sidelobes of the group targets interact with each other, which makes it difficult for group targets detecting and resolving.

To solve this problem, many techniques have been studied. The existing group targets detecting and resolving algorithms can be approximately grouped into two sorts: micro-Doppler (m-D) effect analysis and waveform design method. The m-D

B. Li et al. (Eds.): IoTaaS 2018, LNICST 271, pp. 315–323, 2019.
https://doi.org/10.1007/978-3-030-14657-3_31

characteristics can be regarded as a unique signature of the target and provides additional information for target recognition applications [2–4]. To decrease the mutual interference between the group targets in adjacent range cell, low correlation peak sidelobe level (PSL) deign methods are considered. In [5], the simulated annealing algorithm is used in the polyphase code design. The linear programming method and the reiterative minimum mean-square error algorithm are discussed in [6] and [7], respectively.

In this paper, a group targets detecting and resolving approach using ultra-low sidelobe filtering is proposed. Based on the criterion that minimizes the PSL with a constraint of signal-to-noise ratio (SNR) loss with respect to the conventional matched filtering, the convex method is introduced in the pulse compression process to suppress the sidelobe level, which is quite helpful to decrease the interactions between the group targets. After that, different group targets can be easily detected and resolved from the ultra-low sidelobe filtering results. Numerical and experimental results verify the effectiveness of the proposed algorithm.

2 Signal Model

High range resolution can be achieved by the pulse compression technique, however, high sidelobes always occur. Windowing may be an effective tool to suppress the sidelobe, but it is at the expense of broading the mainbeam and SNR loss [1]. In this section, we will introduce an alternative approach to suppress the sidelobes.

Assuming that the transmitted signal sequence can be expressed as $\mathbf{x} = (x_0, x_1, \ldots, x_{N-1})^T$, where $(\cdot)^T$ denotes the transpose operator, N is the length of transmitted signal sequence.

The ultra-low sidelobe filter is given as $\mathbf{w} = (w_1, w_2, \ldots, w_M)^T$, where M is the length of the ultra-low sidelobe filter, and usually $M \geq N$. To simply the analysis, we usually add zeros to the beginning and end of the transmitted signal to make it extend to signal $\mathbf{s}$, then we can get the new signal $\mathbf{s}$ as

$$\mathbf{s} = (0, \ldots, 0, x_0, x_1, \ldots, x_{N-1}, 0, \ldots, 0,)^T \tag{1}$$

Let the signal $\mathbf{s}$ pass through the ultra-low sidelobe filter $\mathbf{w}$, then the output results can be expressed as

$$y_n = \sum_{i=k_1}^{k_2} w_i^* s_{n-i}, n = 0, 1, 2, \ldots, 2M - 2 \tag{2}$$

Where y_n denotes the filtered results, and $k_1 = \max(0, n - M + 1), k_2 = n - k_1$, $(\cdot)^*$ denotes conjugate operation.

To facilitate derivations, we define the a matrix S as

$$S = \begin{bmatrix} s_0 & \cdots & s_{M-2} & s_{M-1} & 0 & \cdots & 0 \\ 0 & \cdots & s_{M-3} & s_{M-2} & s_{M-1} & \cdots & 0 \\ 0 & \cdots & s_{M-4} & s_{M-3} & s_{M-2} & \cdots & 0 \\ \vdots & \vdots & \vdots & \vdots & \vdots & \vdots & \vdots \\ 0 & 0 & 0 & s_0 & s_1 & \cdots & s_{M-1} \end{bmatrix} \tag{3}$$

Then Eq. (2) can be rewritten in the following form:

$$\mathbf{y} = \mathbf{w}^H S = [y_1, y_2, \ldots, y_M, \ldots, y_{2M-1}] \tag{4}$$

When $\mathbf{w} = \mathbf{s}^*$, the ultra-low sidelobe filter becomes the classical matched filter. The matched filter is optimal to the peak response, but not optimal in terms of the sidelobe level. High sidelobe may deduce the mask phenomenon, where the mainlobe level of the weak targets may be blurred in the high sidelobe level of the strong ones among the group targets. In order to suppress the sidelobes to decrease the mutual interference among the group targets, ultra-low sidelobe filter is quite essential.

There are two famous criteria can be used to decrease the sidelobes, the peak to sidelobe ratio (PSLR) and the integrated sidelobe level (ISL). The first criteria to minimize the maximum value of the sidelobe while the second one to minimizing the integrated sidelobe level of filtered results. In the proposed algorithm, minimizing the PSL after the ultra-low sidelobe filtering is used to design the ultra-low sidelobe filter.

For the further analysis, the PSLR the of the signal $\mathbf{s}$ after the ultra-low sidelobe filtering can be defined as

$$PSL = -20 \log 10 \left(\max_{k \neq M} (|y_M|) \right) \tag{5}$$

Assume that the signal $\mathbf{s}$ is filtered by the ultra-low sidelobe filter, then the SNR_{loss} of the signal $\mathbf{s}$ will occur compared with the matched filter. Then the SNR_{loss} is defined as the ratio between the SNR provided by the ultra-low sidelobe filter and the optimal SNR provided by the matched filter, which can be defined as

$$SNR_{loss} = 10 \log 10 \left(\frac{|\mathbf{w}^H \mathbf{s}|^2}{(\mathbf{w}^H \mathbf{w})(\mathbf{s}^H \mathbf{s})} \right) \tag{6}$$

In order to suppress the sidelobe, we can construct the following optimization

$$\begin{aligned} &\min_{w} \ PSL \\ &s.t. \ \mathbf{w}^H \mathbf{s} = \mathbf{s}^H \mathbf{s} \end{aligned} \tag{7}$$

It should be noticed that the constraint $\mathbf{w}^H \mathbf{s} = \mathbf{s}^H \mathbf{s}$ guarantees that the trivial solution $\mathbf{w} = 0$ will be discard

Based on the definition of PSL, Eq. (7) can be transformed as

$$\min_{w} \max |y|_i, i = 0, 1, 2, M - 2, M, M + 1, \ldots, 2M - 2, i \neq M - 1$$
$$s.t.\, \mathbf{w}^H \mathbf{s} = \mathbf{s}^H \mathbf{s} \tag{8}$$

If the constrained matrix T is defined as

$$T = \begin{bmatrix} s_0 & \cdots & s_{M-2} & 0 & \cdots & 0 \\ 0 & \cdots & s_{M-3} & s_{M-1} & \cdots & 0 \\ 0 & \cdots & s_{M-4} & s_{M-2} & \cdots & 0 \\ \vdots & \ddots & \vdots & \vdots & \ddots & \vdots \\ 0 & \cdots & 0 & s_1 & \cdots & s_{M-1} \end{bmatrix} \tag{9}$$

Then the ultra-low sidelobe problem can be further expressed as

$$\min_{w} \max |\mathbf{w}^H T|_{\infty}$$
$$s.t.\, \mathbf{w}^H \mathbf{s} = \mathbf{s}^H \mathbf{s} \tag{10}$$

Where $|\cdot|_{\infty}$ denotes the infinity norm.

The above constrained infinity norm problem cannot be solved analytically. Many existing solutions consist of weight ISL method or L_p-norm optimization problem with p sufficiently large [6, 7].

In order to solve the constrained infinity norm problem, we convert Eq. (10) as follows

$$\min_{w} t$$
$$s.t.\, \mathbf{w}^H \mathbf{s} = \mathbf{s}^H \mathbf{s}$$
$$\max |\mathbf{w}^H T|_{\infty} \leq t \tag{11}$$

Obviously, Eq. (11) can be considered as a quadratically constrained quadratic program (QCQP). The QCQP problems can be solved with the interior point methods [8–10].

3 Numerical Results

In this section, we will present the first example to demonstrate the ultra-low sidelobe performance of the proposed algorithm. Baker code waveform is considered, and the ultra-low sidelobe filter length is 13, 39 and 65, respectively. Meanwhile, the single target velocity is set as 0 m/s and 10 m/s, respectively. In the simulation, the radar works in the S band.

3.1   Stationary Target Scenery

Figure 1 gives the compared results of matched filter and ultra-low sidelobe filter in the case of stationary target scenery.

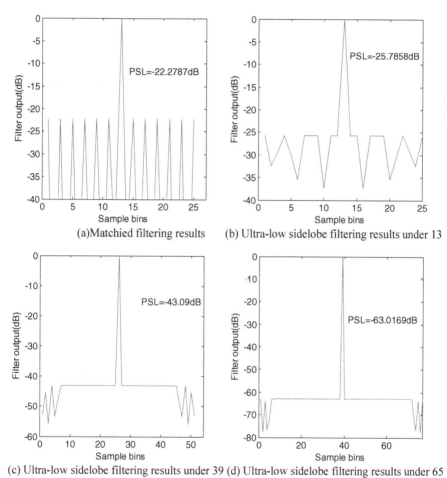

(a)Matchied filtering results

(b) Ultra-low sidelobe filtering results under 13

(c) Ultra-low sidelobe filtering results under 39 (d) Ultra-low sidelobe filtering results under 65

Fig. 1. Results of matched filter and ultra-low sidelobe filter for stationary target

In Fig. 1, matched filter and ultra-low sidelobe filter with the length of 13, 39 and 65 is introduced. From Fig. 1, we can know that the PSL of the matched filter is −22 dB, while the ultra-low sidelobe filtering results are −25.79 dB, −43.09 dB and −63.02 dB, respectively. We can find that there is a substantial gain improvement in PSLR with the increase of the ultra-low sidelobe filter length.

3.2 Moving Target Scenery

For further analysis, the moving target scenery is considered. Assume that the velocity of the target is 10 m/s, and the sample frequency is 1 MHz, the simulation result is illustrated in Fig. 2.

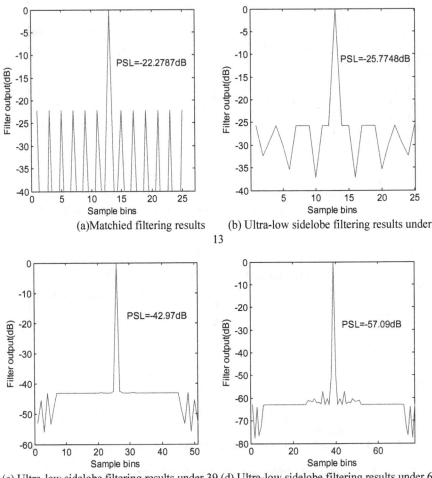

(a)Matchied filtering results   (b) Ultra-low sidelobe filtering results under 13

(c) Ultra-low sidelobe filtering results under 39 (d) Ultra-low sidelobe filtering results under 65

Fig. 2. Results of matched filter and ultra-low sidelobe filter for moving target

From Fig. 2, we can know that sidelobe level of the ultra-low sidelobe filtering results are −25.77 dB, −42.97 dB and −57.09 dB, respectively. And, an interesting phenomenon can be seen that the sidelobe level of the moving target is a little high than the case of stationary target, which may result from the Doppler effect of the moving target. Therefore, the Doppler compensation is necessary in the ultra-low sidelobe filtering algorithm. It will be further investigated in our future work.

4   Experimental Results

To demonstrate the effectiveness of the proposed algorithm in the group targets scenery, in this section, experimental data collected from real radar is utilized to verify the proposed algorithm. Part of the radar system parameters are listed in Table 1.

Table 1. Parameters used in the experiments

Parameters	Value
Time width	50us
Band width	18 MHz
Platform velocity	8 m/s
Range gate number	512
Coherent pulses	1024
Scanning area	−45°–45°
Coherent pulses	1024
Pitching angle	5°

As is shown in Fig. 3, the targets to be studied can be classified as 2 cases. Case 1 is for the first three group targets (T1, T2 and T3) with the same beamwidth in the azimuth and range closed spaced cells. Case 2 is for the single target with the same beamwidth as the first group targets but different range cells. The SNR for both targets is about −19.5 dB, and the velocity is about 152 m/s. To better testify the improvement of the proposed algorithm, the conventional matched filtering algorithm in [1] is performed for comparison.

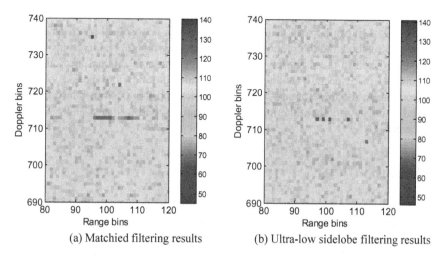

(a) Matchied filtering results   (b) Ultra-low sidelobe filtering results

Fig. 3. Results of matched filter and ultra-low sidelobe filter in the RD domain

In the experiment, the length of ultra-low sidelobe filter is 5 times longer than transmitted signal sequence. The results with the conventional matched filtering algorithm and the proposed algorithm in the Range-Doppler (RD) domain are shown in Fig. 3.

From Fig. 3, we can find that conventional matched filtering results can hardly resolve and distinguish the first group targets, and all the targets in group 1 mutually interact with each other. In this case, false alarm will occur. However, the group targets in Fig. 3(b) can be well distinguished, and it is quite useful for further target detecting and resolving if CFAR detector is followed. To further illustrate the effectiveness of the proposed algorithm, the range profile at the 713 Doppler cell is shown is Fig. 4.

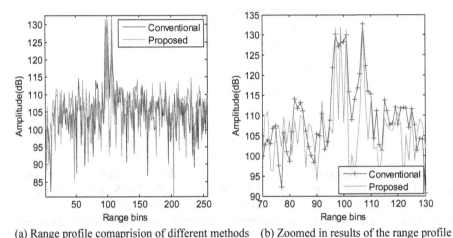

(a) Range profile comaprision of different methods   (b) Zoomed in results of the range profile

Fig. 4. Results of matched filter and ultra-low sidelobe filter in the range profile

From Fig. 4, it can be seen that the two groups can be resolved based on both the conventional method and the proposed one. The three group targets in group 1 can be well distinguished in the proposed algorithm, while the conventional method fails to distinguish them. Especially in Fig. 4(b), there are three obvious peaks, which are corresponding to the three targets in group 1. Based on the experimental results, the proposed ultra-low sidelobe filtering algorithm can be used to detect and resolve the group targets.

5   Conclusion

In this paper, a space group targets detecting and resolving algorithm using the ultra-low sidelobe filtering is proposed. Firstly, the problem of ultra-low sidelobe filtering is converted into the problem of convex optimization by exploiting the optimal PSL criteria, then the ultra-low sidelobe filtering results are used to detect and resolve the space group targets in the main beam. Since the ultra-low sidelobe filter can decrease the mutual interact between group targets, it is useful to distinguish space group targets. Numerical and experimental results demonstrate the effectiveness of the proposed algorithm.

Acknowledgements. The authors would like to thank all the anonymous reviewers for their valuable comments to improve the quality of this paper. This work was supported by the Postdoctoral Science Foundation of China under grant 2017M610966.

References

1. Richards, M.A.: Fundamentals of Radar Signal Processing. McGraw-Hill, New York (2005)
2. Chen, V.C., Li, F., Ho, S.-S., Wechsler, H.: Micro-Doppler effect in radar: Phenomenon, model, and simulation study. IEEE Trans. Aerosp. Electron. Syst. **42**(1), 2–21 (2006)
3. Persico, A.R., Clemente, C., Ilioudis, C., Gaglione, D., Cao, J., Soraghan, J.: Micro-doppler based recognition of ballistic targets using 2D Gabor filters. In: Proceedings of IEEE Sensor Signal Processing for Defence, 1–5 September 2015
4. Zhao, M., Zhang, Q., Luo, Y., Sun, L.: Micromotion feature extraction and distinguishing of space group targets. IEEE Geosci. Rem. Sens. Lett. **14**(2), 174–178 (2017)
5. Deng, H.: Polyphase code design for orthogonal netted radar systems. IEEE Trans. Signal Process. **52**(11), 3126–3135 (2004)
6. Karmarkar, N.: A new polynomial-time algorithm for linear programming. Combinatorica **4**, 373–395 (1984)
7. Cilliers, J.E., Smit, J.C.: Pulse compression sidelobe reduction by minimization of Lp-norms. IEEE Trans. Aerosp. Electron. Syst. **43**(3), 1238–1247 (2007)
8. Boyd, S., Vandenberghe, L.: Convex Optimization. Cambridge University Press, Cambridge (2004)
9. Vandenberghe, L., Boyd, S.: Semidefinite programming. SIAM Rev. **38**(1), 49–95 (1996)
10. Alizadeh, F.: Second-order cone programming. Math. Program. **95**(1), 40–54 (2003)

An Energy Sequencing Based Partial Maximum Likelihood Detection Method for Space-Frequency Joint Index Modulation System

Xiaoke Niu[1], Xingle Feng[1(✉)], Kun Hua[2], Guobin Duan[1], and Shizhe Gao[1]

[1] Chang'an University, Xi'an 710064, China
xlfeng@chd.edu.cn
[2] Lawrence Technological University, Southfield, MI, USA

Abstract. In this paper, an energy sequencing based partial Maximum Likelihood (ML) detection algorithm is proposed for the complex characteristics of receiver detection in space-frequency joint index modulation system. This algorithm can solve the problems of high complexity from ML detection and poor Bit Error Rate (BER) performances by Minimum Mean Square Error (MMSE) detection. The major idea of the proposed algorithm is to demodulate the activated sub-carrier sequence number, antenna sequence number and constellation symbol step by step, where the sub-carrier sequence number is equalized with MMSE and the energy value of each sub-carrier is calculated and sorted. And the P value is set as the number of candidate sub-carriers. Finally, the sequence numbers of alternative sub-carriers, antenna serial numbers and constellation symbols are detected by ML. Simulation results show that the proposed algorithm can reduce both search range of traditional ML methods and the complexity according to the selection of P value. For example, when $P = 3$, the BER can be reduced to 10^{-4} at the SNR of 20 dB in the proposed algorithm.

Keywords: Maximum Likelihood (ML) ·
Space-Frequency joint index modulation · Bit Error Rate (BER) ·
Minimum Mean Square Error (MMSE)

1 Introduction

Traditional information transmission resources in spatial domain, frequency domain and time domain can no longer satisfy the growing demand for high-speed data transmission in 5G. Therefore, the combination of both spatial and frequency domain information sounds like a better idea, however, the dramatically increased numbers of

This work was supported in part by the National Natural Science Foundation of China under Grant 61271262, and in part by the Natural Science Basic Research Plan in Shaanxi Province of China under Grant 2017JM6099, and in part by the Fundamental Research Funds for the Central Universities of China under Grant 300102248307.

B. Li et al. (Eds.): IoTaaS 2018, LNICST 271, pp. 324–332, 2019.
https://doi.org/10.1007/978-3-030-14657-3_32

antennas and sub-carriers have been causing new challenges at the receiver side. In order to solve aforementioned problems, Index Modulation (IM) technology makes a good use of the indexes of transmission medium, such as transmit antennas, sub-carriers, time slots or linear block codes, to modulate information bits by some kinds of mapping rules [1]. Due to the small power consumption of the transmission index bits, IM exploits a feasible trade-off between Spectral Efficiency (SE) and Energy Efficiency (EE), or in other words, the diversity gain and multiplexing gain. IM technology mainly includes Spatial Modulation (SM) and sub-carrier index modulation, both of which use IM to diminish interference and introduce index bits to compensate for the loss of SE. The difference in between is that SM is applied to select antennas, while sub-carrier index modulation is used for selecting sub-carriers. SM is a Multiple-Input Multiple-Output (MIMO) transmission method, which considers the transmitting antennas as a spatial constellation point to carry additional information bits [2]. By activating only one antenna at each time instant, SM system is insensitive to Inter-Channel Interference (ICI) and can suppress the requirement of Inter-Antenna Synchronization (IAS) [3]. Meanwhile, both the detection complexity and the cost of Radio-Frequency (RF) chain are reduced [4]. Thus, SM is regarded as a competitive successor to conventional MIMO techniques. Orthogonal Frequency Division Multiplexing with Index Modulation (OFDM-IM) [5] is an extension of sub-carrier index in SM concept in multi-carrier systems. Under certain conditions, the OFDM-IM scheme has also been proposed to provide higher throughput and better BER performance than plain OFDM. The OFDM-IM scheme also provides an interesting trade-off between complexity, SE and performance by changing the number of active sub-carriers [6].

In this paper, a new space-frequency joint index modulation system, SM-OFDM-IM, that combines SM with OFDM-IM is created [7]. This system combines antenna index in spatial domain and sub-carrier index in frequency domain. However, the receiver has much more strict requirements for channel independence, synchronization and poor real-time performance, and the symbol information that needs to be detected changes with it. In this system, the receiver requires to detect the antenna index bits, sub-carrier index bits and modulation bits. Therefore, it is an important content for index modulation to design a detection algorithm with low complexity and excellent BER performance. At present, there are very few detection algorithms like this proposed SM-OFDM-IM system. In [8], based on Sequential Monte Carlo (SMC) theory and extracts samples at sub-block level, a low complexity detection algorithm is proposed to achieve approximately optimal BER performance. The computational complexity is greatly reduced. But the algorithm only detects the sequence number of the active sub-carriers and constellation symbols, not the sequence number of the active antenna. A low complexity LLR detection algorithm for OFDM-IM system is proposed in [9]. Because the computational complexity of calculating the log likelihood ratio of each bit is extremely high due to the dependence between the sub-carriers caused by the sub-carrier index modulation. Simulation results show that the proposed algorithm achieves near-optimal BER performance with low computational complexity. However, the algorithm can not be directly applied to the proposed system.

Based on the research of signal detection algorithm in SM-OFDM-IM system, an energy sequencing based partial ML detection algorithm is proposed for the characteristics of the receiver. This algorithm reduces the search range of ML and can achieve the compromise between complexity and BER performance. The main contribution of this algorithm is to demodulate the active sub-carrier serial number, antenna sequence number and constellation symbol in sequence.

2 System Model

SM-OFDM-IM system is a combination, that not only SM, but also sub-carrier index modulation are applied to OFDM. The transmitted information is mapped to different combinations of transmit antennas, sub-carriers and modulation respectively, which reflects the flexibility of the implementation of SM-OFDM-IM system. The system model of SM-OFDM-IM is shown in Fig. 1. The active antenna is selected according to the antenna index bits, and the OFDM-IM modulation is carried out on the activated antennas. According to the sub-carrier index bits, some dedicated sub-carriers are chosen and activated. And corresponding constellation symbols are modulated on these sub-carriers. The reverse operation will be implemented at the receiver. Consider a SM-OFDM-IM system with n_t transmit antennas, one transmit RF chain and n_r receive antennas. The N sub-carriers are divided into G sub-carrier blocks of $n = N/G$. Binary bit stream is divided into three parts. The first part is used for sub-carrier index, which modulates $\lfloor \log_2 C_n^k \rfloor$ bits to determine a set of k active sub-carriers for transmission out of n available ones, and the sub-carrier configuration is (n, k), where $\lfloor x \rfloor$ is the greatest integer smaller than x and C_n^k represents the binomial coefficient. In second part, every $\lfloor \log_2 n_t \rfloor$ bits are used for choose 1 out of n_t transmit antennas. In third part, $k \log_2 M$ bits are modulated using $M - QAM$ modulation to form the constellation symbol. Therefore, $\lfloor \log_2 C_n^k \rfloor + \lfloor \log_2 n_t \rfloor + k \log_2 M$ bits are transport at the transmitter, and the SE of SM-OFDM-IM in terms of bits per channel use (bpcu) is

$$R = \left\lfloor \log_2 C_n^k \right\rfloor + \left\lfloor \log_2 n_t \right\rfloor + k \log_2 M \ (bpcu) \tag{1}$$

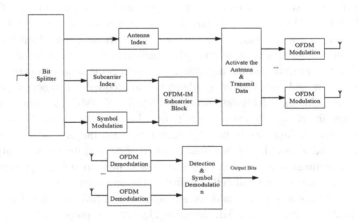

Fig. 1. System model of SM-OFDM-IM

According to the above description, the frequency domain form of the transmit signal $S_g \in C^{n_t \times n}$ of the g-th sub-block can be given by

$$S_g = [s_1^g, s_2^g, \cdots, s_n^g]$$
$$= \begin{bmatrix} s_{1,1}^g \cdots s_{1,n}^g \\ \vdots \\ s_{n_t,1}^g \cdots s_{n_t,n}^g \end{bmatrix}, g = 1, \cdots, G \tag{2}$$

where its elements $S_{i,j}^g \in \{0, S\}, i = 1, 2, \cdots, n_t, j = 1, 2, \cdots, n$ represents the transmitted symbol at the j-th frequency in group g on antenna i, and S is the constellation set of $M - QAM$.

It is assumed that the wireless channels remain constant in the transmission of an SM-OFDM-IM symbol, the frequency domain received signal of SM-OFDM-IM is obtained as

$$y_j^g = H_j^g s_j^g + n_j^g \tag{3}$$

where $H_j^g \in C^{n_r \times n_t}$ is the frequency domain channel matrix at the j-th sub-carrier frequency of group g, whose elements follow i.i.d $CN(0, 1), y_j^g, n_j^g \in C^{n_r \times 1}$ represents the received signal and frequency domain additive white Gaussian noise samples with variance σ^2 at the j-th sub-carrier frequency of group g. The system signal-to-noise ratio (SNR) is defined as $P_s/N_0 = 1/N_0$, where $P_s = 1$ is the normalized transmit power.

At the receiver, using the ML principle decodes the symbol, formula can be written as

$$[\hat{i}, \hat{n}, \hat{m}] = \underset{X \in \Lambda}{\arg\min} \|y - Hs\|_F^2 \tag{4}$$

where Λ denotes all possible transmit signal vector combinations.

3   A Partial Maximum Likelihood Detection Method

In the SM-OFDM-IM system, active antenna sequence numbers, sub-carrier serial numbers and constellation symbols should be detected at the receiver. However, the traditional detection methods of sole SM and OFDM-IM systems are not applicable. Therefore, both the joint detection and stepwise detection are proposed in the dedicated detection algorithms of SM-OFDM-IM system. In [10], the proposed Energy Detection (ED) algorithm for frequency-domain index modulation systems can achieve a bit error rate of 10^{-4} when SNR is 40 dB. But the performance becomes worse when it is applied to SM-OFDM-IM system directly, and the BER is always around 10^{-2}. On this basis, an energy order based partial ML detection algorithm is proposed in this paper.

The detection algorithm is listed as follows:

The received signal is detected in units of each sub-block. It is assumed that the g-th received sub-block is y_g, with the dimension of $n_t \times n$.

First, MMSE equalization. MMSE belongs to the linear detection algorithm. More precisely, it is based on the improved result of Zero Forcing (ZF) detection algorithm. It considers the effect of noise on the detection, so the weight matrix is designed as:

$$G_{MMSE} = (H^H H + \sigma^2 I)^{-1} H^H \tag{5}$$

Where σ^2 is the noise variance, and I is the $n_t \times n$ identity diagonal matrix. According to (5), the signal after the MMSE equilibrium is

$$
\begin{aligned}
\tilde{X}_{MMSE} &= G_{MMSE} \, y_g \\
&= (H^H H + \sigma^2 I)^{-1} H^H \times y_g \\
&= \tilde{S} + (H^H H + \sigma^2 I)^{-1} H^H \times N \\
&= \tilde{S} + \tilde{N}_{MMSE}
\end{aligned}
\tag{6}
$$

where $\tilde{N}_{MMSE} = N \times G_{MMSE} = (H^H H + \sigma^2 I)^{-1} H^H \times N$.

Second, calculating the energy value of the signal after equalization, that is $E_{\tilde{X}_{MMSE}} = \left\| \tilde{X}_{MMSE} \right\|_2^2$, $\hat{E}_{\tilde{X}_{MMSE}}$ is obtained by summing the energy value of the estimated signal on each sub-carrier. And then was sorted by the size in terms of $\hat{E}_{\tilde{X}_{MMSE}}$, that is $[e_1, e_2, \cdots, e_N] = \arg sort(\hat{E}_{\tilde{X}_{MMSE}})$. It is believed that the sub-carrier with the highest energy is most likely to be activated sub-carrier.

Thirdly, the complexity of energy detection is low, but its detection performance is poor due to interference between multiple antennas and Gaussian noise. Therefore, we need to select P $(P = 1, 2, \cdots, n)$ sub-carrier sequence number with higher energy as candidate sub-carrier sequence number. If $P = 2$, select the two sub-carriers with top 2 highest energy value as candidate sub-carriers.

Fourth, the candidate sub-carrier sequence numbers, antenna sequence numbers and constellation symbols are used for ML detection, $D = \left\| y_g - H_g \tilde{x}_g \right\|_F^2 \cdot \tilde{x}_g$ is the transmission symbol for the g-th sub-block estimated according to above steps. Considered that the combination of minimum Euclidean distance is the final detection result. The complexity of the algorithm varies with the value of P.

Finally, $g = g + 1$, repeat above steps and get the detection results of all sub-blocks.

In this detection algorithm, the energy detection is applied to reduce the traversal range of sub-carriers, and then the partial ML detection is carried out on this basis, which narrows the traversal range of ML and reduces the complexity. For the above algorithm, it is very important to determine the order of energy value of the estimated signal. In order to further reduce the complexity, the cost function can be changed to ZF criterion, that is ZF-ML, instead of the MMSE that used in above paper. The simulation results will be compared in Sect. 4.

4 Numerical Results

In this section, the BER performance of the proposed SM-OFDM-IM system, the OFDM and the OFDM-IM scheme are compared by computer simulations under Gaussian noise, as described below.

In order to ensure that OFDM, OFDM-IM and SM-OFDM-IM have the same SE, the simulation parameters of OFDM-IM and SM-OFDM-IM systems are presented in Table 1. The BER performance is given in Fig. 2. Figure 3 is a comparison of Peak Average Power Ratios (PAPR) between OFDM and OFDM-IM systems.

Table 1. OFDM-IM and SM-OFDM-IM systems parameters

Parameters	Vaules of OFDM-IM	Vaules of SM-OFDM-IM
Number of transmit antennas (n_t)	1	2
Number of receive antennas (n_r)	1	2
Number of sub-carriers (N)	128	64
Length of cyclic prefix	32	16
Number of sub-block (G)	64	32
Sub-carrier configuration	$(2, 1)$	$(2, 1)$
Simulation channel	Rayleigh fading channel	Rayleigh fading channel
Modulation mode	BPSK	QPSK

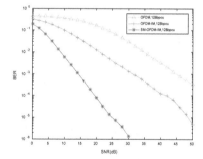

Fig. 2. BER of OFDM, OFDM-IM, SM-OFDM-IM systems

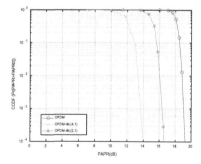

Fig. 3. PAPR of OFDM and OFDM-IM systems

From Fig. 2, at a BER value of 10^{-3}, it can be observed that the SM-OFDM-IM system achieves approximately 30 dB and 15 dB better BER performance than the OFDM and OFDM-IM at the same spectrum efficiency, respectively. This is due to the fact that SM-OFDM-IM adds sub-carrier index and antenna index to transmit data. While the other sub-carriers and antennas remain silent, the sparsity of frequency domain data reduces the sensitivity of the system to frequency offset and the influence of inter-carrier interference on transmission performance. In spatial domain, only one antenna is activated and inter-antenna interference is thereby avoided. SM-OFDM-IM can reduce the SE because of the silence of some sub-carriers and some antennas, and the introduction of index bit information can make up for the transmitting rate. As the result, SM-OFDM-IM is a more promising multi-carrier system than OFDM because of its unique system setting and more flexible parameter configuration.

From Fig. 3, it can be observed that OFDM-IM system have lower PAPR values than conventional OFDM systems under the same number of sub-carriers and the same modulation order. The smaller the number of active sub-carriers is, the lower the PAPR value of OFDM-IM system is.

On the basis of SM-OFDM-IM system model, the impact of different detection methods on BER are compared in this section. The BER performance of the proposed detector, the ML detector and ZF-ML detector with $n_t = 4$, $n_r = 4$, $N = 128$ and the sub-carrier configuration is (4, 1) for BPSK constellations are presented in Fig. 4. And the BER performance with different sizes of P is given in Fig. 5.

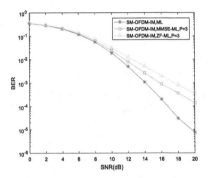 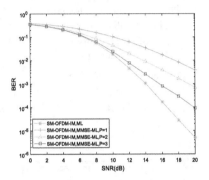

Fig. 4. BER of SM-OFDM-IM system of different detection methods

Fig. 5. BER of SM-OFDM-IM system of different P values

As shown in Fig. 4, the proposed detection algorithm achieves almost 2 dB better BER performance than ZF-ML at the BER value of 10^{-3}. This is because MMSE takes the effect of noise into account. But compared with the ML detection method, the proposed method reduced the complexity at the cost of a lost of about 4 dB BER performance at the value of 10^{-4} due to the smaller search range.

According to the introduction in Sect. 3, the BER performance of SM-OFDM-IM system varies with P values. As shown in Fig. 5, the higher of P values, the larger the search ranges, the higher the complexity and the better the BER performance will be achieved. When P = 4, that is the ML detection, the BER performance is the best and the complexity is highest.

The computational complexity in terms of the complex multiplications performed per sub-block for different detectors is given in Table 2.

Table 2. A total multiplicative computational complexity of different detectors

Different type of decoders	Computational complexity
ML	$(\frac{(n+n_r)n_t}{2}\log_2\frac{n}{2}+n_r n_t^2)M^{nn_t}\sim o(M^{nn_t})$
Proposed partial ML	$10n_t^2 n_r+11nn_t+n_t+(\frac{(n+n_t)n_r}{2}\log_2\frac{n_t}{2}+n_r n_t^2)M^{Pn_t}\sim o(M^{Pn_t})$
ZF-ML	$10n_t^2 n_r+11nn_t+(\frac{(n+n_r)n_t}{2}\log_2\frac{n_t}{2}+n_r n_t^2)M^{Pn_t}\sim o(M^{Pn_t})$

As expected, the calculation complexity of the proposed energy sequencing based partial ML is rather low compared with that of the ML detector. This can be explained that the proposed detector narrows the range of ML by introducing energy ordering. The complexity of the algorithm is slightly higher than that of ZF-ML due to the difference of the weight matrix.

5 Conclusion

In this paper, simulation results has proved that the BER performance of SM-OFDM-IM system outperforms OFDM-IM and OFDM. Then, an energy sequencing based partial ML detection algorithm is proposed for the complex characteristics of the receiver in SM-OFDM-IM system and solved problems of traditional detection methods. The major idea of the algorithm is to demodulate the activated sub-carrier serial numbers, antenna sequence numbers and constellation symbols step by step. The sequence number of sub-carriers is equalized with MMSE, and the energy value of each sub-carrier is obtained, and the number of candidate sub-carriers is set as P value. Then the sequence numbers of partial sub-carriers, antenna serial numbers and constellation symbols are detected by ML. The simulation results show that the proposed algorithm can reduce the search range of traditional ML and reduce the complexity of the algorithm according to the selection of P.

References

1. Basar, E., Wen, M., Mesleh, R., et al.: Index modulation techniques for next-generation wireless networks. Journal **5**(99), 16693–16746 (2017)
2. Mesleh, R.Y., Haas, H., Sinanovic, S., et al.: Spatial modulation. Journal **57**(4), 2228–2241 (2008)
3. Liu, Y., Zhang, M., Wang, H., et al.: Spatial modulation orthogonal frequency division multiplexing with subcarrier index modulation for V2X communications. In: International Conference on Computing, pp. 1–5. IEEE (2016)
4. Yang, P., Di Renzo, M., Xiao, Y., et al.: Design guidelines for spatial modulation. Journal **17**(1), 6–26 (2015)
5. Basar, E., Aygolu, U., Panayirci, E., et al.: Orthogonal frequency division multiplexing with index modulation. Journal **61**(22), 5536–5549 (2013)
6. Mrkic, J., Kocan, E., Pejanovic-Djurisic, M.: Index modulation techniques in OFDM relay systems for 5G wireless networks. In: International Conference on Telecommunications and Signal Processing, pp. 208–211. IEEE (2017)

7. Bai, H.: Research on Multi-Carrier System with Index Modulation. University of Electronic Science and Technology (2016)
8. Zheng, B., Wen, M., Basar, E., et al.: Low-complexity near-optimal detector for multiple-input multiple-output OFDM with index modulation. In: IEEE International Conference on Communication, pp. 1–6. IEEE (2017)
9. Hu, Z., Chen, F., Wen, M., et al.: Low-complexity LLR calculation for OFDM with index modulation. Journal **7**(4), 618–621 (2018)
10. Xu, B.: Research on broadband MIMO-OFDM with index modulation in wireless transmission technology. University of Electronic Science and Technology (2015)

Joint Energy-Efficient Optimization of Downlink and Uplink with eICIC in HetNet

Jie Zheng[1], Ling Gao[2]($\boxtimes$), Dongxiao Zhu[3]($\boxtimes$), Hai Wang[1], Qian Sun[1], Jinping Niu[1], Xiaoya Li[1], and Jianfeng Yang[1]

[1] State-Province Joint Engineering and Research Center of Advanced Networking and Intelligent Information Services, School of Information Science and Technology, Northwest University, Xian 710127, China
jzheng@nwu.edu.cn

[2] State-Province Joint Engineering and Research Center of Advanced Networking and Intelligent Information Services, Xi'an Polytechnic University, Xi'an, China
gl@nwu.edu.cn

[3] Department of Computer Science, Wayne State University, Detroit, MI 48202, USA
dzhu@wayne.edu

Abstract. In this work, we propose to configure uplink access to the macro in almost blank subframes (ABSs) improving the its utilization. The ABSs based on Downlink and Uplink Decoupled (DUDe) are modeled as an energy-efficient optimization problem in Heterogeneous Networks (HetNets). The formulation is a mixed integer programming problem so that we provide a suboptimal algorithm. Simulation study show our proposed algorithm demonstrates better performance according to energy efficiency and capacity.

Keywords: Heterogeneous Network · Energy efficiency ·
Downlink/uplink decoupling (DUDe) · Joint uplink and downlink ·
Interference coordination

1 Introduction

The LTE-A have specified the eICIC that macrocell can keep its downlink in silence for almost blank subframe (ABS) [1]. The users accessing to small cell obtain higher data rate for ABS on account of much less interference.

Moreover, the Downlink and Uplink Decoupling (DUDe) leads to lower energy consumption of HetNets [2–4]. Joint uplink (UL) and downlink (DL) has been mainly studied in the single-tier cellular network. The radio planning is investigated to minimize power consumption of joint uplink and downlink in single-tier network [6]. The association algorithm is proposed to maximize the downlink rate and minimize the power of uplink, but they assumed that it has the same channel gain [7]. To maximize the utility of data rate of joint downlink

B. Li et al. (Eds.): IoTaaS 2018, LNICST 271, pp. 333–339, 2019.
https://doi.org/10.1007/978-3-030-14657-3_33

and uplink, [8] designs the algorithm of joint user association and resource allocation with QoS, however, without consider the eICIC [8].

We formulate EE-eICIC [5] to the corresponding DUDe scenario. We investigate joint energy efficiency optimization of uplink and downlink for eICIC in HetNet. As a result, we use UM-ABS [9], which is to configure uplink transmission for macro in almost blank subframe (ABS) improving the utilization of ABS. And the UM-ABS with downlink and uplink Decoupled (DUDe) can be adopted into an energy efficiency optimization problem.

2   Systems Model

The TDD technology is considered in HetNet, where the subframes of uplink and downlink can be configured dynamically with eICIC. The user association is divided into two types: smallcell-accessed and macrocell-accessed, so we can obtain the SINR expression.

Table 1. The notations of variables

Variable	Definition
N_{sf}	The number of ABS frames
N_m	non-ABS for small cell
A_p	ABS for small cell
$a_u(x_{u,A})$	Time in ABS (non-ABS) used by user
$b_{u,A(nA)}$	Time in ABS (non-ABS) used by user
p_u^{BS}	The power of BS (macrocell or smallcell)
p_{ref}^{macro}	The reference signals power from macrocell over ABS subframes
$P_{Rx}(u)$	The received power in user u for downlink
$P_{Rx}(m)$	The received power in macrocell m from uplink of user u
$P_{Rx}(s)$	The received power in smallcell s from uplink of user u
$P_{small}^{Down}(u)$	The suffered interference from smallcells for downlink
$P_{u \in pico}^{Up}(u)$	The suffered interference from smallcells for uplink
$P_{macro}^{Down}(u)$	The suffered interference from macrocells for downlink
$P_{u \in macro}^{Up}(u)$	The suffered interference from macrocells for uplink
$P_{small}^{Down}(I)$	The suffered interference from smallcells for downlink
$P_{u \in pico}^{Up}(I)$	The suffered interference from smallcells for uplink
$P_{macro}^{Down}(I)$	The suffered interference from macrocells for downlink
$P_{u \in macro}^{Up}(I)$	The suffered interference from macrocells for uplink

The SINR of downlink for one user accessed to smallcell is

$$SINR_{small}^{Down}(u) = \begin{cases} \dfrac{P_{Rx}(u)}{P_{small}^{Down}(u)+P_{pico}^{Up}(u)+P_{macro}^{Up}(u)+N_0} & \text{ABS} \\ \dfrac{P_{Rx}(u)}{P_{small}^{Down}(u)+P_{macro}^{Down}(u)+P_{small}^{Up}(u)+P_{macro}^{Up}(u)+N_0} & \text{non-ABS.} \end{cases} \quad (1)$$

The SINR of uplink for one user accessed to smallcell is

$$SINR_{small}^{Up}(u) = \begin{cases} \frac{P_{Rx}(p)}{P_{small}^{Down}(I)+P_{u\in pico}^{Up}(I)+P_{u\in macro}^{Up}(I)+N_0} & \text{ABS} \\ \frac{P_{Rx}(p)}{P_{small}^{Down}(I)+P_{macro}^{Down}(I)+P_{u\in small}^{Up}(I)+P_{u\in macro}^{Up}(I)+N_0} & \text{non-ABS.} \end{cases} \quad (2)$$

The SINR of downlink for one user accessed to macrocell is

$$SINR_{macro}^{Down}(u) = \frac{P_{Rx}(u)}{P_{small}^{Down}(u) + P_{macro}^{Down}(u) + P_{small}^{Up}(u) + P_{macro}^{Up}(u) + N_0}\text{non-ABS.} \quad (3)$$

The SINR of uplink for one user accessed to macrocell is

$$SINR_{macro}^{Up}(u) = \begin{cases} \frac{P_{Rx}(m)}{P_{small}^{Down}(I)+P_{macro}^{Down}(I)+P_{u\in small}^{Up}(I)+P_{u\in macro}^{Up}(I)+N_0} & \text{ABS} \\ \frac{P_{Rx}(m)}{P_{small}^{Down}(Int)+P_{macro}^{Down}(I)+P_{u\in small}^{Up}(I)+P_{u\in macro}^{Up}(I)+N_0} & \text{non-ABS.} \end{cases} \quad (4)$$

Then, the data rate of uplink and downlink are computed by Shannon capacity.

3 Problem Formulation

The goal in our work is to maximize the energy efficiency of joint uplink and downlink. These variables $\psi = \{R_u, P_u, a_u^{up}, a_{u,A}^{up}, a_u^{down}, b_{u,A}^{up}, b_{u,A}^{down}, b_{u,A}^{up}, b_{u,nA}^{down}, A_p, N_m\}$ are jointly optimized to design the EE-DL-UL-eICIC algorithm by the optimization problem (P1). These variables have been shown in Table 1.

$$P1 : \max_{\psi} \sum_u \frac{R_u}{P_u} \quad (5)$$

$$R_u = R_u^{up} + R_u^{down} \quad (6)$$

$$P_u = P_u^{up} + P_u^{down} \quad (7)$$

$$R_u^{up} \leq a_u^{up} \cdot r_{u,macro}^{up} + a_{u,A}^{up} \cdot r_{u,A,macro}^{up} + b_{u,A}^{up} \cdot r_{u,A,pico}^{up} + b_{u,nA}^{up} \cdot r_{u,nA,pico}^{up} \quad (8)$$

$$R_u^{down} \leq a_u^{down} \cdot r_{u,macro}^{down} + b_{u,A}^{down} \cdot r_{u,A,pico}^{down} + b_{u,nA}^{down} \cdot r_{u,nA,pico}^{down} \quad (9)$$

$$P_u^{down} \leq p_u^{macro} \cdot a_u^{down} + (p_u^{pico} + P_{ref}^{macro}) \cdot y_{u,A}^{down} + p_u^{pico} \cdot b_{u,nA}^{down} \quad (10)$$

$$P_u^{up} \leq p_u \cdot a_u^{up} + p_u \cdot a_{u,A}^{up} + p_u \cdot b_{u,A}^{up} + p_u \cdot b_{u,nA}^{up} \quad (11)$$

$$a_u^{up}(a_{u,A}^{up} + b_{u,nA}^{up}) = 0, \forall u \in U \quad (12)$$

$$a_u^{down}(b_{u,A}^{down} + b_{u,nA}^{down}) = 0, \forall u \in U \quad (13)$$

$$A_p + N_m \leq N_{sf}, \forall p, m \in I_p \quad (14)$$

$$\sum_{u \in U_m} (a_u^{up} + a_u^{down}) \leq N_m, m \in M \quad (15)$$

$$\sum_{u \in U_m} a_{u,A}^{up} \leq A_p, m \in M, p \in P \quad (16)$$

$$\sum_{u \in U_p} (b_{u,A}^{up} + b_{u,A}^{down}) \le A_p, p \in P \tag{17}$$

$$\sum_{u \in U_p} (b_{u,A}^{up} + b_{u,A}^{down} + b_{u,nA}^{up} + b_{u,nA}^{down}) \le N_{sf}, p \in P \tag{18}$$

$$a_u^{up} \ge 0, a_{u,A}^{up} \ge 0, a_u^{down} \ge 0 \tag{19}$$

$$b_{u,A}^{up} \ge 0, b_{u,A}^{down} \ge 0, b_{u,nA}^{up} \ge 0, b_{u,nA}^{down} \ge 0 \tag{20}$$

$$A_p, N_m \le N^+, \forall p, m \in I_{BS} \tag{21}$$

here N^+ represents nonnegative integer number.

The (6) and (7) give the sum of rate and power of downlink and uplink for a user. The (8), (9) and (10), (11) give the maximum data rate and power in uplink or downlink for one user. The (12) and (13) show the uplink or downlink of a user can only associate with one BS (e.g. macrocell or smallcell). The constraint (14) gives the number of ABS subframes provided by macrocells due to the interference with smallcell. The (15), (16) and (17)ensure that the subframes allocated to the user are not larger than the available ABS. The (18) shows that the subframes allocated to user are limited by the ABS period N_{sf}.

It is obvious that P1 belongs to mixed integer programming, which is usually NP-hard [10]. However, the structure of (5) can be reformulated into fractional programming [11]. Thus, the following problem (P2) can be computed with Algorithm 1 for a given ρ (e.g., ρ_n at iteration n).

$$\max_\psi \sum_u (R_u^{up} - \rho_u^{up} P_u^{up}) + (R_u^{down} - \rho_u^{down} P_u^{down}) \tag{22}$$
$$\text{s.t.} \quad (6)\text{–}(21)$$

Algorithm 1. EE-DU-UL-eICIC

1: Choose the iteration accuracy $\varepsilon^{up} > 0, \varepsilon^{down} > 0$ and set the limit of iteration number N_{max}.

2: **while** Quit = 0 and $n \le N_{max}$ **do**

3:     Solving the problem P2 for a certain η^n

4:     **if** $|\rho_u^{up,n}| = |(R_u^{up,n} - \rho_u^{up,n} P_u^{up,n})| < \varepsilon^{up}$ *and* $|\rho_u^{down,n}| = |(R_u^{down,n} - \rho_u^{down,n} P_u^{down,n})| < \varepsilon^{down}$ **then**

5:         Quit = 1;

6:         **return** obtain the optimum configuration of EE-DU-UL-eICIC ψ^{opt} and the result ρ_u^{opt}

7:     **else**

8:         compute $\rho_u = \frac{R_u}{P_u}$ and $n = n + 1$, Quit = 0.

9:     **end if**

10: **end while**

---

4   Algorithm for Nonlinear Programming

For a given ρ, we relax the P2 into P3. The P3 is obtained via ignoring the (12)–(13) and making the (21) on N_m and A_p into positive real numbers. The P3 is expressed with:

$$\max_\psi \sum_u (R_u^{up} - \rho_u^{up} P_u^{up}) + (R_u^{down} - \rho_u^{down} P_u^{down})$$
$$\text{s.t.} \quad (6)\text{--}(11) \ and \ (14)\text{--}(20) \tag{23}$$
$$A_p, N_m \in R^+, \forall p, m \in I_{BS}.$$

where R^+ gives the nonnegative real number. Since the P3 is the convex programming, it is easy to solve by the CVX tools [12]. Then the result computed by CVX are rounded to obtain the feasible solution with (21).

$$Inter(x) = \begin{cases} round\,down\,(x) & x < \frac{N_{sf}}{2} \\ round\,up\,(x) & x \geq \frac{N_{sf}}{2} \end{cases} \tag{24}$$

The DL-accessed, UL-accessed and ABSs allocation schemes are displayed in Algorithm 2.

Algorithm 2. The DL/UL accessed and ABSs allocation algorithm

---

1: To compute the feasible solution N_m^* and A_p^*: $N_m^* = Inter(N_m')$ and $A_p^* = Inter(A_p')$, where N_m' and A_p' are results from CVX tools.

2: To ensure the DL-association or UL-association for user:

$$R_{u,macro}^{up} = r_{u,A,macro}^{up} \cdot \tilde{a}_{u,A}^{up} + r_{u,macro}^{up} \cdot \tilde{a}_u^{up}, R_{u,small}^{up} = r_{u,A,pico}^{up} \cdot \tilde{b}_{u,A}^{up} + r_{u,nA,small}^{up} \cdot \tilde{b}_{u,nA}^{up} \tag{25}$$

$$R_{u,macro}^{down} = r_{u,macro}^{down} \cdot \tilde{a}_u^{down}, R_{u,small}^{down} = r_{u,A,small}^{down} \cdot \tilde{b}_{u,A}^{down} + r_{u,nA,small}^{down} \cdot \tilde{b}_{u,nA}^{down} \tag{26}$$

$$P_{u,macro}^{up} = p_{u,A,macro}^{up} \cdot \tilde{a}_{u,A}^{up} + p_{u,macro}^{up} \cdot \tilde{a}_u^{up}, P_{u,small}^{up} = p_{u,A,small}^{up} \cdot \tilde{b}_{u,A}^{up} + p_{u,nA,small}^{up} \cdot \tilde{b}_{u,nA}^{up} \tag{27}$$

$$P_{u,macro}^{down} = p_{u,macro}^{down} \cdot \tilde{a}_u^{down}, P_{u,small}^{down} = p_{u,A,small}^{down} \cdot \tilde{b}_{u,A}^{down} + p_{u,nA,small}^{down} \cdot \tilde{b}_{u,nA}^{down} \tag{28}$$

where $\tilde{a}_u^{up}$, $\tilde{a}_{u,A}^{up}$, $\tilde{b}_{u,A}^{up}$, $\tilde{b}_{u,nA}^{up}$, $\tilde{a}_u^{down}$, $\tilde{b}_{u,A}^{down}$, $\tilde{b}_{u,nA}^{down}$ is output of CVX tools.

Computing $\rho_u^{up,small} = \frac{R_u^{up,small}}{P_u^{small}}$, $\rho_u^{up,macro} = \frac{R_u^{up,macro}}{P_u^{up,macro}}$. If $\rho_u^{up,small} > \rho_u^{up,small}$, the uplink of user can access to macro cell, or with small cell.

Computing $\rho_u^{down,small} = \frac{R_u^{down,small}}{P_u^{down,small}}$, $\rho_u^{down,macro} = \frac{R_u^{down,macro}}{P_u^{down,macro}}$. If $\rho_u^{down,macro} > \rho_u^{down,small}$, the downlink of user can access to macro cell, or with small cell.

3: Compute the time ratio of frame of every user

$$\hat{a}_u^{down} = \frac{\tilde{a}_u^{down} \cdot N_m^*}{A_m}, \hat{b}_{u,A}^{down} = \frac{\tilde{b}_{u,A}^{down} \cdot A_p^*}{B_{p,A}}, \hat{b}_{u,nA}^{down} = \frac{\tilde{b}_{u,nA}^{down} \cdot (N_{sf} - A_p^*)}{B_{p,nA}} \tag{29}$$

$$\hat{a}_u^{up} = \frac{\tilde{a}_u^{up} \cdot N_m^*}{A_m}, \hat{a}_{u,A}^{up} = \frac{\tilde{a}_{u,A}^{up} \cdot (N_{sf} - N_m^*)}{A_{m,A}^{up}}, \hat{b}_{u,A}^{up} = \frac{\tilde{y}_{u,A}^{up} \cdot A_p^*}{B_{p,A}}, \hat{b}_{u,nA}^{up} = \frac{\tilde{b}_{u,nA}^{up} \cdot (N_{sf} - A_p^*)}{B_{p,nA}}$$
$$\tag{30}$$

Finally, the EE of joint uplink and downlink for one user are computed in macrocell or smallcell.

---

5   Numerical Results

In order to verify our proposed method, we set transmission power of macrocell and small cell to 36 dBm and 30 dBm, and the reference power from macro for ABS to 23 dBm. Each result in following figure is computed by averaging 100 tries.

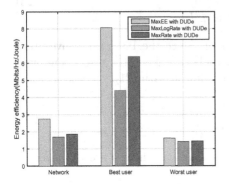

Fig. 1. EE-DU-UL-eICIC algorithm

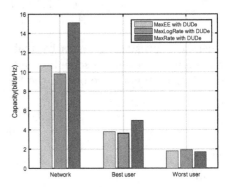

Fig. 2. Capacity vs. users

In Fig. 1, we compared the energy efficiency and capacity among these methods from three aspects: the energy efficiency for network, the best user, and the worst user corresponding to one macrocell, four smallcells, and thirty users. We observe that there is a significant gain in energy efficiency among the best user and the network achieved by MaxEE with DUDe (our proposed) compared with MaxRate with DUDe and MaxlogRate with DUDe. It is shown that the ABS configuration with DUDe need to study from energy efficiency perspective.

The Fig. 2 illustrates the capacity of MaxEE with DUDe compared with the other two algorithms following the same scenarios of Fig. 1. It can be observed that the MaxRate with DUDe achieves the largest rate of the network, the best user and the worst user, but the rates of MaxEE with DUDe is not the lowest. This is due to the fact that MaxRate with DUDe is to maximize the capacity of the network, and the MaxLogRate with DUDe is to strike a trade off between network capacity and user rate.

6   Conclusion

In the paper, we propose an energy efficiency optimization based on downlink and uplink decoupled with eICIC in HetNets. We design a joint downlink and uplink association and ABSs allocation algorithm in HetNet. Numerical results demonstrate that the proposed algorithm achieves superior energy efficiency performance of network.

Acknowledgement. This work was supported in part by the National Natural Science Foundation of China (Grants nos. 61701400, 61501372, 61572401, 61373176, and 61672426), by the Postdoctoral Foundation of China (Grants nos. 2017M613188 and 2017M613186), by the Natural Science Special Foundation of Shaanxi (Grant no. 2017JQ6052 and 17JK0783).

References

1. Deb, S., Monogioudis, P., Miernik, J., Seymour, J.P.: Algorithms for enhanced inter-cell interference coordination (eICIC) in LTE HetNets. IEEE/ACM Trans. Netw. **22**, 137–150 (2014)
2. Boccardi, F., et al.: Why to decouple the uplink and downlink in cellular networks and how to do it. IEEE Commun. Mag. **54**, 110–117 (2015)
3. Lema, M.A., Pardo, E., Galinina, O., Andreev, S., Dohler, M.: Flexible dual-connectivity spectrum aggregation for decoupled uplink and downlink access in 5G heterogeneous systems. IEEE J. Sel. Areas Commun. **34**(11), 2851–2865 (2016)
4. Bacha, M., Wu, Y., Clerckx, B.: Downlink and uplink decoupling in two-tier heterogeneous networks with multi-antenna base stations. IEEE Trans. Wirel. Commun. **16**(5), 2760–2775 (2017)
5. Zheng, J., Gao, L., Wang, H., Niu, J., Li, X., Ren, J.: EE-eICIC: energy-efficient optimization of joint user association and abs for eICIC in heterogeneous cellular networks. Wirel. Commun. Mob. Comput. **2017**, 1–11 (2017)
6. Khalek, A.A., Alkanj, L., Dawy, Z., Turkiyyah, G.: Optimization models and algorithms for joint uplink/downlink UMTS radio network planning with SIR-based power control. IEEE Trans. Veh. Technol. **60**, 1612–1625 (2011)
7. Chen, X., Hu, R.Q.: Joint uplink and downlink optimal mobile association in a wireless heterogeneous network. In: Proceedings of the IEEE GLOBECOM, pp. 4131–4137 (2012)
8. Boostanimehr, H., Bhargava, V.K.: Joint downlink and uplink aware cell association in HetNets with QoS provisioning. IEEE Trans. Wirel. Commun. **14**, 5388–5401 (2015)
9. Zheng, J., Li, J., Wang, N., Yang, X.: Joint load balancing of downlink and uplink for eICIC in heterogeneous networkD. IEEE Trans. Veh. Technol. **66**(7), 6388–6398 (2017)
10. Vanderbei, R.J.: Linear Programming: Foundations and Extensions, 3rd edn. Springer, Heidelberg (2008). https://doi.org/10.1007/978-0-387-74388-2
11. Crouzeix, J.P., Ferland, J.A.: Algorithms for generalized fractional programming. Math. Program. **52**(2), 191–207 (1991)
12. Grant, B., Boyd, S.: CVX: Matlab software for disciplined convex programming (2011). http://cvxr.com/cvx/, Apr. Stanford

Multi-objective Optimization for IoT Devices Association in Fog-Computing Based RAN

Hua Shi[1]([⊠]), Yinbin Feng[2], Ronghua Luo[3], and Jie Zheng[3]

[1] JinLing Institute of Technology, Nanjing 210000, China
shihuawindy@jit.edu.cn
[2] CRRC Nanjing Puzhen Co., Ltd, Nanjing 210000, China
[3] Northwestern University, Xi'an 710000, China

Abstract. The fog-computing based radio access network (F-RAN) is proposed in 5G systems facilitating the deployment of IoT, where fog-computing based access points (FAPs) provide both computational and radio resource closer to IoT devices (IoTDs). On one side, IoTDs try to associate with the FAPs to minimize the power consumption. On the other side, the concentration of IoTDs leads to the long execution delay which consists of transmission time and processing time, where we assume an equal share of computing resource for co-FAP IoTDs. As a result, we investigate multi-objective optimization (MOP) for IoTDs association in F-RAN considering both radio and computing resource. The objects involve minimizing the power consumption and the execution delay of IoTDs. Then we apply quantum-behaved particle swarm optimization with low complexity to solve the MOP. Simulation results show the proposed algorithm achieves a tradeoff between the two objects. It consumes a little more power consumption and brings a big improvement of the average execution delay.

Keywords: Internet of Things · Fog computing · Device association · Multi-objective optimization

1 Introduction

Internet of Things (IoT) is a worldwide network that connects ubiquitous smart devices with little or no human intervention [1], and it can support a wide range of applications, such as smart cities [2] and intelligent transportation [3]. However, a huge volume of data will be generated, and some applications have poor performance due to the limits in terms of power, storage, and computing ability of IoT devices (IoTDs). Fog computing is a promising opportunity to provide shared computing and storage resources to the close proximity of IoTDs and overcome these limitations, which is proposed first by CISCO as "cloud at the edge" [4]. Meanwhile, most of the data exchange in IoT makes use of wireless communication. The fog-computing based radio access network (F-RAN) is proposed in 5G systems facilitating the deployment of IoT, where wide-coverage and fog-computing based access points (FAPs) brings both computational and radio resource closer to IoTDs [5].

© ICST Institute for Computer Sciences, Social Informatics and Telecommunications Engineering 2019
Published by Springer Nature Switzerland AG 2019. All Rights Reserved
B. Li et al. (Eds.): IoTaaS 2018, LNICST 271, pp. 340–347, 2019.
https://doi.org/10.1007/978-3-030-14657-3_34

There are many research efforts for allocating radio and computing resources in IoT. A device association algorithm is proposed considering downlink rate for human-to-human communications and uplink transmit power for coexisted IoTDs [6]. Small-cell assisted traffic offloading in IoT is investigated to minimize the total power consumption with secrecy requirement [7]. [8] proposes joint distributed computing and content sharing among numbers of cooperation F-APs achieving low latency. [9] studies matching between IoT users and resources with the object of cost performance. These works are with single objective by the ignorance to the degradations of other performance. In [10], allocating computing resource of service providers for IoTDs in cloud computing is considered for maximizing the profit of the broker while minimizing the response time and the energy consumptions. [11] addresses the spectrum allocation problem with respect to both spectrum utilization and network throughput in the cognitive radio-based IoT. [12] optimizes the offloading probability and transmission power with one FAP to jointly minimize the energy consumption, execution delay and payment cost. However, they just consider either computing resource or radio resource.

All the above analysis motivates us to investigate multi-objective optimization for IoTDs association in F-RAN considering both radio and computing resource. On one side, IoTDs try to associate with the FAP to minimize the power consumption. On the other side, the concentration of IoTDs sharing the computing resource of one FAP leads to long execution delay which consists of the transmission time and the processing time. To achieve this, we formulate it as an MOP involving minimizing both the power consumption and the execution delay of IoTDs. Then, the MOP is solved based on quantum-behaved particle swarm optimization with low complexity.

2   System Model and Problem Formation

2.1   System Model

We consider a F-RAN network. As shown in Fig. 1, the macro base station (MBS) connects with the center cloud, while small base stations act as FAPs. To facilitate IoTD association, the dual connectivity technology is adopted. MBS provides communicating and computing the control data for IoTDs, and L FAPs (FAP$_1$, FAP$_2$, ..., FAP$_L$) provide services for the traffic data of K IoTDs (IoTD$_1$, IoTD$_2$, ..., IoTD$_K$). We focus on the association control in this paper which decides the binary association indicator x_{kl}. Let $x_{kl} = 1$ when IoTD$_k$ associating with FAP$_l$.

Power Consumption of IoTDs. Since most of IoTDs are battery operated and require a long battery life, one of the most important metrics is power consumption of IoTDs. However, the transmission quality between the IoTDs and FAPs should be met first. In detail, the signal to interference noise ratio from IoTD$_k$ at FAP$_l$ is defined as follows.

$$\text{SINR}_{kl} = g_{kl}p_l/(I_{kl} + \sigma^2) \approx g_{kl}p_k/\sigma^2 \tag{1}$$

where p_k and g_{kl} are the transmission power and channel gain from IoTD$_k$ at FAP$_l$.

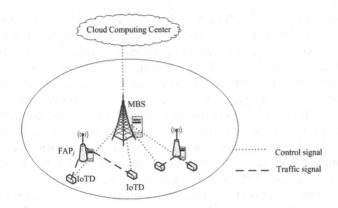

Fig. 1. The model of the F-RAN network

σ^2 represents the channel noise. Considering the relative small transmitting power of IoTDs and the propagation of the signal, the interference is ignored. It is required that $\text{SINR}_{kl} \geq \Gamma_k^{\min}$ in order to ensure a successful transmission. The minimum power consumption of IoTD_k associating with FAP_l is

$$p_{kl} = \Gamma_k^{\min}\sigma^2 / g_{kl}. \tag{2}$$

The corresponding transmitting rate is

$$r_{kl} = W\log_2(1 + \Gamma_k^{\min}), \tag{3}$$

where W is the channel bandwidth.

Execution Delay of IoTDs. We assume the computing rate of FAP_l is C_l and an equal share of C_l for co-FAP IoTDs. The execution delay of IoTD_k consists of the transmission time and the processing time. The time and power consumption for IoTDs to receive the results can be ignored, due to the fact that the size of the outcome for many applications is in general much smaller. Thus, the execution delay for the data size D_k when IoTD_k associating with FAP_l is

$$t_{kl} = D_k/r_{kl} + \kappa D_k \Big/ \Big[(1\Big/\sum\nolimits_{k=1}^{K} x_{kl})C_l\Big], \tag{4}$$

where the required computing rate κD_k is linear with D_k.

2.2   Problem Formulation

Based the two metrics discussed in the previous subsection, the association control problem can be formulated as a multi-objective optimization as follows.

$$\min\{\sum_{k=1}^{K}\sum_{l=1}^{L}x_{kl}p_{kl}, \sum_{k=1}^{K}\sum_{l=1}^{L}x_{kl}t_{kl}\}, \tag{5}$$

s.t.

$$\sum_{l=1}^{L}x_{kl} \leq 1, \forall k = 1, 2, \cdots, K, \tag{6}$$

$$\sum_{l=1}^{L}x_{kl}p_{kl} < P_{k}^{\max}, \forall k = 1, 2, \cdots, K, \tag{7}$$

$$\sum_{k=1}^{K}x_{kl} \leq B_{l}, \forall l = 1, 2, \cdots, L, \tag{8}$$

$$x_{kl} \in \{0, 1\}, \forall k = 1, 2, \cdots, K, l = 1, 2, \cdots, L. \tag{9}$$

Constraint (6) enforces that each IoTD can only associate with one FAP. (7) satisfies the maximum power constraint for each IoTD. (8) ensures that the association do not exceed the capacity constraint for each FAP, where B_l is the number of the channel for FAP$_l$ and each IoTD accesses one channel.

3   QPSO-Based Algorithm for Multi-objective Association

This section presents the algorithm for the multi-objective association problem. It includes two steps as follows.

First Step is Problem Transformation. The previous multi-objective optimization with constraints is transformed to a single-objective and unconstrained optimization problem in the use of the weighted method and the penalty method as follows,

$$\min \alpha_1 (\sum_{k=1}^{K}\sum_{l=1}^{L}x_{kl}\frac{p_{kl}}{P_k^{\max}}) + \alpha_2 (\sum_{k=1}^{K}\sum_{l=1}^{L}x_{kl}\frac{t_{kl}}{t_k^{\max}}\}) $$
$$+ \lambda * \{\sum_{k=1}^{K}\max(\sum_{l=1}^{L}x_{kl}p_{kl} - P_k^{\max}, 0) + \sum_{l=1}^{L}\max(\sum_{k=1}^{K}x_{kl} - B_l, 0)\} \tag{10}$$

In the transformation, a dimensionless quality is firstly obtained corresponding to each objective in order to maintain the balance between the two objectives. In addition, α_1 and α_2 are weight factors reflecting the relative importance of the power consumption and the execution delay, where $\alpha_1 + \alpha_2 = 1$. λ is the penalty coefficient which is multiplied by the violation of the constraints in (7)–(8) as the penalty function. The remaining constraint in (6) will be processed in the next step.

Second Step is QPSO-based Algorithm. The transformed problem in (10) is still computationally difficult. This problem can be considered as a combinational problem of (IoTD$_k$, FAP$_l$) pairs. Its solution is a binary association matrix, the size of which is K rows and L columns. So the computational complexity of this problem is $O(2^{L \times K})$, which shows that this problem is an NP-hard problem. Thus, the heuristic algorithm with low complexity should be considered for this problem. Quantum-behaved particle swarm optimization (QPSO) performs well in terms of computation cost and solution quality [13], so we apply the QPSO algorithm to solve this problem.

QPSO is a population-based optimization tool. The population or swarm represents the set of potential solutions, and each particle in the population represents a solution position. For the problem in (10), the ith particle is represented by a K-dimensions vector $\vec{y}_i = (y_{i1}, y_{i2}, \cdots, y_{iK})$. In $\vec{y}_i$, each y_{ik} corresponds to IoTD$_k$, and its value is the special index of the associated FAP. Since each element has only one value, the constraint in (6) that each IoTD can only associate with one FAP is satisfied.

At the beginning of the QPSO algorithm, an initial swarm consisted of M particles is randomly generated. Particles in the swarm move by iteration through the search space to find a new position with the best function value. After each movement, the best position of each particle and the best position of the swarm are recorded by $\vec{y}_{ibest}$ and $\vec{y}_{gbest}$ respectively. At each iteration, each particle moves its position by the following equations

$$\vec{y}_i = \begin{cases} \vec{z}_i - b * \left| \vec{y}_{mbest} - \vec{y}_i \right| * \ln(1/\mu), & \mu \geq 0.5 \\ \vec{z}_i + b * \left| \vec{y}_{mbest} - \vec{y}_i \right| * \ln(1/\mu), & \mu < 0.5 \end{cases} \tag{11}$$

$$\vec{y}_{mbest} = \frac{1}{M} \sum_{i=1}^{M} \vec{y}_{ibest} \tag{12}$$

$$\vec{z}_i = \varphi \vec{y}_{ibest} + (1 - \varphi) \vec{y}_{gbest} \tag{13}$$

$$b = \omega_{max} - (\omega_{max} - \omega_{min})(iter/iter_{max}) \tag{14}$$

where φ and μ are random numbers in the range [0, 1], $\vec{y}_{mbest}$ is the mean of $\vec{y}_{ibest}$, b is the iterating coefficient reducing linearly. The algorithm ends when the number of the iterations reaches the maximum or the function value error is satisfied.

The QPSO algorithm for the association problem is performed at MBS as the control process, after which IoTDs associate with FAPs accordingly for the traffic data.

4 Performance Evaluation

In this section, the performance of the proposed algorithm is evaluated by simulation. We consider a network with 10 FAPs distributed randomly within a cell with the radius 1000 m. Each FAP has the capacity limit $B_l = 5$, the computing rate C_l in the range [500, 600] Mcycles/s. A number of IoTDs [25, 50] are also distributed randomly within the cell. Each IoTD has the minimum SINR requirement Γ_k^{min} in the range [8, 10] dB, the maximum power as 23 dBm, the data size D_k in the range [2, 8] kbits, and the corresponding required computing cycles as $10^{4*}D_k$. For the wireless propagation between IoTDs and FAPs, we set the pass loss constant as 10^{-2}, the path loss exponent as 4, the multipath fading gain as the Rayleigh distribution with unit mean, and the shadowing gain as the log-normal distribution with 4 dB deviation. As for the QPSO algorithm, we set the size of the swarm as 150, and the number of the iterations as 150. The number of simulation snapshots is set as 200.

The convergence of the proposed QPSO-based MOP IoTDs association algorithm is firstly testified. The values of the two objects which are the average power consumption and the average execution delay of each IoTD at the end of each iteration are shown in Fig. 2(a) and (b) respectively. It can be seen that the values of the two objects achieve convergence when the number of the iterations ends.

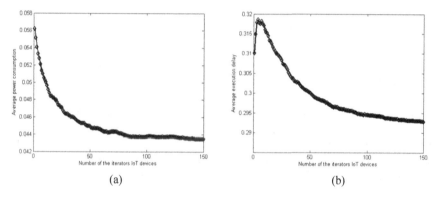

(a)   (b)

Fig. 2. The convergence of the average power consumption (a) and the average execution delay (b) of each IoTD in the proposed QPSO-based MOP IoTDs association algorithm

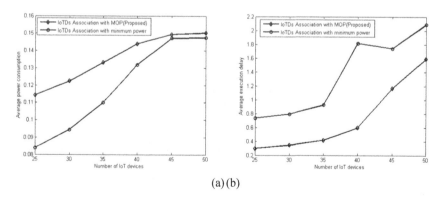

(a)(b)

Fig. 3. The performance of the proposed MOP IoTDs association algorithm in terms of the average power consumption and the average execution delay compared with the IoTDs association algorithm minimizing power consumption in [8]

The performance of the proposed MOP IoTDs association algorithm is compared with the IoTDs association algorithm minimizing power consumption in [8]. The results of the two objects with the number of IoTDs are shown in Fig. 3(a) and (b) respectively. As the capacity constraint of each FAP, part of IoTDs have to associate with the further FAP and the average power consumption increases with the number of IoTDs for both algorithms. Meanwhile, more co-FAP IoTDs share the computing resource of each FAP, so the average execution delay also increases with the number of IoTDs for both algorithms. The compared algorithm optimizes the power consumption, so numbers of IoTDs maybe crowding into one FAP causes longer execution delay. The proposed algorithm achieves a tradeoff between the two objects. It consumes a little more power consumption and brings a big improvement of the average execution delay.

5   Conclusion

In this paper, we investigate the IoTDs association problem in F-RAN where FAPs provide both computational and radio resource closer to IoTDs. We first analyze the power consumption and the execution delay for IoTDs which consists of the transmission time and the processing time. Then we formulate the association as an MOP minimizing both the power consumption and the execution delay. Since the MOP is computationally different, we apply quantum-behaved particle swarm optimization with low complexity to solve the MOP. Simulation results show the proposed algorithm achieves a tradeoff between the two objects. It consumes a little more power consumption and brings a big improvement of the average execution delay.

Acknowledgements. The authors acknowledge the partial economic support from JinLing Institute of Technology through Project No. jit-fhxm-201605 and No. jit-b-201633.

References

1. Perera, C., Zaslavsky, A., Christen, P., et al.: Context aware computing for the Internet of Things: a survey. IEEE Commun. Surv. Tutor. **16**(1), 414–454 (2014)
2. Gaur, A., Scotney, B., Parr, G., et al.: Smart city architecture and its applications based on IoT. Proc. Comput. Sci. **52**(1), 1089–1094 (2015)
3. Kong, L., Khan, M.K., Wu, F., et al.: Millimeter-wave wireless communications for IoT-cloud supported autonomous vehicles: overview, design, and challenges. IEEE Commun. Mag. **55**(1), 62–68 (2017)
4. CISCO: Fog computing and the Internet of Things: extend the cloud to where the things are white paper (2015)
5. Peng, M., Yan, S., Zhang, K., et al.: Fog-computing-based radio access networks: issues and challenges. IEEE Netw. **30**(4), 46–53 (2016)
6. Elhattab, M., Elmesalawy, M., Ibrahim, I.: Opportunistic device association for heterogeneous cellular networks with H2H/IoT co-existence under QoS guarantee. IEEE Internet Things J. **4**(5), 1360–1369 (2017)

7. Yang, X., Wang, X., Wu, Y., et al.: Small-cell assisted secure traffic offloading for narrow-band Internet of Thing (NB-IoT) systems. IEEE Internet Things J. **5**(3), 1516–1526 (2018)
8. Rahman, G.S., Peng, M., Zhang, K., et al.: Radio resource allocation for achieving ultra-low latency in fog radio access networks. IEEE Access **6**, 17442–17454 (2018)
9. Gu, Y., Chang, Z., Pan, M., et al.: Joint radio and computational resource allocation in IoT Fog Computing. IEEE Trans. Veh. Technol. Early Access (2018)
10. Kumrai, T., Ota, K., Dong, M., et al.: Multiobjective optimization in cloud brokering systems for connected Internet of Things. IEEE Internet Things J. **4**(2), 404–413 (2017)
11. Han, R., Gao, Y., Wu, C., et al.: An effective multi-objective optimization algorithm for spectrum allocations in the cognitive-radio-based Internet of Things. IEEE Access **6**, 12858–12867 (2018)
12. Liu, L., Chang, Z., Guo, X., et al.: Multi-objective optimization for computation offloading in fog computing. IEEE Internet Things J. **5**(1), 283–294 (2018)
13. Liu, J., Sun, J., Xu, W.: Quantum-behaved particle swarm optimization for integer programming. In: King, I., Wang, J., Chan, L.-W., Wang, D. (eds.) ICONIP 2006. LNCS, vol. 4233, pp. 1042–1050. Springer, Heidelberg (2006). https://doi.org/10.1007/11893257_114

A Proposed Language Model Based on LSTM

Yumeng Zhang[✉], Xuanmin Lu, Bei Quan, and Yuanyuan Wei

Northwestern Polytechnical University, Xi'an, China
zhangyumeng@mail.nwpu.edu.cn, luxuanmin@nwpu.edu.cn

Abstract. In view of the shortcomings of language model N-gram, this paper presents a Long Short-Term Memory (LSTM)-based language model based on the advantage that LSTM can theoretically utilize any long sequence of information. It's an improved RNN model. Experimental results show that the perplexity of the LSTM language model in the PBT corpus is only one-half that of the N-gram language model.

Keywords: Language model · N-gram · RNN · LSTM · Perplexity

1 Introduction

The language model is widely used in natural language processing, machine translation and speech recognition, and it can directly affect the performance of the corresponding system. Language models are usually divided into two categories: one is a statistical language model based on large-scale corpus. In actual use, the model can only reflect the adjacent constraint relationship of the language due to the limitation of space and time of the system, and cannot solve the recursion problem when the processing language distance is long. The other is a rule-based language model. The model is suitable for the processing of closed corpus, which can effectively reflect the long-distance constraint relationship and recursion phenomenon in language, but the stability of this model is poor, and it is open in processing. Sexual corpus does not have a good consistency in knowledge expression. The commonly used language model is the statistical model N-gram, but it has two disadvantages. First, the semantics of word vectors are not satisfactory in the continuous space. When the system adjusts the parameters of a word or phrase, the word or phrase with similar meanings will also be changed accordingly. Second, the appearing probability of a word depends only on the first few words, and it does not perform well in a particularly long context. In view of the above drawbacks, based on the theory that LSTM can theoretically take advantage of arbitrary long sequence information, a LSTM-based language model is proposed. The model converts each word in the sentence into a word vector and trains it through the LSTM network. Then, it obtains the probability distribution of the word. The experimental results show that the perplexity of the LSTM-based language model in the PBT corpus is only half of that of the N-gram language model.

B. Li et al. (Eds.): IoTaaS 2018, LNICST 271, pp. 348–355, 2019.
https://doi.org/10.1007/978-3-030-14657-3_35

2 N-Gram Language Model

N-gram refers to a sequence of N items in a given piece of text or speech. Items can be syllables, letters, words, or base pairs. The N-gram model is a statistical language model. Usually N-grams are taken from texts or corpus [1]. When N = 1, it is called unigram. When N = 2, it is called bigram. When N = 3, it is called trigram, and so on.

For a sequence of m words (sentences), the probability $P(\omega_1, \ldots, \omega_m)$ can be calculated according to the chain rule:

$$P(\omega_1, \ldots, \omega_m) = P(\omega_1)P(\omega_2|\omega_1)P(\omega_3|\omega_1, \omega_2)\ldots P(\omega_m|\omega_1, \ldots, \omega_{m-1}) \quad (1)$$

In order to reduce the computational difficulty, Markov approximation is used for the above formula, i.e., the probability that the current word appears is only related to the appears the current word. In this way, it is unnecessary to trace back to the initial word of the sentence at the time of calculation, thereby it can greatly reduce the length of the above formula (1):

$$P(\omega_i|\omega_1, \omega_2, \ldots, \omega_{i-1}) = P(\omega_i|\omega_{i-n+1}, \ldots, \omega_{i-1}) \quad (2)$$

When N = 1, a unigram model is:

$$P(\omega_1, \omega_2, \ldots, \omega_m) = \prod_{i=1}^{m} P(\omega_i) \quad (3)$$

When N = 2, a bigram model is:

$$P(w_1, w_2, \ldots, w_m) = \prod_{i=1}^{m} P(w_i|w_{i-1}) \quad (4)$$

When N = 3, a trigram model is:

$$P(w_1, w_2, \ldots, w_m) = \prod_{i=1}^{m} P(w_i|w_{i-2}w_{i-1}) \quad (5)$$

A set of parameters can be calculated by the maximum likelihood method to obtain the maximum value of the training sample probability.

For the unigram model, where $C(w_1, \ldots, w_n)$ represents the frequency of occurrence of N-grams $(w_1, \ldots, w_n)$ in the training corpus data, M represents all the words in the corpus:

$$P(w_i) = \frac{C(w_i)}{M} \quad (6)$$

For the bigram model:

$$P(w_i|w_{i-1}) = \frac{C(w_{i-1}w_i)}{C(w_{i-1})} \tag{7}$$

For the N-gram model:

$$\hat{y}_t = softmax(W^{(s)}h_t) \tag{8}$$

In actual use, because the training corpus is very limited, when the test corpus from the same source is used to verify the training results, some N-grams do not appear in the training corpus. This means that the probabilities calculated by the above formula may be zero, which leads to data sparseness. For languages, the existence of data sparseness makes it not satisfy the law of large numbers, and the calculated probability is distorted. The solution called smoothing or discounting is needed.

The N-gram model predicts the follow-up words based on discrete unit words (for example, n prefix words) that do not have a genetic property in the text. In a continuous context, using only n prefix words as the window range may not be sufficient to accurately describe the context information, and thus does not have the semantic advantage that the word vectors in the continuous space satisfy. Similar-meaning words have similar word vectors. When the model adjusts the parameters of a particular word or phrase, similar words or phrases will change accordingly. At the same time, the size of n prefix words' memory information, used for current word prediction during the operation of the system, will increase exponentially, resulting in an inestimable probability. Therefore, in the face of a large continuous space, it is very difficult to capture context information between vocabularies without processing after the memory information is extracted.

3   LSTM-Based Language Model

Long Short Term Memory Networks, also known as LSTM [2], is a recursive time-recurrent neural network that can effectively handle long-term dependency problems in time series. LSTM network structure was proposed by Hochreater and Schmidhuber in 1997 and further improved and promoted by Alex Graves. In the fields of image analysis, machine translation, handwriting recognition, speech recognition, etc., LSTM has a very outstanding performance and has gotten a flourishing development [3].

LSTM solves long-term dependency problems by transforming RNNs. The long-term memory of information is a self-distribution for LSTM. It does not need to be achieved through deliberate learning. The advantage of LSTM is that it can take into account all the predicative vocabulary in corpus when training the language model.

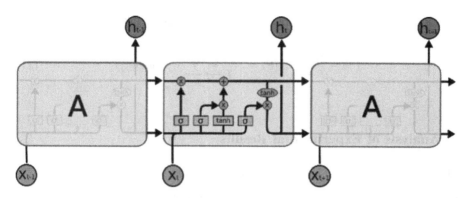

Fig. 1. LSTM structure

Figure 1 shows the structure of the LSTM network, where repeated blocks represent hidden layers in each iteration. Each hidden layer contains several neurons. Each neuron first performs a linear matrix calculation on the input vector and then outputs the corresponding result [4] after the nonlinear function of the activation function. In each iteration, the output of the last iteration interacts with the next word vector of the text to determine whether the information is saved or discarded and the current state is updated. x_t is the input of the hidden layer in this round of iterations. According to the current state information, the predicted output value of the hidden layer y is obtained and the output feature vector h_t is provided for the next hidden layer. Whenever there is a new word vector input in the network, the word vector is combined with the output of the hidden layer at the previous moment to calculate the output of the hidden layer at the next moment, and the hidden layer is used cyclically to maintain the latest state.

Each hidden layer is finally connected with a layer of traditional feedforward network as the output layer. Each node y_i in the output layer corresponds to the unnormalized logarithmic probability of the word i at the next moment, and then the output value y is normalized by the softmax function:

$$\hat{y}_t = softmax(W^{(s)}h_t) \tag{9}$$

$\hat{y}_t$ is the probability distribution calculated by the hidden layer based on all the vocabularies in each round of iterations, that is, when the weights of the entire pre-order vocabulary of the document and the observed word vector x(t) are all determined, the subsequent words can be predicted by the model. The variable |V| of $W^{(s)} \in \mathbb{R}^{|v| \times D_h}$ and $\hat{y} \in \mathbb{R}^{|v|}$ represents the dictionary size of the entire corpus.

In a recurrent neural network, cross-entropy is usually used as a loss function [5]. Equation (10) represents the superposition of cross entropy over the entire corpus when the t-th iteration is performed:

$$J^{(t)}(\theta) = -\sum_{j=1}^{|V|} y_{t,j} \times \log(\hat{y}_{t,j}) \tag{10}$$

For a corpus of size T, the cross entropy is calculated as follows:

$$J = -\frac{1}{T}\sum_{t=1}^{T} J^{(t)}(\theta) = -\frac{1}{T}\sum_{t=1}^{T}\sum_{j=1}^{|V|} y_{t,j} \times \log(\hat{y}_{t,j}) \tag{11}$$

4   Analysis of Experimental Results

4.1   N-Gram Language Model Experiment

The dataset used in the experiment is the Penn Treebank (PTB) corpus. The language model is built by SRILM, and the used statistical data of the corpus file is: file testfile. txt: 3761 sentences, 78669 words and 4794 OOVs. The 1-gram (unigram), 2-gram (bigram) and 3-gram (trigram) models are constructed respectively and the corresponding complexity is calculated. The results are shown in above Table 1, where logprob represents logP(T), ppl represents perplexity of all words, and ppl1 represents perplexity in addition to </s > :

$$ppl = 10 \wedge (-\log P(T)/(sen + word))$$
$$ppl1 = 10 \wedge (-\log P(T)/Word) \tag{12}$$

where Sen and Word represent sentences and words respectively.

Table 1. Three N-gram language models and their perplexity

	LogProb	PPL	PPL1
1-gram	logprob = − 233580.5	ppl = 1020.147	ppl1 = 1451.559
2-gram	logprob = − 185313.2	ppl = 243.7525	ppl1 = 322.4552
3-gram	logprob = − 173834.8	ppl = 173.4205	ppl1 = 225.4726

4.2   LSTM Language Model Experiment

The training model parameters are as follows: the initial value of the relevant parameters is randomly and uniformly distributed, the range is [-init_scale, +init_scale], and init_scale is 0.1; the learning rate is set to 1. When the number of iterations of text training exceeds the maximum value max epoch, it gradually decreases, and the learning rate decay also gradually decreases when the maximum number of iterations is exceeded. The number of LSTM layers is 2 layers, the number of nodes in each hidden layer is 200, the sequence length of a single data is 20, and the dictionary size is 10 K words. At the same time, a batch strategy is adopted, the scale of each batch of data is set to 20, and dropout method is used in order to prevent overfitting.

Firstly, a multi-layer LSTM unit is set up to preprocess the training corpora, and the dictionary id in the training text is converted into a vector format. Text data is sequentially input to the LSTM unit according to the sequence, and the stored state of the previous time period of the LSTM and the current input collectively calculate the output and state of the current unit (the probability of occurrence of the next word). Gradient descent is used to update weights, and gradients can be appropriately pruned (scaled) in order to control gradient explosions. The training corpus is divided into multiple batches, and records such as the cost and state of the network output are recorded, and the results are output according to the number of iterations. Each time a 10% iteration is completed, the current epoch progress, perplexity (natural constant index of the cross-entropy loss function), and training speed (words per second) are displayed. Return the result of the perplexity function at the end of the training.

The operating system is Windows 10, 1709 version, the CPU is Intel i5-6300hq, and the GPU is trained in the Nvidia GTX 960 m environment. Figure 2 shows the decrease curve of Learning Rate. In Fig. 3, the Train Perplexity is the perplexity of the training set, and the Valid Perplexity is the perplexity of the verification set. It can be seen that with the decrease of the Learning Rate, the decrease of perplexity finally tends to be stable. After 13 epochs, the Train Perplexity value was 40.692, the Valid Perplexity value was 119.441, and the Test Set's Test Perplexity value was 115.702.

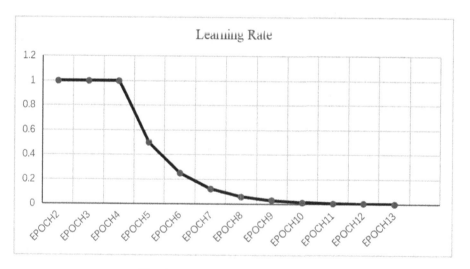

Fig. 2. Decrease curve of learning rate

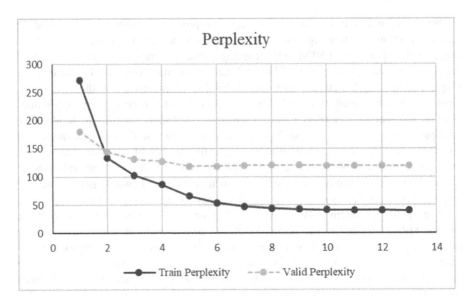

Fig. 3. Train perplexity and valid perplexity trends

4.3   Results and Analysis

The comparison of perplexity can be obtained between the two models from the previous experimental results (Table 2):

Table 2. Perplexity of LSTM model and N-gram model

Model name	Perplexity	
Unigram	ppl = 1020.147	ppl1 = 1451.559
Bigram	ppl = 243.7525	ppl1 = 322.4552
Trigram	ppl = 173.4205	ppl1 = 225.4726
LSTM	115.702	

From the above table, in the N-gram model, when the model is unigram, that is, n = 1, the perplexity degree is the highest; when the model is the bigram, that is, n = 2, the perplexity degree is the second, and when the model is trigram, that is, n = 3, the perplexity degree is the lowest. This is because, for an n-gram model, the larger the value of n, the more constraint information for the result, and the more accurate the model, that is, the trigram has more constraint information on the occurrence of subsequent words, and the reliability is also higher. However, the increase of n causes the model to become more complex, which increases the amount of computations when the model is built. The perplexity of the language model built by LSTM is only 115.702, which is equivalent to half of the trigram. Therefore, this experiment proves that the performance of language model based on LSTM is obviously better than that of N-gram model.

5 Conclusion

At present, the traditional speech recognition technology has not been able to adapt well to the needs of the big data era, and no deeper information can be excavated from the massive data. Deep learning has excellent modeling ability of big data and excellent fitting ability. Using deep learning to study speech recognition has important theoretical significance and practical value. This paper discusses the commonly used traditional statistical language model N-gram, and proposes a language model, based on the improved RNN model—LSTM model, to solve the problem that the N-gram model does not meet the semantic deficiency of word vectors in continuous space, and the LSTM language model has obvious advantages in performance compared with the N-gram.

References

Lin, C.Y., Hovy, E.: Automatic evaluation of summaries using N-gram co-occurrence statistics. In: Conference of the North American Chapter of the Association for Computational Linguistics on Human Language Technology. Association for Computational Linguistics, pp. 71–78 (2003)

Xiong, W., Droppo, J., Huang, X., et al.: Achieving human parity in conversational speech recognition. IEEE/ACM Trans. Audio Speech Lang. Process. PP(99) (2016)

Li, J., Zhang, H., Cai, X.Y., et al.: Towards end-to-end speech recognition for Chinese Mandarin using long short-term memory recurrent neural networks (2015)

Tai, K.S., Socher, R., Manning, C.D.: Improved semantic representations from tree-structured long short-term memory networks. Comput. Sci. 5(1), 36 (2015)

Mikolov, T.A.: Statistical language models based on neural networks (2012)

Impact of Timing Errors on the Performance of Double Iteration Anti-jamming Technology in Physical Layer Security

Qilin Li[1], Lijing Wang[2(✉)], and Zhijiong Cheng[1,2]

[1] The Metering Center of Sichuan Electric Power Corporation of State Grid,
Chengdu 611731, China
li_qi_lin@163.com
[2] National Key Laboratory of Science and Technology on Communications,
University of Electronic Science and Technology of China,
Chengdu 611731, China
wangli_jing@163.com

Abstract. In physical layer security, the double iteration anti-jamming technology based on power cognition can effectively cancel the co-channel interference used in confusing eavesdroppers. This paper analyzes the theoretical performance of double iteration method and the impact of timing errors in receivers on the performance of double iteration anti-jamming technology. Theoretical and simulation results show that, over the AWGN channel, the timing errors of 0.2 chips and 0.4 chips degrade the system performance by about 1 dB and 5 dB respectively, in the interference priority mode. Thus the system performance is sensitive to the timing errors.

Keywords: Physical layer security · Timing errors ·
Double iteration anti-jamming technology · Co-channel interference

1 Introduction

In co-channel interference, for the purpose of jamming the eavesdroppers, the co-channel interference signal and the desired signal occupy the same frequency forming co-channel interference.

However, the co-channel interference also has an effect on the intended receiver. Therefore, cooperative interference cancellation is proposed, which can improve system reliability and significantly increase the security capacity of system. Significant solutions have been made in cooperative interference cancellation, such as time-domain prediction technology [1], transform domain suppression technology [2], code-aided technology [3]. Time-domain prediction technology, considers the interference signal as sinusoidal signal or autoregressive (AR) signal [4, 5] and obtains the optimal estimation of the signal by time-domain filtering. Such an approach is suitable for narrowband interference signal, where its data rate is much lower than the desired signal. The transform domain suppression technology utilizes the difference between the spread spectrum signal and the interference to remove the interference. However, the method is sensitive to the threshold and will damage the desired signal if the

B. Li et al. (Eds.): IoTaaS 2018, LNICST 271, pp. 356–364, 2019.
https://doi.org/10.1007/978-3-030-14657-3_36

interference bandwidth is wide. According to the characteristics of the spreading code, the code-aided technology suppresses the narrowband interference (NBI) and the multiaccess interference (MAI) jointly and meanwhile completes dispreading detection [6]. In fact, it also relies on the prior information of the interference signal and has high computation complexity. Considering the weak self-correlation of broadband signal, the time-domain cancellation is not that acceptable.

In [7–9], the importance of the co-channel interference in physical layer security is emphasized, where the authors propose some cooperative interference power allocation schemes to improve the physical layer security. However, all the schemes are based on ideal interference cancellation, which do not take into account the timing errors. In [5], based on the method of combining power recognition technology and interference cancellation, the iteration mode is determined by sensing the desired signal to interference power ratio, which has a positive effect on broadband interference cancellation and is easy to be implemented. Furthermore, it improves the applications in anti-jamming technique of system. The performance of the schemes mentioned above is under the ideal timing synchronization assumption, but in practice, the timing errors always exist because of the asynchronized clock of the transceiver and the inaccurate estimation of the time delay.

In this paper, we analyze the theoretical performance of double iteration method and the impact of timing errors in receivers on the performance of duo-iteration anti-jamming technology.

The rest of paper is organized as follows. In Sect. 2, the system model is formed. In Sect. 3, we briefly analyze the performance of duo-iteration anti-jamming technology. In Sect. 4, we give the analysis of the system performance under the timing errors. In Sect. 5, the simulation results are presented to verify our theoretical analysis, Sect. 6 concludes this paper.

2 System Model

In Fig. 1, we give the system model. In this model, the transmitted signal consists of the desired signal and interference signal. After modulated and spread, the desired signal is transmitted to the receiver after getting through the shape filter. We consider the transmitted interference signal is a BPSK signal without spreading, and assume that the interference signal and the desired signal have the same bandwidth.

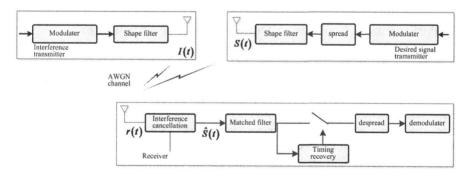

Fig. 1. System model

Denoted the users transmitter signal as $b(t) = \sum_{i=-\infty}^{+\infty} b_i p_T(t - iT_s)$, where $b_i \in \{\pm 1\}$ is the BPSK modulated desired signal. And the spreading sequence is defined as $pn(t) = \sum_{j=-\infty}^{+\infty} pn_j \cdot p_{T_c}(t - jT_c)$, where $p_T(t)$ is a rectangular pulse and its amplitude is 1 and duration is T_s. The spreading chips and the chip period are denoted by $pn_j \in \{\pm 1\}$ and T_c. The spreading gain N satisfies the relationship of $N = T_s/T_c$. And the data symbols are given by

$$S(t) = \sqrt{P_s} b(t) pn(t) g(t) \tag{1}$$

Where P_s is the desired signal power and $g(t)$ is the shaping pulse. And the interference signal takes on the form

$$I(t) = \sqrt{P_I} i(t) g(t) \tag{2}$$

The interference signal is modeled as $i(t) = \sum_{k=-\infty}^{+\infty} i_k \cdot p_{T_c}(t - kT_c)$, a random signal with the same bandwidth of the desired signal, where $i_k \in \{\pm 1\}$ is the BPSK modulated interference signal. And P_I is the energy of the interference signal. And the noise is AWGN.

The filter responses, which has the raised cosine characteristics, can be calculated as

$$h(t) = g(t) \otimes g^*(-t) = \frac{\sin(\pi t/T_c)}{\pi t/T_c} \frac{\cos(\alpha \pi t/T_c)}{1 - (2\alpha t/T_c)^2} \tag{3}$$

Where all the coefficients are normalized. T_c and α represent the duration of chips and the roll-off factor respectively. In the receiver, the received signal is first subjected to the interference cancellation. When the signal-to-interference ratio (SIR) is small, the interference cancellation is performed first. But when the SIR is large, the signal cancellation is performed first until the signal converges or completes the specified iteration times.

3 Performance Analysis of Double Iteration Anti-jamming Technology

In this part, we derive the theoretical expression and analyze the BER performance based on one iteration for the interference cancellation priority mode and the signal cancellation priority mode respectively, which do not take into account the timing errors. And we also compare and analyze the impact of the iteration times. In this section, we assume that the frequency of transceiver is perfectly synchronous. And we define the signal to noise ratio (SNR) and SIR as $SNR = P_s/\sigma_n^2$ and $SIR = P_I/\sigma_n^2$ respectively.

3.1 Performance Comparation

Performance Comparation of the Interference Cancellation Priority Mode
The joint probability distribution of signal and noise, which is conditioned on AWGN, is given by

$$f(x) = \frac{1}{2}\left[\frac{1}{\sigma_n\sqrt{2\pi}}\exp(-\frac{(x-h_1)^2}{2\sigma_n^2}) + \frac{1}{\sigma_n\sqrt{2\pi}}\exp(-\frac{(x+h_1)^2}{2\sigma_n^2})\right] \quad (4)$$

Where $h_1 = \sqrt{P_S}$ is the amplitude of the desired signal.

The cancellation of the interference priority mode is performed as follows: Firstly, the hard-decision of the received signal is given by $\hat{I}(t) = \frac{r(t)}{|r(t)|}$. And the error rate of the transmission signal b_i is calculated as

$$\begin{cases} Pe_2 = \int_{-\infty}^{-b_i \cdot h_1 - h_2} \frac{1}{\sigma_n\sqrt{2\pi}} \cdot \exp\left(-\frac{x^2}{2\sigma_n^2}\right) dx \\ Pe_{-2} = \int_{-b_i \cdot h_1 + h_2}^{+\infty} \frac{1}{\sigma_n\sqrt{2\pi}} \cdot \exp\left(-\frac{x^2}{2\sigma_n^2}\right) dx \end{cases} \quad (5)$$

Where $h_2 = \sqrt{P_I}$ is the amplitude of interference signal. Since the detection of the interference signal is not perfect. After despread, the discrete distribution of the residual interference $I(t) - \hat{I}(t)$ is shown as

$$\zeta(n) = \begin{cases} 0 & 1 - Pe_2 - Pe_{-2} \\ +2h_2 & Pe_2 \\ -2h_2 & Pe_{-2} \end{cases} \quad (6)$$

We can find that it is related not only to the transmission signal, but also to the sampling noise. Therefore, if we use the residual interference distribution as the classification criterion, the noise distribution accuracy will be improved. Denoted the equivalent noise decay factor k_σ, where its exact value can be got from the project experience. After traversing the residual interference, the final expression of BER can be expressed as

$$Pe_s = \sum_{ii=0}^{N} \sum_{jj=0}^{ii} C_N^{ii}(1 - Pe_2 - Pe_{-2})^{N-ii} C_{ii}^{jj} Pe_{-2}^{jj} Pe_2^{ii-jj} \cdot \int_{-\inf}^{N \cdot h_1 - (2 \cdot ii - 4 \cdot jj) \cdot h_2} \frac{1}{\sigma'\sqrt{2\pi}} \exp\left(-\frac{x^2}{2\sigma'^2}\right) dx \quad (7)$$

Where σ' is denoted as $\sigma' = k_\sigma \cdot \sigma_n$. It is assumed that the residual signal and the transmitter signal are unrelated, when the interference power is low and the probability of the interference detection is large. In this case, the equivalent noise power is the sum of the residual interference power and the AWGN signal power. Hence, the signal to noise ratio after dispreading is given by

$$\eta = \frac{N \cdot |h_1|^2}{4(Pe_2 + Pe_{-2})|h_2|^2 + \sigma_n^2} \tag{8}$$

According to (8) and the error rate function, we can derive the statistical error rate formula of the spread spectrum signal

$$Pe_s = Q(\sqrt{\eta}) \tag{9}$$

Performance Derivation of Signal Cancellation Priority Mode

In this mode, we detect the desired signal firstly. The error rate of the desired signal is calculated as

$$Pe_c = Q\left(\sqrt{\frac{N \cdot |h_1|^2}{\sigma_n^2 + |h_2|^2}}\right) \tag{10}$$

Assuming that we transmit a signal b_i, conditioned on whether there is the residual interference, the error rate of its canceled signal is given by

$$\begin{cases} Pe_I|C = \frac{1}{2} \cdot \left(\int_{-\infty}^{-h_2} \frac{1}{\sigma_n \sqrt{2\pi}} \exp\left(-\frac{(x-2h_1)^2}{2\sigma_n^2}\right)dx + \int_{-\infty}^{-h_2} \frac{1}{\sigma_n \sqrt{2\pi}} \exp\left(-\frac{(x+2h_1)^2}{2\sigma_n^2}\right)dx\right) \\ Pe_I|\bar{C} = \int_{-\infty}^{-h_2} \frac{1}{\sigma_n \sqrt{2\pi}} \exp\left(-\frac{(x)^2}{2\sigma_n^2}\right)dx \\ Pe_I = Pe_c \times Pe_I|C + (1 - Pe_c) \times Pe_I|\bar{C} \end{cases} \tag{11}$$

Where, if the interference cancellation is wrong, the error rate of the interference detection is $Pe_I|C$. If the interference cancellation is right, the error rate of the interference detection is $Pe_I|\bar{C}$. The first signal detection not only affects the subsequent interference detection and cancellation, but also affects the final signal detection. Therefore, we use the same method adopted in the interference cancellation priority mode, and get the expressions of the final error rate, which is calculated as

$$\begin{cases} p_{ij} = C_N^{ii}(1 - Pe_{2|C} - Pe_{-2|C})^{N-ii} C_{ii}^{jj} Pe_{-2|C}^{jj} Pe_{2|C}^{ii-jj} \\ Pe|C = \sum_{ii=0}^{N} \sum_{jj=0}^{ii} p_{ij} \cdot \int_{-\inf}^{N \cdot h_1 - (2 \cdot ii - 4 \cdot jj) \cdot h_2} \frac{1}{\sigma' \sqrt{2\pi}} \exp\left(-\frac{x^2}{2\sigma'^2}\right)dx \\ p'_{ij} = C_N^{ii}(1 - Pe_{2|\bar{C}} - Pe_{-2|\bar{C}})^{N-ii} C_{ii}^{jj} Pe_{-2|\bar{C}}^{jj} Pe_{2|\bar{C}}^{ii-jj} \\ Pe|\bar{C} = \sum_{ii=0}^{N} \sum_{jj=0}^{ii} p'_{ij} \cdot \int_{-\inf}^{N \cdot h_1 - (2 \cdot ii - 4 \cdot jj) \cdot h_2} \frac{1}{\sigma_n \sqrt{2\pi}} \exp\left(-\frac{x^2}{2\sigma_n^2}\right)dx \\ Pe = Pe_c \times Pe|C + (1 - Pe_c) \times Pe|\bar{C} \end{cases} \tag{12}$$

If the first signal detection is wrong, the probability of the residual interference $Pe_{2|C}$ equals to $2h_2$. Where, just the same as $Pe_{-2|C}$, $Pe_{2|\bar{C}}$, $Pe_{-2|\bar{C}}$, $Pe_{2|C}$ is the probability after the signal cancellation, so the correlation between the interference detection and the signal declines. In addition, σ' usually can be replaced by σ_n.

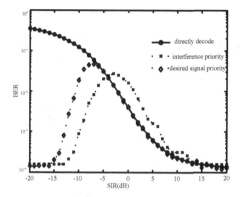

Fig. 2. Comparison of interference cancellation and direct demodulation

The simulation parameters of Fig. 2 are defined as: the spread gain $N = 15$ and $SNR = -0.5$ dB. In Fig. 2, we can see that the performance of the signal cancellation priority is greater than the interference cancellation priority when the SIR is lower than -5 dB. Otherwise the later is greater than the former. In addition, the BER curves of the interference cancellation priority mode and the signal cancellation priority mode are both convex because as the SIR increases, the interference power decreases and the probability of the residual interference signal is increasing. Due to the spread spectrum gain, the intersection of the two curves locates at the negative half-axis of SIR. In practical applications, we can firstly obtain the critical point by numerical method and then select the iteration scheme dynamically by sensing the SIR and SNR until the performance is best.

3.2   Iteration Times Analysis

In Figs. 3 and 4, the BER performance of different iteration times in different interference cancellation modes are compared, where its parameters are defined as $SNR = -0.5$ dB. And we can see that increasing the iteration times can hardly improve the performance. This is because the detection probability of the interference or the desired signal has a lower bound. And the effects of the noise floor and the residual interference cannot be completely eliminated.

In Fig. 3, one iteration can cancel the majority of interference signal in the higher interference power region. Therefore, increasing iteration times almost does not improve the performance significantly. As the interference signal power decreases, the probability and influence of the residual interference signal gradually increase, and it will directly affect the subsequent signal detection. Thus, the performance of signal detection cannot be improved by increasing the iteration times. Meanwhile the probability of interference detection decreases obviously when the interference power is low. Moreover, the impact of the residual interference signal will gradually reach or exceed the interference signal itself. Therefore, increasing the iteration times does not markedly improve the system performance.

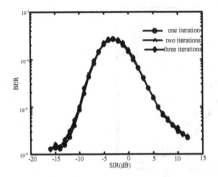
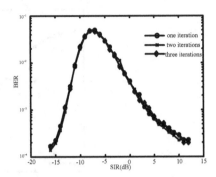

Fig. 3. Interference cancellation priority mode   **Fig. 4.** Signal cancellation priority mode

Similarly, in Fig. 4, the exists of the residual interference in the signal detection makes increasing iteration times useless to improve the system performance.

4   Analysis of Timing Errors

In practice, timing errors always exist. If the sampling point is not at the optimal moment, the energy of sampled signal cannot reach the maximum, and it will introduce the ISI [10]. Ignoring the impact of carrier phase offset, the sampled signal is given by

$$r_n(\tau) = \sum_{i=-K}^{K} \hat{L}_{n-i} h(iT_s + \tau) + v_n \tag{13}$$

Where τ and v_n represent the sampling deviation of symbols and the response of noise over the matched filter respectively. And the length of the forming filter is $2k+1$. In (13), $\hat{L}_0 h(\tau)$ is the sampled signal amplitude and $\sum_{\substack{i=-K \\ i \neq 0}}^{K} \hat{L}_{n-i} h(iT + \tau)$ is the ISI of adjacent symbols at the sampling moment. The ISI has no effect on the output mean due to $\hat{L}_i$ with mean of 0. However, the ISI, at the sampling time, will increase the output noise power [11], which is given by

$$V_I = \sum_{\substack{i=-K \\ i \neq 0}}^{K} |L_i h(iT_s + \tau)|^2 \tag{14}$$

The two major impact on the spread spectrum signal with timing errors are given by

(a) it reduces the integrated signal power, (b) it introduces the ISI power V_I, which reduces the system performance.

5   Simulation Results

The proposed above schemes have been simulated numerically by using MATLAB software, where the spreading gain $N = 15(11.76\,\text{dB})$, and the roll-off factor of match filter $\alpha = 0.22$, and $SNR = -0.5\,\text{dB}$.

In Fig. 5, the BER performance curves of 0.2 chips and 0.4 chips timing errors and ideal timing synchronization are compared. And we give their theoretical derivations and simulation results respectively. We can see that the theoretical derivations and simulation results are the same. In the interference cancellation priority mode, timing errors increase the BER and accelerate the deterioration of the performance. The timing errors of 0.2 chips degrades performance by about 1 dB. In addition, the performance decreases about 5 dB when the timing error is 0.4 chips, so the interference cancellation priority mode is sensitive to timing errors.

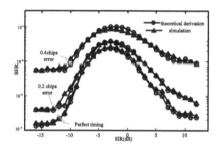
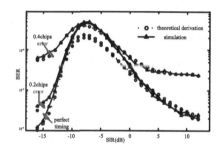

Fig. 5. The interference cancellation prior-   **Fig. 6.** The signal cancellation priority mode
ity mode

In Fig. 6, in the signal cancellation priority mode, the timing errors affect the BER performance significantly when the interference power is large. However, when the interference power is low, the small timing errors has little influence on the BER performance, but the large timing errors enlarge the error platform significantly.

6   Conclusions

In physical layer security, for the double iteration anti-jamming technology based on power recognition, we derive the theoretical expressions for BER with timing errors and analyze the impact of performance. The derivations are verified in simulations. Studies show that timing errors deteriorate the performance of system, but the impacts are different under the different iteration schemes, which requires a sufficiently time delay estimation at the receiver. This paper can provide a theoretical maximum time synchronization error allowed by co-channel interference anti-jamming technology in practical.

References

1. Yan, F., Wang, Y.: A shared frequency interference suppression method based on time-domain signal processing technology. Radar ECM **4**, 21–24 (2015)
2. Guo, L., Wang, B., Sun, Z., Chen, S.: Research on transform domain interference suppression in parallel combinatory spread spectrum system. J. Telemetry Tracking Command **36**(1), 30–35 (2015)
3. Guo, L., Yin, F., Lu, M.: Overview on NBI suppression of DSSS/CDMA systems. Acta Electronica Sinica **37**(10), 2248–2257 (2009)
4. Iltis, R.A., Milstein, L.B.: Performance analysis of narrow-band interference rejection techniques in DS spread-spectrum systems. IEEE Trans. Commun. **32**(11), 1169–1177 (1984)
5. Masry, E.: Closed-form analytical results for the rejection of narrow-band interference in PN spread-spectrum systems - Part II: linear interpolation filters. IEEE Trans. Commun. **33**(1), 10–19 (1985)
6. Buzzi, S., Lops, M., Poor, H.V.: Code-aided interference suppression for DS/CDMA overlay systems. Proc. IEEE **90**(3), 394–435 (2002)
7. Luo, R., Lei, J., Hu, G.: Combating malicious eavesdropper in wireless full-duplex relay networks: cooperative jamming and power allocation. In: Yuan, H., Geng, J., Liu, C., Bian, F., Surapunt, T. (eds.) GSKI 2017. CCIS, vol. 848, pp. 452–463. Springer, Singapore (2018). https://doi.org/10.1007/978-981-13-0893-2_48
8. Kim, J., Choi, J.P.: Cancellation-based friendly jamming for physical layer security. In: Global Communications Conference, pp. 1–6. IEEE (2017)
9. Dong, L., Yousefi'zadeh, H., Jafarkhani, H.: Cooperative jamming and power allocation for wireless relay networks in presence of eavesdropper. In: Proceedings of the IEEE ICC, Kyoto, Japan, 5–9 June 2011, pp. 1–5. IEEE Press (2011)
10. Parkvall, S., Strom, E., Ottersten, B.: The impact of timing errors on the performance of linear DS-CDMA receivers. IEEE J. Sel. Areas Commun. **14**(8), 1660–1668 (2002)
11. Viterbi, A.J.: CDMA: principles of spread spectrum communication. Commun. Technol. **9**(1), 155–213 (1995)

Workshop on Green Communications for Internet of Things

Multi-frequency Large-Scale Channel Modeling and Green Networking Design

Zhenfeng Zhang$^{(\boxtimes)}$, Daosen Zhai, Ruonan Zhang, and Jiaxin Wang

Department of Communication Engineering, Northwestern Polytechnical University, Xi'an 710072, Shaanxi, China
2450182446@mail.nwpu.edu.cn

Abstract. In this paper, we conduct the measurement campaign in the urban microcell (UMi) outdoor-to-indoor (O2I) scenario, with the object to investigate the large-scale fading at 900 MHz, 2.6 GHz, and 3.5 GHz. Specifically, the path loss model is constructed based on the experimental data and the 3GPP TR36.873 channel model standard. Furthermore, according to the channel model, we propose a triple-band frequency-switching (TB-FS) scheme, which can minimize the transmitting power of a base station. Simulation results indicate that considering the propagation characteristics, the low frequency is preferred under small rate requirement and the high frequency is preferred under the large rate requirement. Furthermore, we find that the optimal spectrum is independent from positions of the devices. These results are valuable for the design and optimization of green communications and network deployment.

Keywords: Outdoor-to-indoor (O2I) ·
Channel measurement and modeling · Green communication

1 Introduction

In recent years, the mobile communications have made great progress all over the world. Meanwhile, the communication scenes also become more and more complicated. The fifth generation (5G) network will be available by 2020. There will be massive devices served by the 5G network, which leads to huge energy consumption. Therefore, the energy consumption should be taken into account in the 5G network design. The literature [1] proposed the concept of green communications. In order to improve the energy efficiency, it summarized five technologies including the device-to-device (D2D) communications, ultra dense networks (UDN), massive MIMO, spectrum sharing, and Internet of things (IoT).

This work was supported in part by the National Natural Science Foundation of China (61571370, 61601365, and 61801388), in part by the Fundamental Research Funds for the Central Universities (3102017OQD091 and 3102017GX08003), and in part by the China Postdoctoral Science Foundation (BX20180262).

B. Li et al. (Eds.): IoTaaS 2018, LNICST 271, pp. 367–375, 2019.
https://doi.org/10.1007/978-3-030-14657-3_37

In [2–6], the authors proposed the solutions from different aspects to reduce the communication energy consumption.

In various network deployment environments, the channel features are considerably different, which has a great influence on the wireless network design, such as the power allocation, relay deployment, spectrum utilization, and so on. Channel measurement in different bands can provide useful guidance for the green networking design. Motivated by these factors, abundant channel measurement activities have been conducted. For example, the measurement campaign in [7] investigated the Outdoor-to-Indoor (O2I) path loss in the metropolitan small cell scenario at 3.5 GHz. Based on the two-dimensional (2-D) O2I models given in M.2135 and Winner II, the authors modified the 2-D path loss model by adding a height gain. In [8], the 2.6 GHz O2I wireless link was measured in three metro environments (tunnel, station, and open field). The measurement results indicated that the attenuation in the metro environments were 15 to 20 dB higher than those in free space. It was obtained that the O2I channel followed a Rayleigh distribution and the impact of the window glasses on the signal attenuation was not significant for the overall link budget. The literature [9] characterized the 3-D spatial channels in a gymnasium by field measurement and statistical modeling. It was demonstrated that the average angular spreads were independent from the distance and direction of the microcell base station (BS) in the typical stand coverage scenario.

This article focuses on the outdoor-to-indoor (O2I) channel measurement and modeling at 900 MHz, 2.6 GHz, and 3.5 GHz. The contributions of this paper are two-fold. First, we conducted the urban microcell (UMi) O2I large-scale channel measurement in a hotel. The wireless links from the ground to the high building, where the BS was placed on the ground and the user equipment (UE) distributed on different floors, were measured. This measurement setup is termed as "low-to-high (L2H)" propagation. We then utilized the measurement data to modify the O2I channel model in the 3GPP TR38.673 standard [10]. The O2I path loss model can be used to calculate the transmitting power of the BS. Second, through the analysis of propagation characteristics at the three frequencies, we can obtain that combining hybrid frequencies for network coverage has much greater advantages in energy saving and high-capacity transmission than using a single frequency. We propose a triple-band frequency-switching (TB-FS) scheme, which can minimize the transmitting power of the BS under different rate requirements.

The remainder of this paper is organized as follows. Section 2 describes measurement system setup, and channel measurements. Section 3 describes measurement results. In Sect. 4, the O2I channel model in the 3GPP TR38.673 standard was modified using the measurement data. Section 5 presents triple-Band frequency-switching (TB-FS) scheme. Finally, conclusions are drawn in Sect. 6.

2   Measurement System

The measurement campaign was conducted in a hotel in Chengdu, China, and the environment was a typical UMi O2I scenario. The channel was measured by a tri-band continuous-wave (CW) large-scale channel sounder as shown in Fig. 1. The transmitter (TX) emulated a base station of 2.5 m height placed on the ground, 30 m away from building, as shown in Fig. 1(a). The receiver (RX) was located on different floors, and its height was 1.5 m, as shown in Fig. 1(b). The RX was placed on five different floors, i.e., the 3rd, 4th, 6th, 8th, and 10th floors.

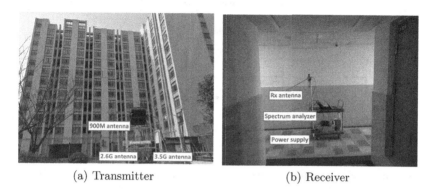

(a) Transmitter   (b) Receiver

Fig. 1. Channel sounder for O2I scenario measurement.

On the TX side, the system is equipped with three RF chains at 900 MHz, 2.6 GHz and 3.5 GHz, and each of them consists of a signal generator, power amplifier (PA), and transmitting antenna. Based on the narrow-band sounding scheme, the TX system generates and transmits a single tone at each target carrier frequency. The RX system adopts a wideband antenna to receive the probing signals at all the three frequencies. The wideband antenna is omni-directional in the azimuth plane and has a half power beamwidth (HPBW) of 40° in the elevation dimension, with the maximum gain of 2.5 dBi. The received single-tone sounding signal is amplified by a low noise amplifier (LNA) and then input into a portable spectrum analyzer (SA). The SA is programmed to measure the received power at the marked frequency points continuously. In this system, we set three markers at the three frequencies. The SA measures the power at the first frequency for 50 times periodically with the interval of 300 ms, and then sends the data to a laptop via Ethernet for storage. Then, the process is repeated for the second and third frequencies.

The system needs to be calibrated before each test, because the transmitting power and gains of the RF equipments such as the amplifiers, change with temperature and time. Then, a path loss value of the specific channel can be obtained by calibrating the measured power attenuation. In what follows, we introduce the detailed calibration and field measurement processes. Firstly, the TX and RX systems are connected directly by radio cables, and an attenuator with −40 dB gain is inserted in the link in calibration, as shown in Fig. 2(a). In field measurements, the RF link is replaced by the transmitting and receiving antennas

and a wireless channel, as shown in Fig. 2(b). By comparing the received power
with the calibration results, the attenuation over the radio channel is obtained.

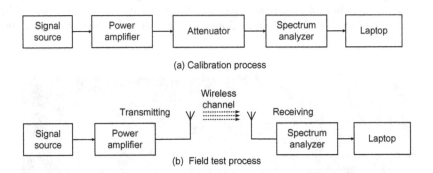

Fig. 2. Calibration and field test processes.

The path loss is calculated according to calibration and measurement procedures by

$$P_{TX} + G_{PA} + G_A - L = P_{RX1}, \tag{1}$$

$$P_{TX} + G_{PA} + G_{TX} - PL + G_{RX} - L = P_{RX2}, \tag{2}$$

where P_{TX} is the transmitting power which can be set manually, $G_{PA} = 10$ dB
is the gain of the power amplifier and $G_A = -40$ dB is the attenuation of
the attenuator. The gains of the transmitting and the receiving antennas are
denoted by G_{TX} and G_{RX}, respectively. The radiation patterns of the antennas
have been measured in an anechoic chamber. P_{RX1} and P_{RX2} are the received
power in calibration and measurement, respectively, which are captured by the
spectrum analyzer. L represents other losses in the measurement system such as
the high-frequency phase-steady cable loss. According to (1) and (2), we can get
path loss value PL by calibration.

3   Measurement Results

The contour diagram in Fig. 3 demonstrates the distributions of the path loss
magnitude at the three frequencies of 900 MHz, 2.6 GHz, and 3.5 GHz. It is
concluded that the penetration capability of the low frequency is better than
that of the high frequency. We analyze the signal attenuation in the room and
corridor as follows.

The rooms on the 3rd floor are representative, so we choose the 3rd floor as
an example to analyze the path loss in the measurement scene. The path loss
of Room 316 is significantly smaller than that of Room 315. Similarly, the path
loss of Room 320 is significantly smaller than that of Room 319. This is because
Rooms 316 and 320 are closer than to the TX, and the signal only needs to
penetrate one wall to reach the rooms in this O2I scenario. In Room 315, the

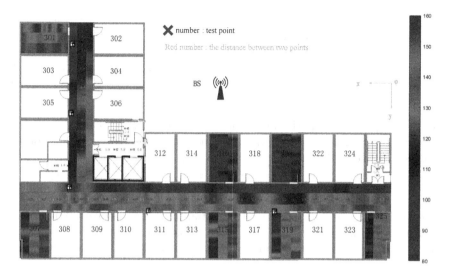

Fig. 3. Path loss on the 3rd floor at 900 MHz, 2.6 GHz, and 3.5 GHz.

signal needs to pass through three walls which causes great fading on the signal. Room 319 is similar to Room 315.

In the long corridor, the path loss at all the three frequencies is obviously larger than that in Rooms 316 and 320. This is because the signal penetrates double walls of the rooms to reach the corridor. We can see that the path loss in the middle of the long corridor is the smallest due to its shortest distance to the TX. From the middle to the right end, the path loss increases gradually with the increment of the distance between the TX to RX. However, the signal can enter the corridor at the right end by the reflection of other buildings outside. Thereby its path loss becomes small again. From the center to the left end, the path loss value increases significantly not only because of the increment of the distance to the TX, but also because the signals need to pass through the double rooms (or an elevator).

Similarly, at the top of the corridor vertical in Fig. 3, the signals have two paths to arrive at the RX. One path is the reflection by other surrounding buildings, and the other path goes through the two walls of Room 302. Thus the path loss is the smallest along the corridor. In the middle part, the path loss becomes low again due to the reduced distance to the TX. From the middle to the bottom part of the corridor, the path loss continuously increases due to the increment of the transmission distance and the obstruction of the rooms.

4   O2I Channel Model

According to the 3GPP TR38.673 channel standard, the O2I path loss of the UMi scenario can be expressed as

$$PL = PL_b + PL_{tw} + PL_{in},\qquad(3)$$

where PL_b, PL_{tw}, and PL_{in} are the outdoor path loss, wall penetration loss, and indoor path loss, respectively. In detail, PL_b, PL_{tw}, and PL_{in} are given by

$$\begin{cases} PL_b = PL_{3D\_UMi}(d_{3d\_in} + d_{3d\_out}), \\ PL_{tw} = 20, \\ PL_{in} = 0.5d_{2d\_in}, \end{cases} \tag{4}$$

where PL_{3D_UMi} is the path loss in the UMi scenario, which can be estimated according to the measurement data in our measurement campaign. The specific expression is given as

$$PL_{3D\_UMi} = 22.0\log_{10}(d_{3D}) + 28 + 20\log_{10}(f_c), \tag{5}$$

where $d_{3D} = d_{3d\_in} + d_{3d\_out}$ denotes the spatial three-dimensional (3D) distance between the BS and UE. d_{2d_in} is the two-dimensional distance from the indoor receiving terminal to the external wall. The O2I model schematic is shown in Fig. 4.

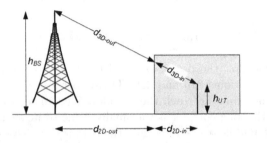

Fig. 4. The schematic of the O2I model [10].

When calculating the path loss with the O2I model, we need to determine the 3D distance between the TX and the RX. To achieve this goal, we need to establish a 3D coordinate system, as shown in Fig. 3. The right of the building is the y axis, and the horizontal position of the receiver is the x axis. The floor height is the z axis. By this way, the 3D coordinates of the BS and the measurement points are defined, and the 3D distance can be obtained. According to the triangular relationship, the 2D distance between the outer wall and the RX can be calculated.

Substituting the 2D and 3D distances and the three carrier frequencies into the O2I model given in (3) to (5), we can obtain PL_b and PL_{in}. Since the signals in the corridor needs to pass through two walls, the penetration loss PL_{tw} is larger than 20 dB given by the O2I model. This part is calculated by subtracting PL_b and PL_{in} from the real measurement data, and then we calculate the average to obtain the penetration loss of the two walls. Finally, the O2I path loss in the measurement scene can be obtained by adding the three parts, PL_b, PL_{in}, and PL_{tw}.

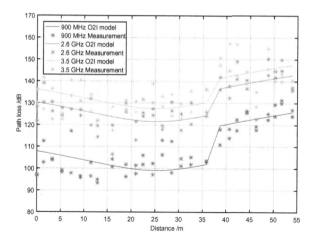

Fig. 5. Measurement results and O2I models.

The measured data and modified channel model are plotted in Fig. 5. As can be seen, the channel model fits the empirical results very well. The path loss when the distance is larger than 39 m becomes larger because the signals need to travel through two rooms or elevator.

5   Triple-Band Frequency-Switching (TB-FS) Scheme

To investigate the energy-saving spectrum allocation scheme, we calculate the transmitting power given data rate requirements according to the proposed channel model in Sect. 4 and the Shannon theorem. The specific parameter settings are as follows. The bandwidth of the 900 MHz carrier frequency is set as 180 KHz. Then according to the relationship of the carrier frequencies, the bandwidths at 2.6 and 3.5 GHz are set as 520 and 700 KHz, respectively. Thus the data rate and transmitting power can be

$$r_i = B_i log_2 \left(1 + \frac{P_i g_i}{n_0 B_i} \right), \tag{6}$$

$$P_i = \left(2^{\frac{r_i}{B_i}} - 1 \right) \frac{n_0 B_i}{g_i}, \tag{7}$$

where r_i refers to the information transmission rate of the i-th carrier frequency, B_i denotes the bandwidth, P_i denotes the transmitting power, g_i denotes the channel power gain which can be obtained by the channel model, and n_0 refers to the noise power spectral density.

The numerical results are evaluated as follows. First, the rate r_i is set as a fixed value, and the transmitting power P_i at the three carrier frequencies is obtained according to the path loss model given in (3) and (7). Thus we take the transmission rate r_i as an independent variable to observe P_i in the band of B_i. It can be observed from Fig. 6 that P_i at the three carry frequencies is different for the transmission rates. It can be seen that two of the three curves intercept with each other. This demonstrates that we can transmit signals at different frequencies and rates to minimize the transmitting power.

We analyze the relationship between the TX-RX distance and the rate intersecting point of the three lines. Interestingly, it is found that when the TX-RX distance is 50, 200 and 500 m, the rate intersecting point remains unchanged. This shows that the transmission rate is independent of the TX-RX distance. This observation indicates that the BS does not need to consider the location of the user when performing frequency switching at various data rates, which may significantly simplify the BS frequency switching operation. At the same time, it can be seen from the three sets of curves in Fig. 6 that if the BS utilizes the frequency switching technology, it can greatly reduce the power consumption.

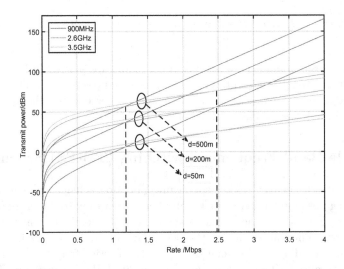

Fig. 6. The effect of the rate requirement on the transmission power.

6   Conclusion

In this work, we have performed a measurement campaign for the UMi O2I channels at 900 MHz, 2.6 GHz, and 3.5 GHz. We have constructed the O2I path loss model of the 3GPP TR36.873 standard based on our channel measurement results. The modified model indicates that the propagation loss through the concrete wall is very large. Furthermore, we have proposed the TB-FS scheme to save the BS power consumption. The numerical results show that the transmitting power at the three frequencies is independent from the TX-RX distance. As such, we can switch the carrier only according to the rate requirements to reduce the BS transmitting power. The proposed channel model and TB-FS scheme can facilitate the development of green networks for the future mobile communication services.

References

1. Gandotra, P., Jha, R., Jain, S.: Green communication in next generation cellular networks: a survey. IEEE Access **5**(99), 11727–11758 (2017)
2. Lin, X., Andrews, J.G., Ghosh, A., Ratasuk, R.: An overview of 3GPP device-to-device proximity services. IEEE Commun. Mag. **52**(4), 40–48 (2014)
3. Chen, L., et al.: Green full-duplex self-backhaul and energy harvesting small cell networks with massive MIMO. IEEE J. Sel. Areas Commun. **34**(12), 3709–3724 (2016)
4. Goldsmith, A., Jafar, S.A., Maric, I., Srinivasa, S.: Breaking spectrum gridlock with cognitive radios: an information theoretic perspective. Proc. IEEE **97**(5), 894–914 (2009)
5. Kaushik, A., et al.: Spectrum sharing for 5G wireless systems (Spectrum sharing challenge). In: IEEE International Symposium on Dynamic Spectrum Access Networks (DySPAN), September 2015, pp. 1–2 (2015)
6. Gao, H., Ejaz, W., Jo, M.: Cooperative wireless energy harvesting and spectrum sharing in 5G networks. IEEE Access **4**, 3647–3658 (2016)
7. Li, C., Zhao, Z., Tian, L., Zhang, J.: Height gain modeling of outdoor-to-indoor path loss in metropolitan small cell based on measurements at 3.5 GHz. In: International Symposium on Wireless Personal Multimedia Communications (WPMC) (2014)
8. Arriola, A., et al.: Characterization of an outdoor-to-indoor wireless link in metro environments at 2.6 GHz. In: IEEE International Conference on ITS Telecommunications (2017)
9. Zhong, Z., Zhang, R., Ren, K., Wang, K., Li, B., Zhang, X.: Measurement and modeling of 3-dimensional radio channels with cross-polarizations in a gymnasium. In: IEEE European Conference on Antennas and Propagation, pp. 2473–2477 (2017)
10. 3GPP TR36.873-V12.4.0(2017–03) Study on 3D channel model for LTE

The Position Relationship for RSU Assisted Vehicular Opportunistic Networks

Yixin He$^{(\boxtimes)}$, Yi Jiang, and Song Pan

School of Electronics and Information, Northwestern Polytechnical University,
Xi'an, China
914416406@qq.com, jiangyiv88@nwpu.edu.cn

Abstract. Vehicular Ad-hoc networks (VANETs) are promising information-sharing technologies for modern intelligent transportation systems (ITS), in which information can be disseminated and shared among a group of moving vehicles through wireless communication devices. Previous data dissemination schemes for VANETs are unable to meet the requirements of effective and efficient data dissemination in VANETs due to the dynamic nature of the framework such as the intermittent connection between vehicles. A position relationship and RSU assistance (PR) data dissemination algorithm was proposed for opportunistic VANETs. Through making use of road side unit (RSU), position information and transfer probability to improve the performance of PR algorithm. Simulation results show that the data delivery ratio of PR protocol is improved, the number of network copy is controlled, and network overhead is reduced.

Keywords: Vehicular Opportunistic Networks · Position relationship · RSU

1 Introduction

Vehicular Opportunistic Networks (VON) are wireless vehicle communication network, in which information can be disseminated and shared among a group of moving vehicles through wireless communication devices. Vehicular Opportunistic Networks is regarded as one of the most promising ad hoc networks in recent years. But some drawbacks like non-homogeneous distribution of nodes, frequent change of topology and unstable communication links, caused by high speed vehicles in the network, have brought challenges to the routing protocol design.

At present, the research on vehicle opportunistic network is at an initial stage. An opportunistic routing protocol based on expected delay was proposed, which optimizes the data transmission delay (EDOR) between source nodes and target nodes when they are both moving vehicles, but this routing protocol is very complex [1]. A traffic distribution based opportunistic routing (TDOR) in urban VANETs was proposed. TDOR protocol was designed as a two-phase algorithm, namely intersection selection phase and next hop selection phase, which improved the path, but this routing protocol was bad in energy consumption [2]. A data dissemination mechanism based on density-sensing (DDMD) for Vehicular Opportunistic Networks was proposed, this protocol

© ICST Institute for Computer Sciences, Social Informatics and Telecommunications Engineering 2019
Published by Springer Nature Switzerland AG 2019. All Rights Reserved
B. Li et al. (Eds.): IoTaaS 2018, LNICST 271, pp. 376–383, 2019.
https://doi.org/10.1007/978-3-030-14657-3_38

was aware of the network environment of node density [3]. But these three agreements focused on strengthening one aspect and ignored the other aspects.

Vehicular Opportunistic Networks is an important component of future wireless networks, therefore, it is an important method building road side unit (RSU) to assist communication for improving the performance of VON. This paper presents a position relationship and RSU assistance (PR) data dissemination algorithm was proposed for opportunistic VANETs. According to the design of the scheme design of RSU can connect directly to the backbone network and communicate with the vehicle for wireless communication. PR algorithm outperforms EDOR algorithm, TDOR algorithm and DDMD algorithm with less network resource consumption in broad scenarios.

2 The Position Relationship for RSU Assisted Vehicular Opportunistic Networks

2.1 System Overview and Network Modeling

It is assumed that the time for some vehicles to establish communication can be ignored, moreover, the vehicle can get the position information through the GPS.

Definition: Published information (PI) and Feedback information (FI)

1. PI means the message to be transmitted, it contains five parts, namely vector, position, source, TTL and message.
2. FI means the message to be returned, it contains three parts, namely vector, position and number.

2.2 Position Information

In PR algorithm, if vehicles have the network copy, they will send published information at set intervals. The feedback information will determine whether they continue to retain the network copy. The user can set the maximum network copies according to the requirement.

PR algorithm makes the above decision by calculating position information.

(1) The position relationship of send PI vehicle and send FI vehicle

When the send PI vehicle received the FI, it can calculate the angle of two vehicles, according to the vector information. The computation can be written as:

$$\zeta = cos^{-1} \frac{\overrightarrow{A_V} * \overrightarrow{B_V}}{\left|\overrightarrow{A_V}\right|\left|\overrightarrow{B_V}\right|} = cos^{-1} \frac{A_{VX} * B_{VX} + A_{VY} * B_{VY}}{\sqrt{(A_{VX}^2 + A_{VY}^2) + (B_{VX}^2 + B_{VY}^2)}} \tag{1}$$

Where $\overrightarrow{A_V}$ is A vehicle's velocity vector, $\overrightarrow{B_V}$ is B vehicle's velocity vector.
If $|\zeta| < 90$, it is thought that the direction of the two vehicles are same.
If $|\zeta| > 90$, it is thought that the direction of the two vehicles are not same.

(2) The position relationship of the vehicle and RSU

PR algorithm defines two relative states of vehicles and RSU: Driving toward RSU and Driving to the back to RSU. PR algorithm can get the two relationships by calculating the vehicles' velocity vector, angle relation and transfer probability.

As shown in Fig. 1, vehicle A travels at a speed of V_A, and the vehicle B travels at a speed of V_B. Both V_A and V_B are velocity vectors, which represent the traveling speed and direction of vehicle A and B. S represents RSU, AS is the distance vector of vehicle A to RSU, and BS is the distance vector of vehicle B to RSU. α is the angle between AS and V_A, and β is the angle between BS and V_B.

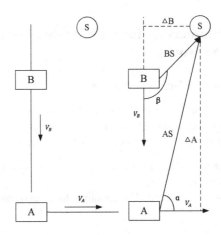

Fig. 1. Diagram of the vehicle and RSU

Obviously, to calculate α and β, we must get information about AS and BS. According to the foregoing assumptions, the AS and BS can be obtained by GPS. The calculation can be written as:

$$\alpha = tan^{-1}(AS_Y/AS_X) - tan^{-1}(A_{VY}/A_{VX}) \tag{2}$$

$$\beta = tan^{-1}(BS_Y/BS_X) - tan^{-1}(B_{VY}/B_{VX}) \tag{3}$$

Where, $\alpha, \beta \in (0, \pi)$.

By calculating α and β, we can get the following four results:

1. If $\alpha < \frac{\pi}{2}$, the vehicle A is considered to be traveling toward the RSU.
2. If $\alpha > \frac{\pi}{2}$, the vehicle A is considered to be far from the RSU.
3. If $\beta < \frac{\pi}{2}$, the vehicle B is considered to be traveling toward the RSU.
4. If $\beta > \frac{\pi}{2}$, then consider vehicle B away from RSU.

2.3   Transfer Probability

An Encounter-based Routing Algorithm for Social Opportunistic Networks proposed that it computes forwarding efficiency of each node based on past encounter information. Meanwhile, it computes the average contact duration based on the past information, considering transmitting messages of varied size differs in time. Thus, a message may directly forwarded to the destination node, or it is forwarded to other nodes if and only if the candidate node may encounter the destination node with a high probability and the average contact duration between them is longer than that of current node [4].

When any node A and B meet in the network, the computation of transfer probability can be written as:

$$P_{A,B} = P_{A,B_{old}} + (1 - P_{A,B_{old}}) * Q_{A,B} \tag{4}$$

$$Q_{A,B} = F_{A,B} * D(A) \tag{5}$$

Where $Q_{A,B} \in [0, 1]$

$Q_{A,B}$ represents the effective forwarding capability of nodes A and B. The effective forwarding capability is proportional to the transmission probability.

D(A) represents the number of nodes that A meets different new nodes at a given time interval accounts for the total number of nodes.

$F_{A,B}$ represents the social attribute of the node, the number of nodes A meeting with a specific node at a certain time interval accounts for the total number of nodes that the current node meets with other nodes.

If the nodes A and B do not meet in a time unit, the transmission probability will be reduced gradually. The computation can be written as:

$$P_{A,B} = P_{A,B_{old}} * \zeta^k \tag{6}$$

Where $\zeta \in [0, 1]$, k means time unit.

Through the analyses above, we can get the following four results.

1. If $\alpha < \frac{\pi}{2}$ and $\beta < \frac{\pi}{2}$, the vehicle A and B are considered to be traveling toward the RSU.
2. If $P_{RSU,B} \geq P_{A,B}$, the vehicle A and B are considered to be traveling toward the RSU.
3. If $\alpha > \frac{\pi}{2}$ and $\beta > \frac{\pi}{2}$, the vehicle A and B are considered to be far from the RSU.
4. If $P_{RSU,B} < P_{A,B}$, the vehicle A and B are considered to be far from the RSU.

2.4   Routing Algorithm

PR algorithm can be explained as follows:

1. Vehicle A carrying the message (PI) periodically broadcasts the trajectory of the target vehicle.
2. One-hop neighboring vehicles and RSU calculate the position relationship and transfer probability.

3. The vehicle calculating the value of α, β and ζ, we can get the following four results.

  • If $|\zeta| < 90$ and vehicle A, B runs away from RSU, pass information to the vehicle B.
  • If $|\zeta| > 90$ and vehicle A, B runs toward RSU, pass information to the RSU.
  • If $|\zeta| < 90$ and vehicle A, B runs toward RSU, it need to calculation transfer probability. If $P_{RSU,B} \geq P_{A,B}$, pass information to the RSU, or pass information to the vehicle B.
  • If $|\zeta| > 90$ and vehicle A, B runs away from RSU, the vehicle A continues to drive.

4. If the message had received, the vehicle B will discard the message.
5. This process is repeated until the target vehicle receives the message or until the message is invalid (Fig. 2).

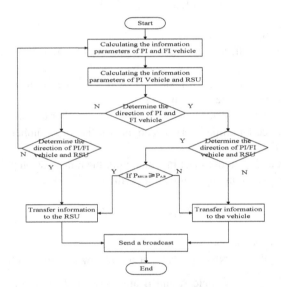

Fig. 2. PR algorithm flowchart

3 Performance Evaluation

In this section, we evaluate the performance of PR algorithm through extensive simulations using the ONE simulator. Since DTN routing is one of the possible solutions for message delivery to moving target without the help of stationary nodes, we have compared the performance of PR algorithm with two alternate DTN routing mechanisms—Epidemic Dissemination (ED) algorithm and Random Choice (RC) algorithm.

3.1 Simulation Parameters

We conducted 101 rounds of simulations with different random seeds for each vehicle number N. The scenario chosen for simulation was the road map of Helsinki, Finland. Each vehicle's movement pattern is determined by Shortest Path Map Based Movement model. In this model, vehicles take the shortest path on the road of the map exactly. For each simulation, a vehicle node was selected randomly as the target vehicle, and the simulation was repeated 10 times with different target vehicles for each random seed. Only a transmission message was considered in the simulation, and the loss of transmission during the communication procedure was not taken into consideration.

The simulations parameters are listed in Table 1.

Table 1. Simulation parameters

Parameter	Value
Size of network area	4500 * 3400 m^2
Simulation time	4500 s
Vehicle/RSU Transmit range	30/100 m
Vehicle/RSU Transmit speed	1/10 Mbps
Vehicle number N	[500:200:1900]
Average node speed	15–80 MPH
Message size	7 kB

3.2 Simulation Results and Analysis

(1) Network Copy

When the size of the simulation region is constant, the number of vehicles determines the average space density of the network.

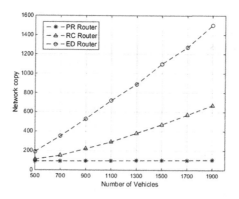

Fig. 3. The network copies comparison for different densities

In Fig. 3, we have plotted the network copies of algorithms with the vehicles number N increases. While the ED algorithm and RC algorithm employs multicast strategy, the overhead of ED algorithm and RC algorithm are obviously higher than the PR algorithm. But the PR algorithm adaptively removes outdated message copies and controls the number of replicas via RSU. From Fig. 3, it is obvious that PR algorithm has the lowest network copy of the three, and the gap between PR algorithm and the others increases as the density increases.

(2) Data Packet Successful Delivery Rate

Data packet successful delivery rate indicates the proportion of the successful transmit of the information to the destination vehicle. Data packet successful delivery rate represent the proportion of the successful transmit of the information to the destination vehicle.

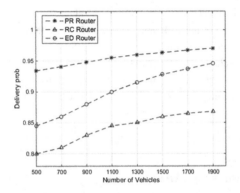

Fig. 4. The rate comparison for different densities

In Fig. 4, because the PR algorithm can choose RSU or vehicles to relay, it has the highest rate of the three. Due to its flooding characteristics, the ED algorithm increases its delivery rate as the number of vehicles increases, but it has been slightly lower than the PR algorithm. The success delivery rate of the RC algorithm has always been in a relatively low state, which is due to the random selection of the RC algorithm when selecting the relay node, so there may be a portion of the forwarding object that may have been forwarded or the useless vehicle nodes are passed on the information.

(3) Network Overhead

Although the PR algorithm can control the number of the network copy, with the increase of simulation time, if the speed of deleting the network copy less quickly than the speed of creating the network copy, network overhead can be well controlled.

Figure 5 demonstrates the transmission overhead (number of message replicas) for different densities. Since both the PR algorithm employ unicast strategy, the overhead of PR algorithm has the same characteristic with the network copy. From Fig. 4, it is obvious that PR algorithm has the lowest overhead of the three, and the gap between PR algorithm and the others increases as the density increases.

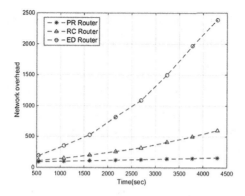

Fig. 5. The network overhead comparison for different densities

4   Conclusion

In this paper, we present and discuss PR algorithm, an opportunistic routing algorithm in vehicular networks. The main idea of PR algorithm is to calculate position relationship and transfer probability to identify a "better" message carrier (Vehicle or RSU). And the transmission procedure of PR algorithm is implemented completely through the message carrying and forwarding across vehicles, with the help of RSU. The evaluation results show that, when compared to the existing algorithms, PR algorithm has a good performance in various vehicle densities in terms of controlling the network copy, increasing the data packet successful delivery rate, and reducing the network overhead. Most notably when the vehicle density is high, PR algorithm had an impressive performance.

Acknowledgments. This work was supported in part by the National Natural Science Foundation of China (61571370, 61601365, and 61801388), in part by the Fundamental Research Funds for the Central Universities (3102017OQD091 and 3102017GX08003), in part by the China Postdoctoral Science Foundation (BX20180262), and in part by the Seed Foundation of Innovation and Creation for Graduate Students in Northwestern Polytechnical University (Grant no. ZZ2018129).

References

1. Liu, L., Pei, J.: Opportunistic routing algorithm for vehicular ad hoc networks based on delay expection. Transduc. Microsyst. Technol. **36**(10), 150–153 (2017)
2. Cai, Q., Niu, J., Liu, C.: Adaptive data dissemination algorithm for vehicular opportunistic networks. Comput. Sci. **38**(16), 58–63 (2016)
3. Zhang, L., Miao, J., Xu, X.: A data dissemination mechanism based on density-sensing for vehicular opportunistic networks. Comput. Eng. Design **38**(12), 36–40 (2017)
4. Yang, Y., Wang, X., Zhang, L., Liu, S., Lin, Y.: An encounter-based routing algorithm for social opportunistic networks. Comput. Technol. Dev. **28**(2), 64–73 (2018)
5. Khabazian, M., Aissa, S., Mehmet-Ali, M.: Performance modeling ofmessage dissemination in vehicular ad hoc networks with priority. IEEEJ. Sel. Areas Commun. **29**(1), 61–71 (2017)

A Multi-channel Based Reader Anti-collision Protocol for Dense RFID Environments

Yi Jiang[✉], Wei Cheng, Yixin He, and Song Pan

School of Electronics and Information,
Northwestern Polytechnical University, Xi'an, China
{jiangyiv88, pupil_119}@nwpu.edu.cn

Abstract. With the recent developments in Internet of things, RFID is getting more and more attention. In the dense RFID environment, reader collision occurs when some readers potentially interfere with the operation of others, and cannot correctly communicate with tags, which will limit the performance of RFID system. As we know, the tag collision has been widely discussed, but the reader collision has not received as much attention in the past few years. In this paper, we propose a novel multi-channel based reader anti-collision protocol (MRAP), which is suitable for dense RFID environments, to solve the reader collision problem. Based on the connected matrix and the distance between any two readers, we design the principle to confirm channel and communication timeslot for each reader. Using multi-channel, a new communication process is generated, in which the irrelevant readers can communicate with tags simultaneously, even in the same channel. The simulation and performance analysis results show that our protocol can achieve better performances than previous protocols in enhancing system throughput and system efficiency.

Keywords: Reader anti-collision protocol · Multi-channel · RFID · Pulse

1 Introduction

Radio frequency identification (RFID) is a perception processing technology in the Internet of things, which is widely used in several trades, such as traffic, logistics, security, etc. The routine RFID system is composed of many readers and tags. The tags are always passive, so they do not have an ability to distinguish the different frequency from the readers. The tag memories can be read in the read range of readers. A reader can choose the number of tags which it plans to communicate by adjusting its read range. In some applications, the dense RFID environment consists of many readers, in which some potentially reader interferences occur and the relevant readers cannot correctly communicate with tags. This is called the reader collision problem. As we know, the tag collision has been widely discussed, but the impact of reader collision has not been fully explored in the past few years. Each reader has the read range and interference range respectively. The reader collision problem can be divided into two cases: reader to reader collision and reader to tag collision. The two cases should be avoided in normal communication.

B. Li et al. (Eds.): IoTaaS 2018, LNICST 271, pp. 384–392, 2019.
https://doi.org/10.1007/978-3-030-14657-3_39

(1) Reader to reader collision: Due to the power limit, the responses of tags are easy to be affected by the close readers at the same time and frequency, which means readers can affect the communications of other readers with tags in their interference range.

(2) Reader to tag collision: If a tag is located in the overlap reading region among more than one reader, it cannot respond the interrogations from several readers when they read it simultaneously. Because it is too simple not to distinguish the frequency from the different readers.

The Pulse [1, 2] has two channels, one of which is used to communicate with readers, another is used to read tags. Song et al. [3] propose a slot occupied probability (SOP) protocol based on Pluse, which can effectively decrease the number of collision timeslot by choosing unused timeslot. The above two protocols do not use multiple data channels, the performances of which need to be improved. In paper [4], the channels are divided into odd and even channels, in which the odd ones is used first. The system throughput is not optimal using this method to allocate channels. Meguerditchian et al. [5] use two control channels, from which distinguishes the same timeslot or the same frequency, but it does not consider the case that the two collisions occur simultaneously.

In this paper, we propose a novel reader anti-collision protocol (MRAP) based on the Pulse protocol, which has one control channel and several data channels and is very suitable for dense RFID environments. The number of the data channels is decided by the number of readers. To allot the multiple channels, we design a connected matrix of readers in which the unconnected readers should be arranged in the same channel. It is better for solving the reader to reader collision. By the distance of any two readers, we define a method to further confirm channels and communication timeslots for every reader, and give the relationship between the number of channels and backoff time delay [1]. Using the above method, we define the detail communication process to avoid the collision in dense reader model, in which the uncorrelated readers can share the same timeslot in the same channel, but the connected readers do not have the same timeslot to communicate with tags, even if they belong to the different channels. To prove it, we take a concrete application for example. The simulation and performance analysis results show that our protocol outperforms previous protocols in the system throughput and system efficiency even in a highly dense network.

2 Assumption and Principle of Using Multiple Channels

2.1 Choosing Channel by the Connected Matrix of Readers

A graph $G = (V, E)$ is used to indicate the reader collision model, where the V expresses the vertices of the graph and the E expresses the connections between any two vertices called edges [7]. In reader collision model, $V = \{R_1, R_2, \ldots, R_n\}$ is the set of readers in which n is the number of readers, and $E = \{e(R_i, R_j)\}, i, j \in 1, 2, \ldots n$ is the set of edges connecting two readers R_i and R_j, if they have a communication overlapped region. If two readers are connected by a direct edge, they will possibly interfere with each other leading to either reader to reader collision or reader to tag collision.

The symmetric matrix $D = \{d_{ij}\}, i, j \in 1, 2, \ldots n$ with size n × n is constructed to describe whether any two readers in RFID system will potentially interfere with each other. In the graph of reader collision model $G = (V, E)$, if $e(R_i, R_j) \in E$, then $d_{ij} = 1$, else $d_{ij} = 0$.

The situation of interference between any two readers is described in Fig. 1(a). Suppose the circle surrounding a reader is its read range, so the situation of connection among readers is showed in Fig. 1(b).

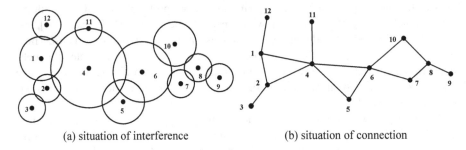

(a) situation of interference   (b) situation of connection

Fig. 1. Reader collision in dense RFID reader environment

Suppose the interference range of reads is not considered, the unconnected readers should be arranged in the same channel or the same timeslot when the value of their corresponding edge is $d_{ij} = 0$, since that case can avoid interfering with each other.

If the reader has many choices for choosing its data channel, it will choose the channel with the biggest frequency distinctions among its connected reader channels, which can reduce the interference range between the two connected readers [4].

2.2 Further Confirming the Channel and Timeslot by the Distance Between Other Readers

Reader to reader collision occurs when a tag responds an interrogation interfered by another reader, which can be solved by using the different channels. Reader to tag collision occurs when one tag is located in the overlapping interrogation region of more than one reader, which can be solved by using the different timeslots. The different timeslots can be represented as the different backoff time delays [1], which prevent the relevant readers from communicating with tags simultaneously. However, the appearance of collisions is different from the variance of the distance between two readers, by which we will further confirm the channels and timeslots for the different readers. Suppose that the readers are expressed as $R_i, i \in 0, 1, 2, \ldots, n$, where r_i is the read range and I_i is the interference range of the reader R_i. Then we define $D(R_i, R_j)$ as the distance between the reader R_i and the reader R_j. Taking Fig. 2 for instance, the concrete conditions can be described as follows:

(a) If $D(R_i, R_j) < r_i + r_j$, the reader R_i and the reader R_j can be arranged in the different channels, then choose the different timeslots. There is a potential for reader to reader collision and reader to tag collision simultaneously.

(b) If $r_i + r_j < D(R_i, R_j) < max\{I_i, I_j\}$, the reader R_i and the reader R_j can be arranged in the different channels, then choose the different timeslots. There is a potential for reader to reader collision and reader to tag collision simultaneously.

(c) If $max\{I_i, I_j\} < D(R_i, R_j) < max\{r_i + I_j, r_j + I_i\}$, the reader R_i and the reader R_j can be arranged in the same channel, then choose the different timeslots. There is only potential for reader to tag collision.

(d) If $D(R_i, R_j) > max\{r_i + I_j, r_j + I_i\}$, the reader R_i and the reader R_j can be arranged in the same channel, then choose the same timeslot. There is no potential for reader collision.

The potential for reader to reader collision and reader to tag collision can also be solved by choosing the different timeslots, but the unconnected readers can use the same timeslot in the different channels or the same channel, which can effectively enhance the system throughput and improve the system performance.

Choosing the different channels can reduce the interference range between two readers. It is because that the interference signal strength of adjacent channels is reduced by spectral mask [5].

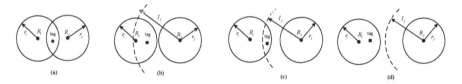

(a)   (b)   (c)   (d)

Fig. 2. The distance between two readers

3 MRAP Protocol

The proposed distributed protocol MRAP, is based on Pulse protocol [1, 2]. There are one control channel and several data channels. A reader will periodically broadcast a beacon signal on the control channel, when it is reading tags. The beacons which belong to the irrelevant readers can transmit on control channel. If a reader wants to communicate with tags, it will sense the control channel for the beacon of its relevant readers first. During a period of time, the reader does not sense any relevant beacon, which means it can start communicating with tags on its data channel.

The communication range of control channel is larger than that of data channels. We can suppose that the maximum of the control communication range is expressed as $max\{I_i + r_j\}, i, j \in 0, 1, \ldots, n$, which can be achieved by enhancing the transmit power. Perhaps two readers can communicate on the control, but they interfere with each other on their data channel. The control channel is special and independent of the RFID spectrum. Any communication on the control channel does not affect the normal communication on data channels.

Our protocol is present at the reader, which can transmit signal on both control channel and data channel simultaneously. The communication between readers is executed on the control channel and the communication between reader and tags is executed on the data channels.

We depict the method of how to confirm data channels and timeslots for readers, which can be used in the process of the MRAP protocol. It is assumed that the power of a beacon signal is boosted enough to be received by all of neighboring readers.

- On the basis of the connected matrix of readers, corresponding $d_{ij} = 0$, the irrelevant readers can be found. Using the principle of choosing channels (2.A), the reader can confirm which channels can be chosen.
- According to the distance information sent from other relevant readers and the principle (2.B), the reader can further confirm an optimal channel from the available channels and a timeslot which should be different among the relevant readers.

The process of MRAP protocol is described as follows:

- *Waiting* State: Every reader which wants to read tags must sense the control channel first. If a reader does not receive *beacon_i* signal from the relevant readers $R_i (i \in$ *relevant readers*), which means other relevant readers are not communicating with tags at that time, it will star its *Waiting State*. We define T_L as the length of timeslot. If a reader does not receive any beacon corresponding to its relevant readers during the thrice T_L time, it will star its *Contend State*. Otherwise it will continue to wait.
- *Contend State*: In this state, the reader must choose a delay of backoff time (*backoff_delay*) to prevent the readers from reading tags simultaneously. If the reader does not receive the information for *backoff_delay* from other relevant readers, and the previous *backoff_delay* is equal to zero, it will choose a random backoff time and broadcast it. Otherwise the reader will avoid choosing this backoff time received from others. When the previous *backoff_delay* is not equal to zero, the reader should resume previous *backoff_delay* and broadcast it. If a relevant beacon signal is received by the reader during the period of backoff time delay, it will return to the *Waiting State*. If the *backoff_delay* is over, the reader will enter the *Communicate State*.
- *Communicate State*: The reader can communicate with tags in this state, and it must periodically send a beacon on the control channel to keep the control of the data channel. If the communication between reader and tags is finished, it will return to the *Waiting State*. When the application queue is empty, it will enter the *Idle State*.
- *Idle State*: In this state, there is no message to transmit in either control channel or data channels, which means the system is idle.

Normally, the value of *backoff_delay* is decreased in the *Contend State*. When the relevant beacon is received, the value of *backoff_delay* will be stopped, and will be restarted as long as the control channel is idle again.

4 Simulation Experiments

4.1 Simulation Scenarios

To simulate the performance of the MRAP protocol, we assume the simulation scene as following. The data channel frequency is defined as the range from 860 MHz to 960 MHz, in which control channel frequency is 930 MHz. The transmission power of readers is −45 dBm. The read range and interference range on data channel is 1.62 m and 7.1 m respectively. The sensing range on control channel is 5.4 m. The data is sent between readers and tags by 2 Mbps. To better display the function of our protocol, it will remove some unnecessary effect, such as interference between two channels, path loss and fade, etc. The collision is the only reason leading to packet loss.

- The simulation field is a 10 m × 10 m area, in which 400 tags is evenly distributed at an interval of 0.5 m.
- All the readers are fixed and randomly distributed in the simulation field. We will run the simulation 100 times in the different cases, then compute the average from them.

We assume the simulation time is 60 s and the interval of sending a packet is 0.5 ms. We use the same readers to take part in the experiment, and the number is 4, 9, 16, 25, 36, 49 and 64 respectively.

4.2 Results of Simulation and Analysis

We assume the number of channels should be varied with the different number of readers. When $n = 4$, the number of channels is 1. When $n = 9$ or 16, the number of channels is 2. When $n = 25$ or 36, the number of channels is 3. Finally, When $n = 49$ or 64, the number of channels is 4. We will compare our protocol with the previous protocols, such as CSMA [8], Colorwave, and Pulse.

The system throughput can be increased as a sign that the read rate is being enhanced. The increased system efficiency indicates that the ability of readers finding and eliminating the collisions is improved. We will show that the MRAP protocol is more effective in both dimensions than previous protocols by simulations.

(1) System Throughput

The system throughput comparing with the different protocols at $n = 25$ is shown in Fig. 3. Figure 4 shows the system throughput with varying number of readers.

In Fig. 3, even though the throughput of pulse is up to a saturation point of $n = 25$, it is still lower than our protocol. It shows the proposed protocol is not affected by the increasing number of readers, because it has more than one channel used to transmit. The number of channels will increase with an increase in the number of readers.

From Fig. 4, we can see that the system throughput of Colorwave is the lowest, because the timeslots randomly chosen to communicate are underutilized. We find the curve rising with an increase of the reader scale, which is due to the increase of utilization with more readers. CSMA suffers from the hidden terminal problem, so the system throughput is lower. As the number of readers increases, the throughput of it

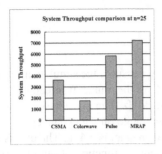
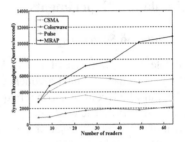

Fig. 3. Throughput comparing with the different protocols at n = 25

Fig. 4. Throughput comparison with varying number of readers

does not rise, so it is unable to fit for dense networks. In Pulse, the beacon signal is sent on the control channel when the data channel is being used, so the unnecessary collision can be avoided. It has the higher throughput compared with CSMA. Our Protocol has many data channels which can be used in transmitting signals to tags simultaneously. It avoids collisions by using the different timeslots between any relevant readers, so the system throughput is highest. With the increasing number of readers, the curve of our protocol keeps rising, the property above is more obvious. The proposed MRAP protocol is very fit to perform in dense reader environment.

(2) System Efficiency

The system efficiency comparing with the different protocols at $n = 25$ is shown in Fig. 5. Figure 6 shows the system efficiency with varying number of readers.

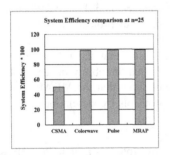
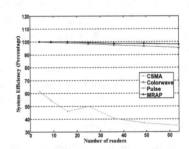

Fig. 5. Efficiency comparing with the different protocols at n = 25

Fig. 6. Efficiency comparison with varying number of readers

In Fig. 5, the number of readers is 25, which is a saturation point for Pulse, but the efficiency of our protocol is still the highest one. It is because that the number of readers is less than the saturation point in a single channel, by which the process of the proposed protocol is more simple and effective. The collisions can be reduced by using the different channels to the relevant readers.

From Fig. 6, we can see that the system efficiency of CSMA is the lowest. When the number of readers increases, it continues decreasing. It is because the increasing number of hidden terminals. In spite of low throughput, the efficiency of Colorwave is higher than CSMA. In Pulse, the beacon is used to remove the effect of the hidden terminal, so its efficiency is in the middle. The successful transmission rate of all protocols decreases with the increasing number of readers. The proposed protocol is highest because it has many data channels which can be used in transmitting signals to tags. The number of readers in a single channel is less than Pulse which takes all readers in one channel, so its efficiency is higher than that of Pulse. With the number of readers increased, the characteristic is more obvious.

5 Conclusion

We describe a novel reader anti-collision protocol MRAP based on pulse by using multiple channels, which is suitable for large scale RFID systems. We propose the principle to confirm the channel and communication timeslot for each reader, which can increase the utilization of channels. During the novel process of communication to avoid collisions, we solve the problem of how to use the different channels and timeslots between the relevant readers, in which the irrelevant readers can enjoy a same timeslot in the same channel, but the relevant readers do not have a same timeslot to communicate with tags, even if they belong to the different channels. It can reduce the reader collision effectively and increase the efficiency of data transmission. The simulation and performance analysis results show that our protocol can achieve better performance than the existing protocols in terms of increasing system throughput and system efficiency. Based on the comprehensive analysis and comparison, our protocol can be used in applications requiring dense RFID environments.

Acknowledgement. This work was supported in part by the National Natural Science Foundation of China (61571370, 61601365, and 61801388), in part by the Fundamental Research Funds for the Central Universities (3102017OQD091 and 3102017GX08003), in part by the China Postdoctoral Science Foundation (BX20180262), and in part by the Seed Foundation of Innovation and Creation for Graduate Students in NWPU (Grant no. ZZ2018129).

References

1. Birari, S.M., Iyer, S.: PULSE: a MAC protocol for RFID networks. In: Enokido, T., Yan, L., Xiao, B., Kim, D., Dai, Y., Yang, L.T. (eds.) EUC 2005. LNCS, vol. 3823, pp. 1036–1046. Springer, Heidelberg (2005). https://doi.org/10.1007/11596042_106
2. Birari, S., Iyer, S.: Mitigating the reader collision problem in RFID networks with mobile readers. In: 2005 13th IEEE International Conference on Networks. Jointly Held with the 2005 IEEE 7th Malaysia International Conference on Communication, vol. 1, p. 6 (2005)
3. Song, I.C., Fan, X., Chang, K.H.: Enhanced pulse protocol RFID reader anti-collision algorithm using slot occupied probability in dense reader environment. KSII Trans. Inf. Syst. 2(6), 299–311 (2008)

4. Shin, K., Song, W.: RAC-multi: reader anti-collision algorithm for multichannel mobile RFID networks. Sensors **10**(1), 84–96 (2010)
5. Meguerditchian, C., Safa, H., El-Hajj, W.: New reader anti-collision algorithm for dense RFID environments. In: 18th IEEE International Conference on Electronics, Circuits and Systems (ICECS), Beirut, Lebanon, pp. 85–88, December 2011
6. Kim, J.G., Lee, W.J., Yu, J.: Effect of localized optimal clustering for reader anti-collision in
7. Jensen, T.R., Toft, B.: Graph Coloring Problems. Wiley-Interscience Publication, Hoboken (1995)
8. ETSI EN 302 208-1 v1.1.1: CTAN: Number footnotes separately in superscripts. Place the actual footnote at the bottom of the column in which it was cited. Do not put footnotes in the reference list. Use letters for table footnotes, September 2004. http://www.etsi.org

UMa Large-Scale O2I Modeling and Base Station Number Optimization

Yutong Wang[✉], Daosen Zhai, Ruonan Zhang, Yi Jiang, and Qi Guo

Department of Communication Engineering, Northwestern Polytechnical University,
Xi'an 710072, Shaanxi, China
wangyutong@mail.nwpu.edu.cn

Abstract. The Large-scale fading model of wireless channels plays an important role in the design and analysis of green communication systems and networks. In this paper, we conduct the propagation measurement in the urban macrocell (UMa) outdoor-to-indoor (O2I) scenario at 39 GHz. Based on the field measurement data, we modify the path loss model specified by the 3GPP TR38.901 standard. Furthermore, using the proposed channel model, we analyze the optimal number of base stations (BSs) in a given area to minimize the total power consumption of all the BSs, including both the statistic power and transmitting power. The proposed channel model and optimization solution can be utilized to design green communication systems especially for the millimeter wave wireless networks.

Keywords: outdoor-to-indoor (O2I) · urban macro (UMa) ·
Green communication

1 Introduction

The outdoor-to-indoor (O2I) coverage is an important application scenario of the mobile communication systems. With the dense deployment of buildings, the O2I channels have become more and more complicated. As a consequence, the previous channel models cannot well depict the propagation properties of the O2I channels especially for the millimeter wave (mmWave) bands. Motivated by the need to improve the channel models, researchers have conducted extensive measurement to investigate and model the O2I channels in different scenarios and frequency bands. For instances, the authors in [1] analyzed the propagation characteristics of the O2I channels at 0.85 and 1.9 GHz, where the transmitter was deployed on the rooftop of several multistory buildings in a university campus. According to the measurement results, it was found that the received

This work was supported in part by the National Natural Science Foundation of China (61571370, 61601365, and 61801388), in part by the Fundamental Research Funds for the Central Universities (3102017OQD091 and 3102017GX08003), and in part by the China Postdoctoral Science Foundation (BX20180262).

B. Li et al. (Eds.): IoTaaS 2018, LNICST 271, pp. 393–400, 2019.
https://doi.org/10.1007/978-3-030-14657-3_40

power at 1.9 GHz was 10 dB smaller that at 0.85 GHz. The authors in [2] studied the wireless channels at 3, 10, 17, and 60 GHz in an O2I scenario. The results revealed that the signal attenuation fluctuated significantly, depending on the materials of the windows. In [3], the 2.6 GHz O2I channel was characterized for the tunnel and open field environments in subways. The results indicated that the attenuation values were 15 to 20 dB higher than those in free space, due to the non-line-of-sight (NLOS) propagation. The penetration capabilities of the linear polarization (LP) and circular polarization (CP) were compared in [4]. The analysis found that the CP had more parallel polarization components and hence it had a stronger penetration capability with respect to the LP in most O2I scenarios.

The fifth generation (5G) mobile communication networks is undergoing standardization. The 5G is expected to meet the user requirement for ultra-high data rates, but at the expense of increased cost [5]. The network will include a plenty of base stations (BSs), mobile devices, and antennas [6]. Therefore, it is necessary to properly deploy BSs to reduce energy consumption. A fairness-aware multiple drone base stations (DBS) deployment algorithm was proposed in [7], which could maximize the proportional fair sum-rate by using the particle swarm optimization (PSO). The results were provided for the downlink coverage probability of a wireless network with predetermined BS locations in [8]. The authors proposed a chance-constrained stochastic formulation for the optimal network deployment.

This paper focuses on the large-scale O2I channel analysis and modeling at 39 GHz in the mmWave band. Specifically, we have performed a channel measurement campaign in a typical urban macrocell (UMa) scenario, and then modified the model parameters in the 3GPP TR38.901 standard to improve the accuracy based on our measurement data. Furthermore, in order to reduce the energy consumption of the mmWave network, we analyze the optimal number of BSs in a given area based on the modified large-scale channel fading model. Simulation results indicate that the total power consumption of all BSs can be minimized through optimizing the BS density.

The rest of the paper is organized as follows. Section 2 introduces the channel measurement scenario and system. Section 3 presents the measurement results and the proposed path loss model. In Sect. 4, we analyze the optimal number of BSs for the network deployment. Finally, Sect. 5 concludes the paper and points out the future research issues.

2   Channel Measurement Scenario and System

2.1   Measurement Scenario

The channel measurement campaign was carried out at the Minhang campus of Shanghai Jiao Tong University, which was a typical UMa O2I scene. As shown in Fig. 1, the transmitter (Tx) was installed on a five-story building with a height of about 24 m, and the antenna height was about 25 m above the ground. The receiver (Rx) was located on the second and third floors in an adjacent building.

The floorplan of the measured positions is shown in Fig. 1(c). The height of the receiving antenna was 1.5 m high a total of 61 positions on the second floor and 58 positions on the third floor were measured.

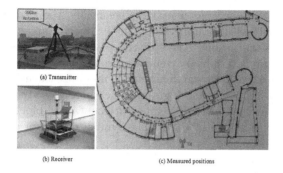

(a) Transmitter

(b) Receiver   (c) Measured positions

Fig. 1. Transceiver and measurement scenarios.

2.2   Measurement System

The block diagram of the large-scale channel sounding system is shown in Fig. 2. The Tx consists of a signal source, a high-frequency power amplifier (PA), a transmitting antenna, and connection cables. The transmitting antenna is sectorial with a half power beam width (HPBW) of 60°. The Tx is powered by the municipal electricity due to the huge power consumption of the amplifier. The Rx is composed of a UPS power supply device, a handheld GPS device, a spectrum analyzer, a receiving antenna, a laptop computer, and connection cables. An omnidirectional cylindrical antenna is used in the Rx.

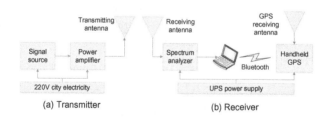

Fig. 2. Block diagram of the large-scale channel measurement system.

The transmitting power of the signal source is 10 dBm, and the gain of the high-frequency power amplifier is 63 dB. The gain of the Tx and Rx antennas are 8.5 and 2.5 dBi, respectively. The spectrum analyzer is connected to the laptop through a Category 5 network cable. The handheld GPS device records the positions of the Rx, and sends the information to the laptop through Bluetooth to calculate the three-dimensional Tx-Rx distance in the data processing.

2.3   Measurement System Calibration

In order to obtain accurate path loss in the measurement, the entire measurement system must be calibrated, including connection cables, antennas, power amplifiers, and so on. The patterns of the Tx and Rx antennas have been measured in a microwave anechoic chamber. Specifically, the system is calibrated by the back-to-back calibration method. The Tx and Rx are connected directly with cables. Since the output power of the power amplifier is very large, an attenuator with a predetermined attenuation coefficient is inserted between the power amplifier and the spectrum analyzer. To minimize the measurement error, the same transmission power and high-frequency phase-steady cables in the calibration process are used in the field measurement. The block diagram of the calibration process is shown in Fig. 3. The symbol definition is summarized in Table 1.

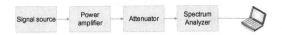

Fig. 3. Calibration operating system block diagram.

Table 1. The symbol definitions.

Symbols	Meanings
P_{TX} (dBm)	The transmitting power of the signal source
G_{PA} (dBm)	The gain of the power amplifier
G_A (dB) > 0	The attenuation coefficient of the attenuator
G_{TX}	The gain of the Tx antenna
G_{RX}	The gain of the Rx antenna
P_{RX1}	The received signal power of the spectrum analyzer in calibration
P_{RX2}	The received signal power of the spectrum analyzer in field measurement
PL	The path loss of the wireless channel
L	The other loss of the measurement system

According to the calibration diagram, we can obtain.

$$P_{TX} + G_{PA} - G_A - L = P_{RX1}. \tag{1}$$

$$P_{TX} + G_{PA} + G_{TX} - PL + G_{RX} - L = P_{RX2}. \tag{2}$$

Substituting (1) into (2) yields

$$PL = P_{RX1} - P_{RX2} + G_A + G_{TX} + G_{RX}. \tag{3}$$

In order to verify the channel sounding system and the calibration scheme, we conducted a field measurement experiment. The experiment was performed in an open space to emulate the free-space scenario. The, Tx-Rx distance was set as 5, 10, and 15 meters for air interface testing. The path loss was obtained by using (3). The experimental results indicate that the errors between the measurement results and the free-space model were between 0.6–1.1 dB, within the allowable range.

3 Measurement Results and Modeling

Path loss is one of the large-scale fading parameters of wireless channels. This paper models the path loss at 39 GHz in the UMa scenario based on the field measurement data and the 3GPP TR 38.901 standard. The path loss model is expressed as

$$PL = PL_b + PL_{tw} + PL_{in} + N(0, \sigma_p^2), \tag{4}$$

$$PL_b = 28 + 23.46 \cdot \log_{10}(d_{3D}) + 21.23 \cdot \log_{10}(f_c), \tag{5}$$

$$PL_{in} = 0.5 d_{2D\_in}, \tag{6}$$

where PL_b is the basic outdoor path loss given in (5), PL_{in} is the indoor loss depending on the depth into the building and given in (6), and d_{2D_in} is 2D distance between Tx and Rx. In (4), $\sigma_p=4.4$ is the standard deviation of the penetration loss. PL_{tw} is the building penetration loss through the external wall given by

$$PL_{tw} = PL_{npi} - 10\log_{10} \sum_{i=1}^{N} P_i \times 10^{\frac{-L_{material\_i}}{10}}$$

$$= 5 - 10\log_{10}(0.9 \cdot 10^{\frac{-L_{concrete}}{10}} + 0.05 \cdot 10^{\frac{-L_{glass}}{10}} + 0.05 \cdot 10^{\frac{-L_{wood}}{10}}), \tag{7}$$

$$L_{glass} = 2 + 0.2 f_c, \tag{8}$$

$$L_{concrete} = 5 + 4 f_c, \tag{9}$$

$$L_{wood} = 4.85 + 0.12 f_c, \tag{10}$$

where PL_{npi} is an additional loss added to the external wall loss to account for non-perpendicular incidence, P_i represents the proportion of the i-th material, where $\sum_{i=1}^{N} P_i = 1$, and N is the number of materials.

The measurement data and our model are depicted in Fig. 4, where (a) represents the path loss in the second floor and (b) represents the path loss in the third floor. As can be seen, our model can well match the measurement data. The positions of 9, 10, 46, and 47 shown in Fig. 1 have line-of-sight (LOS) rays, and thus their path loss is relative small. For these positions, we do not add the penetration loss into the path loss model.

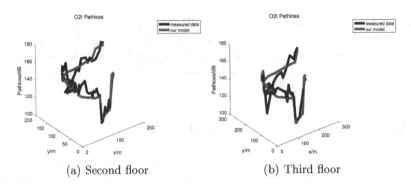

(a) Second floor   (b) Third floor

Fig. 4. Measured data and our model

4   Base Station Deployment OPTIMIZATION

The power consumption of a base station (BS) consists of two parts, i.e., the transmitting power and the statistic power. Given a network coverage area, with the increment of the number of BSs, the radius of the coverage area of a BS decreases. As a result, the transmitting power may be reduced to satisfy the minimum received power requirement P_r. On the contrary, the smaller the number of BSs is the larger the transmitting power is consumed, but the smaller statistic power is required. As such, there is a tradeoff between the transmitting and statistic power by changing the number of BSs. In other words, there is an optimal three-dimensional distance between the cell-edge user and the BSs, which can minimize the total power consumption of all BSs. Suppose that the geographic area is defined as S. The height of the BS is h. We denote P^c and P^t as the statistic and transmitting power, respectively. Specifically, the BS coverage is shown in Fig. 5.

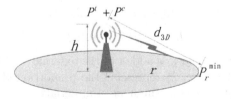

Fig. 5. Base station coverage.

The mathematical analysis for the optimal d_{3D} is given as follows. The BS height h and coverage radius r are known, and thus the three-dimensional distance from the user to the BS d_{3D} is obtained as

$$d_{3D}^2 = h^2 + r^2. \tag{11}$$

The total area is divided by the coverage area of the BSs and the number of BSs, denoted by N_b, is obtained as

$$N_b(d_{3D}) \approx \frac{S}{\pi r^2} = \frac{S}{\pi(d_{3D}^2 - h^2)}. \tag{12}$$

The summation of the minimum received power and the path loss gives the dynamic transmitting power of the i-th BS, denoted by P_i^t, as

$$P_i^t(d_{3D}) = P_r^{\min} + PL(d_{3D}). \tag{13}$$

The total power is the summation of the static and transmitting power of all the N_b BSs and can be obtained as

$$P^{tot}(d_{3D}) = \sum_{i=1}^{N_b(d_{3D})} (P_i^c + P_i^t(d_{3D}))$$

$$= \sum_{i=1}^{\frac{S}{\pi(d_{3D}^2 - h^2)}} (P_i^c + P_r^{\min} + PL(d_{3D})), \tag{14}$$

where P^{tot} is the total power, and P_i^c is the BS static power.

The total power minimization problem can be formulated as

$$\min_{d_{3D}} P^{tot}(d_{3D})$$

$$s.t. \quad d_{3D} > h \tag{15}$$

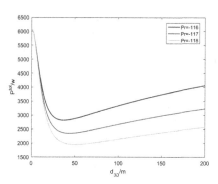

Fig. 6. The effect of the 3D-distance on the total power consumption.

In this work, we solve the problem via numerical results. Specifically, we set $h = 12.5\,\mathrm{m}$, $P^c = 30$ watt, and $S = 10^5\,\mathrm{m}^2$. The relationship between the total power of the BSs and the three-dimensional distance from the edge users to a BS is shown in Fig. 6. The vertical distance between the Tx and the Rx is set by the path loss model. From Fig. 6, we can find that with under different minimum received power (MRP) requirements, the optimal d_{3D} and the optimal number of BSs vary. For example, when MRP is -116 dBm, the optimal d_{3D} is 37 m, and the corresponding number of BSs is 26. When MRP is -118 dBm, the optimal d_{3D} is 53 m, and the corresponding number of BSs is 12.

5   Conclusions

In this paper, we have measured the 39 GHz large-scale fading in the typical UMa O2I scenario. Based on the measurement data, we have modified the parameters in the 3GPP TR38.901 standard to improve the channel model for the O2I mmWave channels. Moreover, by using the proposed channel model, we have evaluated the optimal number of BSs for a given coverage area to minimize the total power consumption. Our study provides guidance for the green networking of the 5G system, especially for the mmWave O2I coverage. In the future works, we will perform more field measurement on the mmWave channels such as in the urban microcell (UMi) and indoor scenarios, to provide more measurement data to refine the channel models for network deployment.

References

1. Calderon Jimenez, M.J., Arana, K., Arias, M.R.: Outdoor-to-indoor propagation mechanisms in multistorey building for 0.85 GHz and 1.9 GHz bands. In: IEEE 37th Central America and Panama Convention (CONCAPAN XXXVII), Managua, pp. 1–6 (2017)
2. Diakhate, C.A.L., Conrat, J.M., Cousin, J.C., Sibille, A.: Millimeter-wave outdoor-to-indoor channel measurements at 3, 10, 17 and 60 GHz. In: 2017 11th European Conference on Antennas and Propagation (EUCAP), Paris, pp. 1798–1802 (2017)
3. Arriola, A., Briso, C., Moreno, J., Echeverria, E.: Characterization of an outdoor-to-indoor wireless link in metro environments at 2.6 GHz. In: 2017 15th International Conference on ITS Telecommunications (ITST), Warsaw, pp. 1–4 (2017)
4. Zhong, Z., Liao, X.: Circular polarization benefits in outdoor to indoor scenarios for MIMO cellular networks. In: 2014 6th International Conference on Wireless Communications and Signal Processing (WCSP), Hefei, pp. 1–5 (2014)
5. Gandotra, P., Jha, R.K., Jain, S.: Green communication in next generation cellular networks: a survey. IEEE Access **5**, 11727–11758 (2017). https://doi.org/10.1109/ACCESS.2017.2711784
6. Andrews, J.G., et al.: What will 5G Be? IEEE J. Sel. Areas Commun. **32**(6), 1065–1082 (2014)
7. Akarsu, A., Girici, T.: Fairness aware multiple drone base station deployment. IET Commun. **12**(4), 425–431 (2018)
8. Chatterjee, S., Abdel-Rahman, M.J., MacKenzie, A.B.: Optimal base station deployment with downlink rate coverage probability constraint. IEEE Wirel. Commun. Lett. **7**(3), 340–343 (2018)

Direction Based Charging
in Rechargeable Wireless Sensor Network

Sadia Batool[1], Fei Tong[1,2(✉)], Songyuan Li[1], and Shibo He[1]

[1] Zhejiang University, Hangzhou, China
s.batool39@yahoo.com, {ftong,s18he}@zju.edu.cn, songyuanli.zju@gmail.com
[2] Key Laboratory of Computer Network and Information Integration,
Southeast University, Ministry of Education, Nanjing, China

Abstract. One of the main issues in rechargeable wireless sensor networks (RWSNs) is the stainability of network operation. Recently, wireless power transmission technology has been applied in RWSNs to transmit wireless power from chargers to sensor nodes. In this paper, we focus on how a charger changes its direction to cover different nodes for energy provisioning and, how to sense maximum data from the area of network. We intend to tackle the challenge that how to cover different nodes from each direction so that the charger can provide maximum power to maintain network performance. In this regard, a direction scheduling algorithm is proposed to fill the charging demand. Furthermore, a proposed Linear Programming based solution is proposed to determine maximum data sensing rate. The simulation results show that the proposed method performs better compared with existing methods.

Keywords: Rechargeable wireless sensor networks ·
Energy provisioning · Wireless power transmission ·
Direction scheduling

1 Introduction

Wireless sensor networks are widely deployed in remote or dangerous areas to sense ambient environment and generate data to reflect the real-world scenarios [1]. In these applications, prolonging the network lifetime and throughput is critical for achieving acceptable quality of services. Recently, researchers consider energy harvesting from different sources [2,3] such as solar power [4], thermal power [5], radio frequency (RF), and wind [6]. Especially, RF power transfer becomes the most popular one because of its reliability, safety and efficiency.

To transmit charging power in RWSN, we use directional charger. A directional charger changes its orientation at different time slot to different nodes to transmit energy according to their energy demands. Different directional chargers are deployed at different positions In RWSN. After fulfilling energy demand, sensor nodes start their sensing tasks. The objective of this study is to transmit

B. Li et al. (Eds.): IoTaaS 2018, LNICST 271, pp. 401–408, 2019.
https://doi.org/10.1007/978-3-030-14657-3_41

maximum power and sense maximum data from different sensor nodes. How to improve the throughput performance in RWSNs is a significant topic.

In most of the existing works, the covering area of a charger is assumed to be omnidirectional, which is not practical for many RF wireless charging platforms, such as WISP and Powercast. The key idea of this study is to maximize an RWSN's throughput by scheduling the directions of chargers optimally. Then, a charger directional scheduling algorithm is help full to provide near-optimal coverage of sensor nodes in each time slot. The Existing algorithm is also helpfull to obtain optimal sensing rate of each sensor node. Extensive simulations are conducted to verify the effectiveness of our design network.

Wireless charger deployment problem has attracted plenty of attention in recent years. In [7], it demonstrated how to minimize charging cost by reducing energy consumption rate and improving recharging efficiency. In [8], it verified the effectiveness of Friis' Free space equation in wireless charging and first investigated the minimum static charger deployment problem. In [9], the authors optimized the basic station deployment in order to improve data collection rate. In [10], the authors divided the sensing areas into grids and placed omnidirectional wireless charger on various grid points. All these studies did not consider a directional charging method to transmit energy.

The main contributions of this paper can be summarized as follows:

- We notice that chargers provide directional charging service in practical systems, which lead to a new problem of optimally scheduling chargers direction, so that the throughput of an RWSN can be maximized.
- We design a near-optimal charger direction scheduling algorithm, and further propose a solution based on linear programming to maximize the network throughput.
- Extensive simulations are conducted to show that our design is effective and robust under different network settings.

The rest of the paper is organized as follows. Section 2 describes the Problem Statement. Section 3 shows the proposed Charging Orientation Scheduling scheme for an RWSN. Simulation results of the proposed scheme are shown in Sect. 4, which are then compared with those of the other schemes in Sect. 5. Finally, Sect. 6 concludes the paper.

2   Problem Statement

In practical hardware settings, antennas of chargers can be directional, which leads to a new design challenge of scheduling orientation of chargers optimally, so that the network throughput is maximized. In this work, we consider an optimization problem as follows. Given a network area, where a set of $N = \{1...n\}$ rechargeable sensor nodes, a set of $P = \{1...p\}$ sink stations and a set of $M = \{1...m\}$ chargers are randomly deployed. Each node is required to sense data and forward it to the sink station. Chargers are deployed to transmit wireless power to sensor nodes. A Charger is directional and the covering area is a

Table 1. Notation definition

Symbol	Definition
r_{it}	Sensing rate of node i in time t
N	Total nodes which are deployed in the area of network
M	Chargers which are deployed in the area of network
$f_{it}^{(in)}$	Data in-flow in node i
$f_{it}^{(out)}$	Data out-flow of node i
$p_o(i)$	Parameter energy consumption for node to sense one bit of data
$p_r(i)$	Parameter energy consumption for node to receive one bit of data
$p_t(i)$	Parameter energy consumption for node to transmit one bit of data
$p_c(ij)$	Charging power that comes from charger j to node i
X_{ij}	Node i coverage area where charger j provide energy at each direction of the node
C_{ij}	Relationship between direction of charger and state of the node
Y_j	Direction of the node where charger j provide energy to the node i
$P_c(i)$	Direction of the node where charger j provide energy to the node i

90 degree sector to transmit charging power to the sensor node and the charger can change its orientation to provide power from different directions. We need to schedule each charger's orientation optimally, so that the network throughput is maximized. To simplify this problem, we divide the time series into slots $T = \{1...t\}$ and suppose that in each slot, the charger can only select one of the four sectors to cover the nodes in the area of RWSNs. In our work we apply a tree topology with single-path routing, where each node has only one link to its next hop, and routing table is fixed. Thus, we have the following model:

$$max \quad \Sigma_{t=1}^{T}\Sigma_{i=1}^{N}r_{it} \tag{1}$$

$$s.t. \quad r_i + f_{it}^{(in)} = f_{it}^{(out)} \tag{2}$$

$$p_o(i)r_i + p_r(i)f_{it}^{(in)} + p_t(i)f_{it}^{(out)} \geq P_c(i) \tag{3}$$

$$P_c(i) = \Sigma_{i=1}^{M}X_{ij}p_c(ij) \tag{4}$$

$$X_{ij} = C_{ij} \times Y_j \tag{5}$$

The notations used in this paper are summarized in Table 1. Equation (1) shows the sensing rate of node i. Equation (2) shows the data conservation constraint, i.e., for each sensor node, the aggregated incoming data flow is equal to the aggregated outgoing data flow. Equation (3) calculates the available power of every sensor node. Equation (4) is the charging power, showing that the consumption power should never exceed the total charging power. Equation (5) shows the sensor node coverage area, where a charger provides energy at each direction (Figs. 1 and 2).

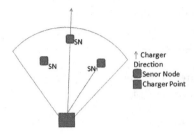

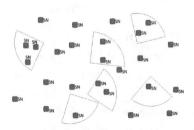

Fig. 1. Coverage area of directional based charger, which changes its orientation at different time slot to cover different nodes.

Fig. 2. Rechargeable wireless sensor network with different sensor nodes and directional chargers.

2.1 Energy Charging Model

We first specify our energy charging model based on the Frii's free space equation:

$$p_r = \frac{GsGr\eta}{Lp}\left(\frac{\lambda}{4\pi(d+b)^2}\right)p_o, \tag{6}$$

where d is the Euclidean distance between sensor node and charger, p_o is the source power, G_s is the source antenna gain, G_r is the receive antenna gain, Lp is polarization loss, η is a parameter to adjust the model for short distance transmission, and b is the parameter to set the rectifier efficiency. For the ease of the presentation, we simplify the energy charging model as:

$$p_r = \frac{a}{(d+b)^2}. \tag{7}$$

This has been proved experimentally to be a valid approximation of energy charging in [8]. We use the powercast wireless charging sensor nodes fabricated in our lab and the $XT91501$ powercast transmission. We fit experimental parameters for energy charging model. Based on the experimental results, we choose $a = 7.593$ and $b = 0.3154$ in our simulation.

2.2 Energy Consumption Model

The energy consumption model for wireless communication in our work is shown as follows:

$$p_t(k, d) = (p_r + \epsilon d^2) \times k, \quad p_r(k) = p_r \times k. \tag{8}$$

The model describes the communication energy, where d is the Euclidean distance between transmitter and receiver, k is the number of transmitted data bits, and p_t, p_r, and ϵ are the constant parameters concerned with the communication environment. Equation (8) shows the power consumption for data transmission and the power consumption of data reception. To obtain feasible values of these parameters, we perform communication experiments with Zigbee. Based on the results, we choose $p_t = p_r = 558\ nJ/bit$, and $\epsilon = 44.66\ pJ/bit/m^2$.

3 Solution of the Problem

To tackle the problem, we propose the following solution. The only constraint comes from the limited energy that in each time slot, the total energy used by sensing and communication should never exceed the sum of remaining energy and charged energy. The consumed energy should be expressed as follows:

$$\phi_i^{(t)} = p_o r_i + p_r f_i^{(t)} + (p_t + \varepsilon d_i^2) g_i^{(t)}, \tag{9}$$

where $f_i^{(t)}$ is the income flow of node i in timeslot t and $g_i^{(t)}$ is outcome flow. The charged energy of node i in timeslot t is as below:

$$\Psi_i^{(t)} = \Sigma_{j=1}^M \frac{a}{d_{ij} + b^2} x_{ij}^{(t)}, \tag{10}$$

where $x_{ij}^{(t)}$ is index denoting whether node i can be charged by charger j in time slot t. For charger j, it has four sectors to select in each timeslot. Thus, we use a vector $y_j^{(t)}$, to denote its directions. For instance, if charger j selects the second sector to cover, then we have $y_j^{(t)} = [0, 1, 1, 1]$, we use coefficient vector a_{ij}, to denote whether node i can be charged by charger j, if charger j selects current sector to cover. For instance, if node i can be charged by charger j when it turns to the first sector, then we have $a_{ij} = [1, 0, 0, 0]$. Therefore, we can build up the relationship between $x_{ij}^{(t)}$ and $y_i^{(t)}$, as follows:

$$x_{ij}^{(t)} = a_{ij} y_j^{(t)}. \tag{11}$$

Then the energy constraint can be expressed as:

$$e_i^t = e_i^{(t-1)} - \Phi_i^{(t)} + \Psi_i^{(t)}, \ 0 \le e_i^t \le B \tag{12}$$

Our optimization objective is to maximize the network throughput. e_i^t is the total energy that node i consumes in time t, $\Phi_i^{(t)}$ is the consumed energy in node i at time t, and $\Psi_i^{(t)}$ is the energy consumed to charge node i at time t. Thus, the optimization problem is as follows:

$$max \ \sum_{t=1}^T \sum_{i=1}^N r_{it} \tag{13}$$

We set score for each node in each iteration to evaluate its demand of being charged. Intuitively, those nodes with low energy should have the charging priority. On the other hand, we adopt a tree based routing protocol in our design. Then node in the upper layer should obtain more energy, since they will take on more relay task.

Therefore, we design such scoring mechanism based on $\frac{g}{e+p+\varepsilon}$, where g is related to the number of node layer (a node close to sink is in a top layer), e denotes the remaining energy after the previous time slot, p is the charging power that node has received in current time slot, and ε is a factor to avoid the denominator being zero. Thus, in each time slot, we run the following mechanism to obtain chargers' orientation scheduling. In each iteration, we update each node's score, then we can calculate each sector's total score, and we select the sector with highest score greedily. This process continues until every charger has decided its covering sector. After the charger's orientation has been decided, it is easy to obtain each node energy level in current time slot. After that we adopt the tree based routing protocol to transmit data. The network throughput is recorded and the nodes energy is updated from time to time.

4   Simulation Results

According to the system model, we conduct simulation using MATLAB to evaluate the performance of the proposed method.

4.1   Simulation Setup

We consider a $100 \times 100 \; m^2$ network area, where 50 sensor nodes are randomly deployed and a sink station is used for data collection. Operation time is divided into different slots to sense data from the network area. One bit of data is sensed in each time slot. We set the data sensing power p_o as 0.00024 mW/bit, data transmission power p_t as 0.00024 mW/bit, and data receiving power p_r as 0.000558 mW/bit. The simulation results have been obtained by running extensive simulations. After providing energy to each sector, the maximum data collection rate of different nodes in different time can be tested.

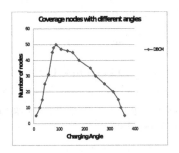

Fig. 3. Covered nodes with different angles.

4.2   Experimental Results

We demonstrate the results using different number of chargers which provide energy to different nodes at different time slot with 90 degree angle. Data sensing rate of different node can increase on the base of charging power. In Fig. 3, different number of nodes are covered by a charger with an orientation angle ranging from 0 to 360 degree. It is observed that maximum nodes coverage is achieved at 90 degree.

5   Performance Comparison

In this section, we compare our results with other methods. The performance evaluation of our proposed Directional Based Charging Method (DBCM) and existing Moveable Charging Based Method (MCBM) [11] is presented in Fig. 4. Both methods cover more than 60 present of the network nodes. Our main purpose is to satisfy charging demand of different nodes in RWSNs. We can see that in DBCM, a charger covers more nodes with a charging angle of 90 degree. Then the charger number is fixed to one, and its charging angle varies from 0 to 360 degree. We compare DBCM with two existing methods, i.e., Power Balance Aware Deployment (PBAD) [12] and Random Position Random Orientation (RPRO) [12]. The results are shown in Fig. 5. As shown in the figure, when the charging angle is around 90 degree, all three methods can reach the largest coverage. Still, DBCM performs better than the other two methods.

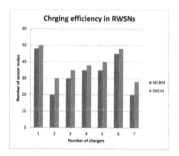

Fig. 4. Coverage *vs.* number of chargers (with charging angle of 90 degree).

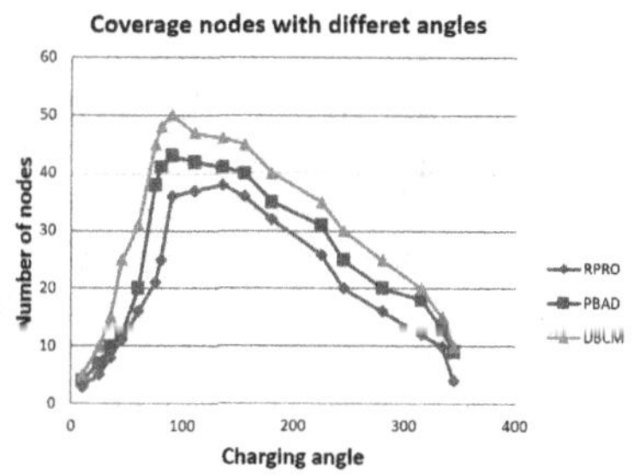

Fig. 5. Coverage *vs.* charging angle (only one charger).

6   Conclusion

In this paper, we consider an RWSN's throughput maximization problem by scheduling the directions of different chargers optimally. We design a charger direction scheduling algorithm to provide near-optimal coverage of sensor nodes in each time slot. It is helpful for obtaining optimal sensing rate of each sensor nodes from routing table. Extensive simulations are conducted to verify the effectiveness of our design.

Acknowledgment. This work was supported in part by the National Natural Science Foundation of China under grant 61702452, in part by the China Post-Doctoral Science Foundation under Grant 2018M630675, and in part by the Ministry of Educations Key Lab for Computer Network and Information Integration, Southeast University, China.

References

1. Hodge, V.J., et al.: Wireless sensor networks for condition monitoring in the railway industry: a survey. IEEE Trans. Intell. Transp. Syst. **16**(3), 1088–1106 (2015)
2. Wan, Z.G., Tan, Y.K., Yuen, C.: Review on energy harvesting and energy management for sustainable wireless sensor networks. In: 2011 IEEE 13th International Conference on Communication Technology (ICCT). IEEE (2011)
3. Kim, S., et al.: Ambient RF energy-harvesting technologies for self-sustainable standalone wireless sensor platforms. Proc. IEEE **102**(11), 1649–1666 (2014)
4. Hsu, J., Kansal, A., Srivastava, M.: Energy Harvesting Support for Sensor Networking (2004)
5. Raghunathan, V., et al.: Design considerations for solar energy harvesting wireless embedded systems. In: 2005 Fourth International Symposium on Information Processing in Sensor Networks, IPSN 2005. IEEE (2005)
6. Park, C., Chou, P.H.: Ambimax: autonomous energy harvesting platform for multi-supply wireless sensor nodes. In: 2006 3rd Annual IEEE Communications Society on Sensor and Ad Hoc Communications and Networks, SECON 2006, vol. 1. IEEE (2006)
7. Tong, B., et al.: How wireless power charging technology affects sensor network deployment and routing. In: 2010 IEEE 30th International Conference on Distributed Computing Systems (ICDCS). IEEE (2010)
8. He, S., et al.: Energy provisioning in wireless rechargeable sensor networks. IEEE Trans. Mobile Comput. **12**(10), 1931–1942 (2013)
9. Bogdanov, A., Maneva, E., Riesenfeld, S.: Power-aware base station positioning for sensor networks. In: INFOCOM 2004, Twenty-Third Annual Joint Conference of the IEEE Computer and Communications Societies, vol. 1. IEEE (2004)
10. Chiu, T.C., et al.: Mobility-aware charger deployment for wireless rechargeable sensor networks. In: 2012 14th Asia-Pacific Network Operations and Management Symposium (APNOMS). IEEE (2012)
11. Jian, W.-J., et al.: Movable-charger-based planning scheme in wireless rechargeable sensor networks. In: 2015 IEEE Conference on Computer Communications Workshops (INFOCOM WKSHPS). IEEE (2015)
12. Li, S.-L., et al.: A power balance aware wireless charger deployment method for complete coverage in wireless rechargeable wireless sensor network. Sensors (Basel, Switzerland) **18**(6), 1–13 (2016). 06/2018

Energy-Efficiency Random Network Coding Scheduling Based on Power Control in IoT Networks

Bin Li[✉], Hong Jiang, and Chao Chen

Department of Communication Engineering, Northwestern Polytechnical University, Xi'an 710072, China
libin@nwpu.edu.cn

Abstract. Random network coding (RNC) is an efficient coding scheme to improve the performance of wireless multicast networks, especially for the IoT network with multiple devices. Meanwhile, energy-efficient transmission is also an insistent demand in IoT network. Therefore, in this paper, we considered the heterogenous wireless channels of the devices caused by the transmitting distances and analyzed the energy consumption of overall network by using adaptive random network coding (ARNC). Then, we proposed a new power control metric that both considered the energy consumption and the network throughput. Based on the new metric, we optimized the transmitting power of the BS and proposed an energy-efficient ARNC scheduling based on power control to improve quality of service. The simulation results also showed the effectiveness of the optimization and proposed methods compared with the traditional methods.

Keywords: IoT networks · ARNC · Power control · Energy-efficiency

1 Introduction

With the rapid development of the mobile communications and demand of the connection among the different kinds of the devices, the Internet of Things (IoT) becomes an important part of the information technology of the future. Technically, IoT is expected to enable people-thing and thing-thing interconnections by combining communication technologies and networks. Nowadays, IoT was included in the fifth Generation (5G) standard through the 3GPP access network, where the more reliable connection, the higher throughput, the lower energy consumption are required in IoT.

This work is partially supported by NSFC (Nos. 61601365, 61571370, 61801388), Key Research and Development Plan in Shaanxi Province (Nos. 2017ZDXM-GY-101), the Fundamental Research Funds for the Central Universities (3102017OQD091 and 3102017GX08003), and in part by the China Postdoctoral Science Foundation (BX20180262).

© ICST Institute for Computer Sciences, Social Informatics and Telecommunications Engineering 2019
Published by Springer Nature Switzerland AG 2019. All Rights Reserved
B. Li et al. (Eds.): IoTaaS 2018, LNICST 271, pp. 409–417, 2019.
https://doi.org/10.1007/978-3-030-14657-3_42

On the other hand, network coding (NC) has drawn significant attention to increase system throughput and reliable connection for the last few years since the pioneering work of Alswede et al. [1]. NC has shown the potential abilities to improve network efficiency for reducing the number of transmitted packets. As for multicast networks, random network coding (RNC) in [2] that is based on the concept of NC has attracted significant research interest. And the sender encodes original packets by combining them using random coefficients, So the receivers can decode the complete information only they get a full set of independent coded packets. It means the lower quality of service (QoS). Therefore, an adaptive random coding is proposed by our team in [3], where the users can decode out more data according to the dynamic coding structure in ARNC and the network throughput remarkably improved when the user number is large. This property is suitable in IoT networks to provide the high throughput and reliable connection.

In our previous work [4], we proposed a novel network coding (adaptive random network coding) with different feedback schemes. But we didn't take energy consumption into account. Consequently, in this paper, we considered a IoT networks in which the devices had different transmitting distances from the base station (BS). We first introduced the ARNC scheduling schemes and analyzed the corresponding energy consumption. Then we proposed the energy-efficient metric called energy-efficiency ratio (EER) that both considered the overall energy consumption and network throughput. Based on the EER, we optimized the transmitting power of the BS and proposed an energy-efficient ARNC scheduling based on power control to improve quality of service. Moreover, the impact of network parameters on the EER was discussed based on the simulation results.

For the related work, [5] investigated the feasibility of improving the energy efficient that was applied to battery-limited IoT networks. From [6], it considered the energy harvesting for mass deployment of IoT devices in heterogenous networks and developed an effective energy-harvesting-aware routing algorithm to improve energy efficiency. To further enhance energy efficiency performance of 5G IoT, in [7], the authors proposed one integrated system structure for better energy efficiency. In addition, there are lots of concepts and techniques dedicated to save energy, mainly focus on reducing transmission, since the energy used for encoding is incomparable smaller than energy for broadcasting. Therefore, the goal of [8] was to develop new coding scheme for data compression to save energy for IoT solution. However, Some papers about NC focus more on the higher throughput, coding latency and transmitting delay. The contribution in [9] was to develop diversity schemes to optimize the throughput of system with RNC. [10] mainly focused on the low latency application of RNC as well as data storage application that use large blocks of data. For [12], it analyzed the delay bounds for transmitting packets from a source node to the destination by introducing RNC, but it ignored the energy consumption.

This work is organized as follows: Sect. 2.1 presents the system model. Then the details of different ARNC transmission schemes are introduced in Sect. 2.3.

And the energy consumption is presented in Sect. 3. While in Sect. 4 shows simulation results and discussion. Finally, we present our conclusion in Sect. 5.

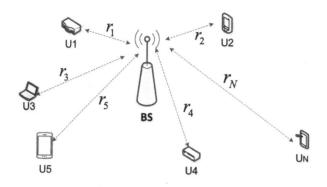

Fig. 1. System model. System with one BS and N users.

2   System Model and Setting

2.1   System Model

As shown in Fig. 1, we consider a single hop wireless network, such as Wi-Fi networks or micro cell network in 5G, which consists one transmitter (AP or micro base station (MBS)) and N devices. For simplicity, we assume the IoT network operates on a single frequency and each time slot only transmits one packet. The information at the transmitter are divided into data batches. Each data batch consists of M source packets (denoted as $\alpha_1, \alpha_2, ..., \alpha_M$) and sends to the N devices within a transmission deadline of T time slots (usually $T > M$). In our system, the devices always have the different transmitting distances, causing the heterogenous wireless channels between the transceivers, as shown in Fig. 1. According to [11], the received signal-to-noise ratio (SNR) γ_n at the device n can be written as:

$$\gamma_n = \frac{GE_{tx}|h_n|^2}{r_n{}^2 N_0} \qquad (1)$$

where N_0 is the white Gauss noise power, E_{tx} denotes the transmitting energy consumption per bit for BS. $G = (G_r G_t \lambda^2) / \left(M_l N_f (4\pi)^2 \right)$, where G_t and G_r indicate the transmitter and receiver antenna gains respectively. λ represents the carrier wavelength, N_f is the noise figure and M_l is the link margin. When the packets are modulated by QPSK, the average bit error rate (BER) for the n can be expressed as

$$p_n^b = \int_0^\infty Q\left(\sqrt{2\gamma_n}\right) f\left(|h_n|^2\right) d|h_n|^2 \qquad (2)$$

where $Q(x)$ is the Q-function, $|h_n|^2$ follows the exponential distribution, i.e.,$|h_n|^2 \sim E\left(1/\sigma_n^2\right)$. We set the packet length as f bits. Similar to [11], the PER of user n (ϵ_n) is defined as:

$$\epsilon_n = 1 - \left(1 - p_n^b\right)^f. \tag{3}$$

2.2   ARNC Encoding and Decoding

Adaptive random networking coding (ARNC) is the coding scheme that suitable for high throughput transmission with lower latency scenarios and proposed by our team in [4]. For easily understanding of paper, we introduce the ARNC briefly in this section below.

As for M prioritized packets, the BS creates M generations. At time slot t, generation G_m ($1 \leq m \leq M$) generates the coding packet $c_{mt} = \sum_{j=1}^{m} \alpha_{tj} p_j$, where α_{tj} is coding coefficient which is randomly chosen from the finite field $\mathbb{F}_q$. For example, when $M = 3$, the coding packets from the three generations are

* G_1: only contains p_1, i.e., $G_1 : c_{1t} = p_1$.
* G_2: combines p_1 and p_2, i.e., $G_2 : c_{2t} = \alpha_{t1} p_1 + \alpha_{t2} p_2$.
* G_3: combines p_1, p_2 and p_3, i.e., $G_3 : c_{3t} = \alpha_{t1} p_1 + \alpha_{t2} p_2 + \alpha_{t3} p_3$

Please note that at each time slot, the scheduler decides which generation to work and controls the type of RNC packets for multicasting. While for the decoding, at the receiver side, each device has a $M \times T$ decoding matrix $\mathbf{s}$, which includes the received ARNC coefficient vectors. For example, when $M = 3$ and $T = 5$, the decoding matrix of n may be shown as

$$s_n^t = \begin{pmatrix} 1 & \alpha_{12} & 0 & 0 \\ 0 & \alpha_{22} & 0 & 0 \\ 0 & 0 & 0 & 0 \end{pmatrix} \tag{4}$$

which indicates the device n successfully receives four ARNC packets (c_{11} and c_{22}) until the tth time slot. Apparently, although n doesn't collect a full set of the coded packets, it can still recover the partial transmitted packets (p_1 and $p2$). After that, if n receives one more coding packet from G_3, it can decode all the information in one block.

2.3   Transmitting Scheduling Strategy

To find an optimal scheduling strategy that can maximize the network average throughput, the BS needs to make the action that decides the transmitting power and the coding packet in each time slot. To be specific, the optimal action a_t at time slot t depends on the current network status $\mathbf{S}_t$ that is gotten from the device feedback information. When all the users receive coded packets successfully or the hard deadline, the BS will turn to the next information block. To fully understand the strategy, we firstly specify the network dynamics by ($\mathbf{S}_t$, $\mathbf{A}$, E_{tx}, r, T), in which boldface letters refer to vectors or matrices.

1. T is the time slots associated with deadline, and the time slot index is t $(0 \leq t \leq T - 1)$.
2. Network state $\mathbf{S}_t$: As for BS, $\mathbf{S}_t$ denotes all the devices status from the feedback information, which is defined by $\mathbf{S}_t = \boldsymbol{s}_1^t \cup \boldsymbol{s}_2^t \cup ... \cup \boldsymbol{s}_n^t \cup ... \cup \boldsymbol{s}_N^t$ where $\boldsymbol{s}_n^t$ shows the network status of n, according which the throughput of the n can be calculated. Therefore, the network throughput can be calculated depending on $\mathbf{S}_t$.
3. Action set $\mathbf{A}$: During the delivery process, the BS takes action a_t $(a_t \in \mathbf{A})$ to decide which coded packet will be sent at the next time slot according to $\mathbf{S}_t$. For instance, by taking the action $a_t = \{$transmit c_{2t} from G_2 at $t\}$, the BS transmits $c_{2t} = \alpha_{t1}p_1 + \alpha_{t2}p_2$ at t.
4. Transmitting energy consumption per packets E_{tx}: The BS broadcasts the coded packets with power E_{tx}, which is selected from $[E_{min}, E_{max}]$.
5. The immediate network throughput $r(\mathbf{S}_t, a_t, E_{tx})$. This denotes the network throughput associated with a_t according to $\mathbf{S}_t$. It can be written as:

$$r(\mathbf{S}_t, a_t, E_{tx}) = \mathbb{E}[r(\mathbf{S}_{t+1}|\mathbf{S}_t, a_t, E_{tx})] = \sum_{n=1}^{N} \mathbb{E}[r(\boldsymbol{s}_n^{t+1}|\boldsymbol{s}_n^t, a_t, E_{tx})] \quad (5)$$

where $\mathbb{E}[\cdot]$ means the expectation function of $\mathbf{S}_{t+1}$. $r(\boldsymbol{s}_n^{t+1}|\boldsymbol{s}_n^t, a_t, E_{tx})$ is the future network throughput of n when $\boldsymbol{s}_n^{t+1}$ is updated from $\boldsymbol{s}_n^t$ under P_{tx} and a_t. It is noticed that when the coded packet is correctly received by n, $\boldsymbol{s}_n^t$ will changed to $\boldsymbol{s}_n^{t+1}$. Otherwise, do nothing. Thus, we have

$$\mathbb{E}[r(\boldsymbol{s}_n^{t+1}|\boldsymbol{s}_n^t, a_t, E_{tx})] = (1 - \epsilon_n(E_{tx}))r(\boldsymbol{s}_n^{t+1}) \quad (6)$$

where $r(\boldsymbol{s}_n^{t+1})$ shows the throughput of n under $\boldsymbol{s}_n^{t+1}$ and can be calculated according to Sect. 2.2.

3   Energy Consumption Analysis with Perfect Feedback

In this scheduling, BS exploits the feedback information from the devices to indicate whether the previous transmitted packet has been received successfully. Based on that, the BS updates $\mathbf{S}_t$ and decides the optimal action a_t depending on S_t. Thus, the overall network throughput is shown as

$$\Gamma(E_{tx}, T) = \sum_{t=0}^{T-1} r(\mathbf{S}_t, a_t, E_{tx}) + r(\mathbf{S}_T) \quad (7)$$

Correspondingly, the total energy consumption for each data block is

$$E_{total}(E_{tx}, T) = E_c + TLE_{tx} + TNLE_{cr} + TNE_{feed}. \quad (8)$$

Here E_c shows the circuit energy consumption for ARNC, and the detail is given by [13]. E_{cr} is the receiving and decoding energy consumption per bit for the devices. E_{feed} represents feedback energy consumption per packet. To

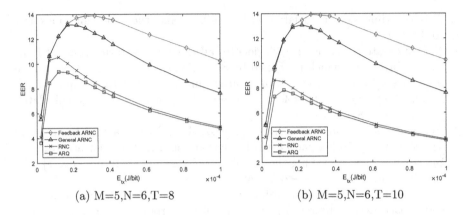

(a) M=5,N=6,T=8   (b) M=5,N=6,T=10

Fig. 2. Transmitting power versus EER.

fully evaluate the performance of the ARNC scheduling, we propose a new energy-efficient metric called energy-efficiency ratio (EER) that both consider the network throughput and total energy consumption. EER reflects the average throughput per time per device for unit energy consumption and is shown as:

$$\varepsilon(E_{tx}, T) = \frac{\Gamma(E_{tx}, T)}{NTE_{total}(E_{tx}, T)} \tag{9}$$

Our goal is to find an optimal transmitting power that maximizes the EER:

$$E_{tx}^* = \arg\max_{E_{tx} \in \mathbf{E}} \{\varepsilon\} \tag{10}$$

To solve the problem (10), we must first get the overall network throughput $\Gamma(E_{tx}, T)$. Here we adopt greedy scheduling technique (GST) during the scheduling decision with given $E_{tx}(E_{tx} \in \mathbf{E})$. In the GST, the BS finds the appropriate $a_t^\dagger \in \Omega$ (here Ω means the set of optimal action) in t to maximize $r(\mathbf{S}_t, a_t, E_{tx})$ until completion time slot or deadline, It is denoted as

$$\left\{a_t^\dagger | E_{tx}\right\} = \arg\max_{a_t \in \mathbf{A}} \left\{r(\mathbf{S}_t, a_t, E_{tx}) | E_{tx}\right\} \tag{11}$$

Accordingly, for given E_{tx}, we can get the overall network throughput $\Gamma_\Omega(E_{tx}, T)$ under the optimal transmission action in each time slot. That is

$$\Omega | E_{tx} = \left(a_0^\dagger, a_1^\dagger, \cdots, a_{T-1}^\dagger\right) | E_{tx} \tag{12}$$

Then, the Eq. (10) can be rewritten as

$$E_{tx}^* = \arg\max_{E_{tx} \in \mathbf{E}} \{\frac{\Gamma_\Omega(E_{tx}, T)}{NTE_{total}(E_{tx}, T)}\} \tag{13}$$

By traversing E_{tx} in the domain of definition, we finally get the optimal E_{tx}^*.

4   Simulation Results and Discussions

In this simulation, we assume each data block is divided into M packets and the BS needs to deliver these packets to N users within a deadline of $T > M$ time slots. Meanwhile, we assume all the fading channels between transceiver are independent and let $\sigma_n^2 = 1$. According to [11], the other system parameters are shown as follows: $1/\lambda = 2.4\,GHz$, $G_r G_t = 5\,dB$, $N_f = 10\,dB$, $M_l = 38\,dB$, $N_0/2 = -174 dBm$, $f = 1000\,bit$, $E_{feed} = 6 \times 10^{-5}\,(J/bit)$, $E_{cr} = 1 \times 10^{-5}\,(J/bit)$. $E_c = 4 \times 10^{-5}\,(J/bit)$.

Figure 2 shows how E_{tx} affects the network performance using our proposed mechanisms under our metric. We incorporate the traditional automatic repeat request(ARQ), RNC, and general ARNC proposed in [4] for comparison. Here we set $M = 5$, $N = 6$, and the distance of the devices in $r_1 = 280\,m$, $r_2 = 250\,m$, $r_3 = 220\,m$, $r_4 = 190\,m$, $r_5 = 160\,m$, $r_6 = 130\,m$. As we discussed earlier, there exists an optimal E_{tx} for the transmit scheduling. It is clear that our ARNC scheduling scheme offers the best performance gain among all schemes, and the optimal transmission energy of the ARNC achieves the greatest energy efficiency.

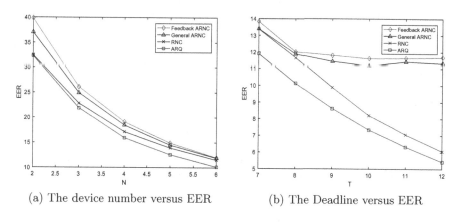

(a) The device number versus EER   (b) The Deadline versus EER

Fig. 3. Transmitting power versus EER.

Figure 3(a) depicts the number of devices versus the EER. Here we set $M = 6$, $T = 7$. The distance between BS and devices are 130 m, 160 m, 190 m, 220 m, 250 m, 280 m, respectively. From the Fig. 3(a), it is worth noticed that the performance curve declines as the number of users increasing in all cases. In order to satisfy the needs of more users, the average network throughput of system will inevitably decrease and consume more energy, simultaneously. However, our methods also achieve the better performance compared with other schemes.

In Fig. 3(b), the system parameters are set as follows: $M = 6$, $N = 6$, other parameters are the same to Fig. 3(a). It can be seen that when T increases, the EER of ARQ and RNC declines, correspondingly. This is because, when T increases, ARQ and RNC has more time to collect the coded packet to achieve more original packets. However, they also cost more energy consumption for transmission, leading to the lower EER. While for the general ARNC and our method, when the $T < 9$, the EERs decrease because we cost more time and energy consumption for decoding. However, when $T \geq 9$, the EER tend to stable because the general ARNC and our method already has enough time slots for decoding. Thus, the left time slots are the reluctant and the BS does not need to allocate the transmitting power in these time slots. Meanwhile, we can also find that our method has the best performance gain among the compared schemes, and this advantage is more obvious when T is large.

5 Conclusion

In this paper, we investigate the scenario that the BS multicasts prioritized data to the different transmitting distance users in the heterogenous wireless networks. Unlike [4], we consider the effects of total energy consumption and average network throughput by using ARNC. Then we propose a new power control to find an optimal transmitting power to improve the overall system performance by relying on the indicator of EER maximization. Finally, simulation results shows that the energy efficiency of proposed scheme is better than other traditional methods while meeting the target requirement of transmission.

References

1. Ahlswede, R., Cai, N., Li, S.-Y., Yeung, R.: Network information flow. IEEE Trans. Inf. Theory **46**(4), 1204–1216 (2000)
2. Ho, T., et al.: A random linear network coding approach to multicast. IEEE Trans. Inf. Theory **52**(10), 4413–4430 (2006)
3. Alshaheen, H., Rizk, H.T.: Improving the energy efficiency for biosensor nodes in the WBSN bottleneck zone based on a random linear network coding. In: 2017 11th International Symposium on Medical Information and Communication Technology (ISMICT), Lisbon, pp. 59–63 (2017)
4. Li, B., Li, H., Zhang, R.: Adaptive random network coding for multicasting hard-deadline-constrained prioritized data. IEEE Trans. Veh. Technol. **65**(10), 8739–8744 (2016)
5. Lee, B.M.: Improved energy efficiency of massive MIMO-OFDM in battery-limited IoT networks. IEEE Access **6**, 38147–38160 (2018)
6. Nguyen, T.D., Khan, J.Y., Ngo, D.T.: A distributed energy-harvesting-aware routing algorithm for heterogeneous IoT networks. IEEE Trans. Green Commun. Netw. **2**(4), 1115–1127 (2018)
7. Zhang, D., Zhou, Z., Mumtaz, S., Rodriguez, J., Sato, T.: One integrated energy efficiency proposal for 5G IoT communications. IEEE Internet Things J. **3**(6), 1346–1354 (2016)
8. Stojkoska, B.R., Nikolovski, Z.: Data compression for energy efficient IoT solutions. In: 2017 25th Telecommunication Forum (TELFOR), Belgrade, pp. 1–4 (2017)
9. Hu, G., Xu, K., Xu, Y.: Throughput optimization with random network coding in cooperative wireless network. In: 2017 First International Conference on Electronics Instrumentation and Information Systems (EIIS), Harbin, pp. 1–6 (2017)

10. Nielsen, L., Rydhof Hansen, R., Lucani, D.E.: Latency performance of encoding with random linear network coding. In: 2018 24th European Wireless Conference European Wireless, Catania, Italy, pp. 1–5 (2018)
11. Li, B., Li, H., Wang, W., Yin, Q., Liu, H.: Performance analysis and optimization for energy-efficient cooperative transmission in random wireless sensor network. IEEE Trans. Wirel. Commun. **12**(9), 4647–4657 (2013)
12. Cogill, R., Shrader, B.: Delay bounds for random linear coding in parallel relay networks. IEEE Trans. Mobile Comput. **14**(5), 964–974 (2015)
13. Angelopoulos, G., Médard, M., Chandrakasan, A.P.: Energy-aware hardware implementation of network coding. In: Casares-Giner, V., Manzoni, P., Pont, A. (eds.) NETWORKING 2011. LNCS, vol. 6827, pp. 137–144. Springer, Heidelberg (2011). https://doi.org/10.1007/978-3-642-23041-7_14

Energy-Efficient Trajectory Optimization in UAV-Based Internet of Things (IoT) Network with Delay Tolerance

Bin Li$^{(\boxtimes)}$, Xianzhen Guo, and Zhou Zhou

Department of Communication Engineering, Northwestern Polytechnical University,
Xi'an 710072, China
libin@nwpu.edu.cn

Abstract. With high flexibility and the ability to achieve better wireless channels, the utilization of unmanned aerial vehicles (UAVs) in internet-of-things (IoT) network has gain great popularity. However, the limitation of the power supply and the movement of the UAV make it necessary to optimize the UAV's trajectory to maximize the ground coverage and prolong the communication duration. In this paper, we consider a scenario where the UAV flies in a circular trajectory in the air and equips the transceiver to collect the data with delay tolerance from a sets of the devices in a certain region. In this way, we formed a UAV-based IoT network. Our aim is to achieve the largest communication coverage by optimizing the altitude and radius of UAV's trajectory under the given transmitting power, so as to achieve the energy-efficient communication coverage. Besides, the situation that the directional antenna is also deployed on the UAV to improve the energy utilization. Then, we model the optimization problem as a joint 2-dimensional optimization problem and propose an exhaustive search (ES) over a 1-D parameter in a certain range. Numerical results are presented showing the optimal altitude and radius in the different cases and the antenna beam angles.

Keywords: Unmanned aerial vehicle · Internet-of-things ·
Energy-efficiency · Trajectory optimization · Coverage optimization

1 Introduction

With the high mobility and the lower cost compared with the manned aircraft, unmanned aerial vehicles (UAVs) have attracted a lot of attentions and found a wide range of applications during the past few decades [1]. Along with the

This work is partially supported by NSFC (Nos. 61601365, 61571370, 61801388), Key Research and Development Plan in Shaanxi Province (Nos. 2017ZDXM-GY-101), the Fundamental Research Funds for the Central Universities (3102017OQD091 and 3102017GX08003), and in part by the China Postdoctoral Science Foundation (BX20180262).

© ICST Institute for Computer Sciences, Social Informatics and Telecommunications Engineering 2019
Published by Springer Nature Switzerland AG 2019. All Rights Reserved
B. Li et al. (Eds.): IoTaaS 2018, LNICST 271, pp. 418–425, 2019.
https://doi.org/10.1007/978-3-030-14657-3_43

maturity of the technology and the relevant regulations, a worldwide deployment of the UAVs is expected. A good example is that the UAVs can act as mobile aerial base stations (BSs) to provide the downlink and uplink communications for the ground users and boost the capacity of wireless networks [7]. Compared with the traditional network that largely dependent on the fixed infrastructure (BSs) and could be severally disrupted in the case of natural disasters such as floods or earthquakes [2], the UAV-based BSs are easily for deployment and can quickly provide the emergency communications in the disaster area.

One the other hand, Internet of Things (IoT) utilizes the intelligent interfaces to connect devices, vehicles, and other smart objects, in order to provide the smart environments. For some IoT devices, the information generated from them is not delay sensitive (such as temperature sensed by the sensor nodes), and it is not necessary to connect with UAV constantly. Under this delay tolerance network, the devices just communicate with the BS opportunistically.

Fortunately, the UAV provide the ideal platform for such application. Because of the movement of the UAVs, they can dynamically communicate with the target devices in the large area when necessary. However, the UAV's work performance is constrained due to the limited energy supply. In order to fully use the UAV's endurance, we hope that the UAV can work in an energy-efficient way and covers as many devices as possible. Therefore, we are expected to design an optimal trajectory to maximize the effective coverage area under the given UAV endurance.

Accordingly, in this letter, we consider a the delay tolerance IoT network where a UAV flies in a circular trajectory and collects the data from a sets of devices in a certain region (Fig. 1). Our aim is to achieve the largest effective coverage area by optimizing the flying altitude and radius of UAV's trajectory. Besides, the situation that the directional antenna is also deployed on the UAV to improve the energy utilization. Finally, we model the optimization problem as a joint 2-D optimization problem and propose an exhaustive search (ES) over a 1-D parameter in a certain range. Numerical results are presented to show the optimal height and radius in different cases.

For the related work, Zeng et al. [4] provided us an overview of UAV-assisted communications with emphasis on the use cases, network architecture, channel characteristics, and UAV design considerations. The work in [5] considered a scenario where an UAV collects information from a set of sensors located on a straight line and reformulated their design as a dynamic programming (DP) problem to minimize the aviation time of UAV for data collection. In [6], the total amount of data that can be transmitted from the UAV to the uses over the flying duration was formulated and calculated, considering the time-varying channel between UAV and users as LoS Link. The work in [3] took different QoS requirements into consideration and proposed a maximal weighted area (MWA) algorithm for UAV-BSs that maximized the number of covered users using the minimum transmit power. Yang et al. [9] considered the optimal path planning for UAV, which collect data over IoT sensor network. [10] proposed an approach to connect UAVs with IoT to stream sensor data to cloud services was presented, providing an efficient solution for data transmissions. Unlike our paper, these work above not considered the delay tolerance of the data and the effect of the directional antennas on the UAVs.

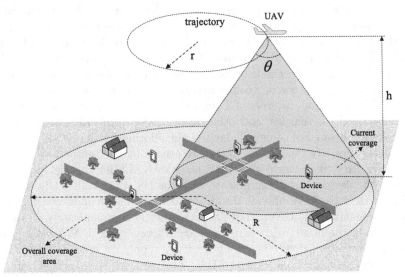

Fig. 1. System model. The UAV-based IoT network. h denotes the UAV flying altitude. θ shows antenna beam angle. R and r are the radius of the efficient overall coverage area and the UAV trajectory, correspondingly.

This work is organized as follows: In Sect. 2, the system model is presented, followed by the problem formulation in Sect. 3. Simulation and discuss are given in Sect. 4. Conclusions are drawn in Sect. 5.

2   System Model and Settings

Here we consider a scenario where a UAV flies in a circular path above the target region and serves as a data gateway over a set of the devices in IoT network. These devices distribute uniformly on the ground and they are expected to transmit the collecting information to the UAV in this system. Let μ denote the set of devices that need to transmit information to the UAV. Without loss of generality, a three-dimensional (3D) cartesian coordinate is considered and (x_i, y_i, z_i) denotes the location of device i in the set of μ. While for the UAV, since its flying trajectory is a circle and the location is time-varying, we denote $(r \cos \frac{v}{r}t, r \sin \frac{v}{r}t, h)$ as the 3D location of the UAV, where r, v and h represent the radius of trajectory, the UAV's speed, and UAV's flight altitude, respectively.

On the other hand, a directional antenna is equipped at the UAV and its beam angle is denoted by θ. According to the directional antenna property, the antenna gain will increase with the decrease of the antenna's beam angle [8]. The relationship between these two parameters can be approximately expressed as:

$$G_a(dBi) = 10Lg(9.7/(\theta_{E,3dB} \times \theta_{H,3dB})) \tag{1}$$

where $\theta_{E,3dB}$ and $\theta_{H,3dB}$ are the elevational angle and horizontal angle of the 3dB gain [11]. What's more, for easily calculation, we make the assumption that the devices can only get the reliable links with the UAV when they are inside the antenna's coverage.

According to the reference [12], the air-to-ground (A2G) links are either line-of-sight (LoS) or non line-of-sight (NLoS) with some probability. Note that in practice, epically for rural macro-cell (RMa) scenario, the UAV-ground channel is more likely to have the LoS link as compared to the terrestrial ground to ground channel. For simplicity, we assume that the communication link from each device to the UAV is dominated by the LoS channel. The extension to the non-LoS and multi-path channel will be discussed in the urban scenarios in our future work. In addition, for the lower flying speed of UAV, the Doppler shift caused by the mobility of the UAV is also ignored. Therefore, we can conclude that the time-varying channel follows the free-space path model, and the received signal power P_r can be expressed as:

$$P_r(t) = P_T \frac{G_a G_R (\frac{\lambda}{4\pi})^2}{d(t,i)} \qquad (2)$$

where $d(t,i) = \left(r\cos\frac{v}{r}t - x_i\right)^2 + \left(r\sin\frac{v}{r}t - y_i\right)^2 + (h - z_i)^2$ denotes the transmitting distance between the UAV and device i at t time slot. G_R shows the device's antenna gain while G_a is the gain of the antenna equipped on the UAV. P_T is the transmitting power of the devices. Thus, the instantaneous channel capacity in bit/second can be expressed as:

$$R_i(t) = \begin{cases} B\log_2(1 + \frac{P_r(t)}{\sigma^2}) = B\log_2(1 + \frac{\gamma_0}{d(t,i)}) & 0 \leq t \leq \frac{\alpha r}{v} \\ 0 & \frac{\alpha r}{v} < t \leq \frac{\pi r}{v} \end{cases} \qquad (3)$$

where $\alpha_{\max} \in (0, \pi)$ is the angle between the line that connects the UAV and the center of its trajectory and the line parallel to the x coordinate in the trajectory. Thus, $\alpha = arctan(\frac{h}{x_i^2 + y_i^2})$. B denotes the channel bandwidth. $\sigma^2 = N_0 B$ is the white Gaussian noise power at the UAV receiver. γ_0 shows the reference received signal-to-noise ratio (SNR) at σ^2. According to the LoS channel property, the SNR will decline with the increase of the communication distance. In some cases, the SNR is too poor to support the data collection and the communication link is outage. Accordingly, in order to guarantee the quality of service (QoS), we set a power threshold P_{th} and the received power P_r must exceed the threshold for the successful transmission. Therefore, the distance between a device and the UAV cannot exceed a distance threshold according to Eq. (3), which is denoted by d_{th}. From the analysis above, we can model the total quantity of information bit uploaded to the UAV from each device as

$$Q_i^T = 2 \int_0^T R_i(t)dt = 2 \int_0^{\frac{r\alpha}{v}} R_i(t)dt \qquad (4)$$

where T is the time period for one hover circle.

3   Problem Formulation and Proposed Solution

In this UAV-based IoT network, there are a set of devices uploading the collecting data to the UAV. However, the total amount of information collected by UAV from different devices is varied because of their different locations. For the movement of the UAV, some devices may be disconnected with UAV intermittently when they are moved out of the coverage of the antenna beam in UAV. To evaluate the effective connection for each device, we set a throughput threshold Q_{th} for the total amount of the transmission information in T, to determine whether the devices are in the UAV's effective coverage. Let $\varepsilon_i \in \{0,1\}$ be a binary decision variable such that $\varepsilon_i = 1$ when the throughout of device i is more than Q_{th} and $\varepsilon_i = 0$ otherwise. This condition can be written as

$$Q_i^T = \int_0^{\frac{r\alpha}{v}} R(t)dt \geq Q_{th}\varepsilon_i \tag{5}$$

The optimization problem is then formulated as

$$(h^*, r^*) = \arg\max_{h,r} \sum_{i \in \mu} \varepsilon_i$$
$$subject\ to$$
$$Q_i^T \geq \varepsilon_i Q_{th}, \forall i \in \mu \tag{6}$$
$$\varepsilon_i \in \{0,1\}, \forall i \in \mu$$
$$d(t,i) \geq d_{th}^2$$

In this network, since the UAV flies in a circular trajectory, the effective overall coverage area is also expected to be a circle. For the given flying altitude, when the radius of trajectory r is too large, the communication hole in which the devices can not connect to the UAV at any time will be formed in the center of the coverage area, causing the communication blind area in IoT network. When r is too small, the effective coverage area is also limited by the antenna beam. On the other hand, if we increase the flying altitude, the communication coverage will be enhanced obviously. However, for the given transmitting power, the higher altitude also causes the large path loss, leading to the lower received SNR and capacity. This phenomenon also restricts the coverage area. Therefore, there exists the optimal flying altitude and hovering circle under the constraint of the throughput in (6). As we can see that (6) is a joint 2-D optimization problem and difficult to solve. The difficulty arises from the coupling between the optimization of height and radius for the large quantity of devices.

Accordingly, we propose a solution to simplify this optimization problem. To find the largest coverage area, we only need to find the furthest device that can satisfy the required capacity. Here we utilize exhaustive search (ES) over the altitude and radius of trajectory respectively to find the furthest communication point from the region center. The distance between this point and the center of coverage is regarded as the radius of the largest coverage R. What's more, considering the symmetry of circular region, it is reasonable to just consider

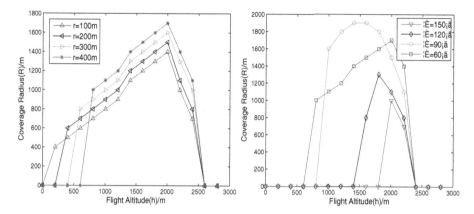

Fig. 2. Maximal coverage vs flight altitude

Fig. 3. The effect of beam angle

the device on the x-coordinate whose location is $(x_i, 0, 0)$ and only calculate the capacity when the UAV's location is $(r\cos(vt/r), r\sin(vt/r), h)$.

The process for implementing the ES is as follows:

(1) Initialize the distribution of a set of devices distributed on directivity of the x-coordinate, like $(x_i, 0, 0)$.
(2) Initialize the flight height h and the flight radius r.
(3) Calculate the total capacity (Q_i) of the device i using (5).
(4) Determine whether the device i is in the UAV's coverage by comparing the value of Q_i and Q_{th}.
(5) Update the location of the device until find the furthest device which satisfy the required QoS.
(6) Calculate the coverage radius by $R = x_i$.
7) Update the value of h or r and repeat (1) to (6).

4   Simulation Results and Discussions

In the simulation, we focus on the UAV's maximal coverage and discuss how the parameters affect the size of the coverage area. In the communication system, the communication bandwidth is $B = 1\,\mathrm{Mhz}$ and the noise power spectrum density at the UAV receiver is assumed to be $N_0 = -160\,\mathrm{dBm/hz}$. Therefore, the noise power $\sigma^2 = N_0 B = -100\,\mathrm{dBm}$. The device's transmission power is $P = 0.1\,w(or -10\,\mathrm{dB})$. We assume that the carrier frequency is $f_0 = 2.4\,\mathrm{Ghz}$ and the wave length is $\lambda = c/f_0 = 0.125\,\mathrm{m}$.

Figure 2 shows how the optimal coverage area changes with the flight altitude h when the trajectory radius $r = 100\,\mathrm{m}$, $200\,\mathrm{m}$, $300\,\mathrm{m}$, $400\,\mathrm{m}$ and $\theta = 60°$ respectively. As we can see that the coverage radius R first increases with h gradually and then falls to 0 rapidly from the maximum value. This is because that the altitude of the UAV has a direct affect on the path loss. In the first

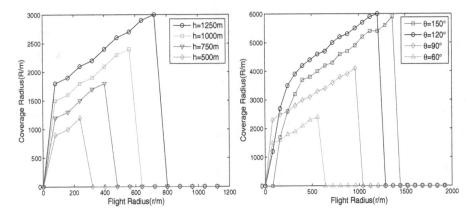

Fig. 4. Maximal coverage vs flight radius

Fig. 5. The effect of beam angle

period, according to $R = h\cos(\frac{\pi-\theta}{2})$, R is in direct proportion to h. However, when the UAV flies too high, the path loss caused by the distance dominate the communication performance and few devices can connect to the UAV. As a result, the coverage will decline to 0 ultimately.

Figure 3 presents the effects of the beam angle when $r = 100$ m. By comparing the maximum of different line for $\theta = 60°, 90°, 120°, 150°$, We can conclude that the optimal radius is the largest when $\theta = 60°$ and there must be an optimal beam angle for a largest covered region. What's more, from the difference between the line for $r = 100$ m, $\theta = 60°$ in Fig. 2 and the line for $r = 100$ m, $\theta = 90°$ in Fig. 3, we can find that an optimal beam angle enables the UAV to achieve a bigger coverage with lower flight altitude, which matters a lot on UAV's energy efficiency.

Figure 4 depicts the relationship between the maximal coverage and the flight radius when the flight radius $h = 500$ m, 750 m, 1000 m, 1500 m, $\theta = 60°$ respectively. As shown in the figure, in a certain range, the increase of the radius can enlarge the coverage. However, there is a limitation for the increase of the flight radius, beyond which the coverage falls to 0. It is because the coverage in this paper is defines as the circular region that can be covered all the time, which is modeled by $(r\cos\alpha - x_0)^2 + (r\sin\alpha - y_0)^2 + (h - z_0)^2 \geq d_{th}^2$.

Figure 5 describes the how the beam able affects the coverage radius when $h = 1000$ m, $\theta = 60°, 90°, 120°, 150°$. Just like what we do in analysing the beam angle's effects in Fig. 3, we find that there is also an optimal value for the angle. The optimal beam angle can also be determined by ES over a 1-D parameter.

5   Conclusion

In this paper, we consider the scenario that the UAV collect the delay tolerance data from the devices with different transmitting distance and then we formed a UAV-based IoT network. To evaluate the communications performance, we considered the ergodic capacity from UAV to the devices in each UAV flight cycle in order to find the largest coverage area. During the process, we modeled the relationship between the coverage area, the flight altitude, and the flight radius to find the optimal flying altitude and hover radius respectively. What's more, the effects of the directive antenna's beam angle are discussed and presented in the simulations.

References

1. Gupta, L., Jain, R., Vaszkun, G.: Survey of important issues in UAV communication networks. IEEE Commun. Surv. Tutorials **18**(2), 1123–1152 (2016). Secondquarter 2016
2. Say, S., Inata, H., Liu, J., Shimamoto, S.: Priority-based data gathering framework in UAV-assisted wireless sensor networks. IEEE Sens. J. **16**(14), 5785–5794 (2016)
3. Alzenad, M., El-Keyi, A., Lagum, F., Yanikomeroglu, H.: 3-D Placement of an unmanned aerial vehicle base station (UAV-BS) for energy-efficient maximal coverage. IEEE Wirel. Commun. Lett. **6**(4), 434–437 (2017)
4. Zeng, Y., Zhang, R., Lim, T.J.: Wireless communications with unmanned aerial vehicles: opportunities and challenges. IEEE Commun. Mag. **54**(5), 36–42 (2016)
5. Rashed, S., Soyturk, M.: Effects of UAV mobility patterns on data collection in wireless sensor networks. In: 2015 IEEE International Conference on Communication, Networks and Satellite (COMNESTAT), Bandung, pp. 74–79 (2015)
6. Zeng, Y., Zhang, R.: Energy-efficient UAV communication with trajectory optimization. IEEE Trans. Wirel. Commun. **16**(6), 3747–3760 (2017)
7. Zhan, C., Zeng, Y., Zhang, R.: Energy-efficient data collection in UAV enabled wireless sensor network. IEEE Wirel. Commun. Lett. **7**(3), 328–331 (2018)
8. Goldsmith, A.: Wireless Communications. Cambridge University Press, Cambridge (2005)
9. Yang, Q., Yoo, S.J.: Optimal UAV path planning: sensing data acquisition over IoT sensor networks using multi-objective bio-inspired algorithms. IEEE Access **6**, 13671–13684 (2018)
10. Gaur, A.S., Budakoti, J., Lung, C.H., Redmond, A.: IoT-equipped UAV communications with seamless vertical handover. In: 2017 IEEE Conference on Dependable and Secure Computing, Taipei, pp. 459–465 (2017)
11. Cheng, D.K.: Field and Wave Electromagnetics. Pearson New International Edition, Pearson Schweiz AG (2013)
12. Mozaffari, M., Saad, W., Bennis, M., Debbah, M.: Unmanned aerial vehicle with underlaid device-to-device communications: performance and tradeoffs. IEEE Trans. Wirel. Commun. **15**(6), 3949–3963 (2016)

Robust Power Allocation for Cognitive Radio System in Underlay Mode

Rongfei Fan[1(✉)], Qi Gu[2], Xuming An[1], and Yu Zhang[1]

[1] Beijing Institute of Technology, Beijing 100081, China
{fanrongfei,yuzhang}@bit.edu.cn, 1952590139@qq.com
[2] Beijing Jiaotong University, Beijing 100044, China
15112094@bjtu.edu.cn

Abstract. In this paper, the power allocation problem in cognitive underlay system is studied given a number of samples of interference channel's gain. System throughput of secondary users is aimed to be maximized while limiting the worst-case interference outage probability. With only a number of samples of interference channel's gain, we define an uncertain region of possible probability density function (PDF) of interference channel's gain and further derive the closed-form expression for the interference outage probability constraint. By replacing the formulated constraint on interference outage probability with closed-form one, we show the originally formulated problem is a convex optimization problem.

Keywords: Cognitive radio · Underlay · Robust power allocation

1 Introduction

Underlay cognitive communication is a popular technique in face of spectrum shortage problem, which permits the licensed user (primary user) and unlicensed user (secondary user) to coexist on the same band while limiting the interference to primary user [1]. Thus how to limit the interference is an important issue. In literatures, the interference channel's gain is usually assumed to be a random variable. Thus a probabilistic constraint should be imposed on the event of interference outage, in which the interference to the licensed receiver is above a given threshold [2]. In this case, the distribution of interference channel's gain should be known perfectly. This assumption is hard to realize and some literatures suppose the distribution of interference channel' gain to be uncertain.

In this paper, we only assume a number of samples of interference channel's gain and define the an uncertain region of probability density function (PDF) of interference channel's gain. Then over the uncertain region of PDF, closed-from interference outage probability is derived. With this transformation, the power allocation problem aiming at system throughput maximization is shown to be convex.

Supported by National Natural Science Foundation of China with Grant No. 61601025, and China Scholarship Council.

B. Li et al. (Eds.): IoTaaS 2018, LNICST 271, pp. 426–430, 2019.
https://doi.org/10.1007/978-3-030-14657-3_44

2   System Model and Problem Formulation

Consider a cognitive radio system in underlay mode. There are N primary users, each of which occupy a spectrum with bandwidth B. These N primary users constitute the set $\mathcal{N} \triangleq \{1, 2, \ldots, N\}$. There are M secondary transmitters communicating to a secondary base station. These M secondary transmitters constitute the set $\mathcal{M} \triangleq \{1, 2, \ldots, M\}$.

These M secondary transmitters will access into the secondary base station over N channels via non-orthogonal multiple access (NOMA). On nth channel, the mth secondary transmitter will transmit with power $p_{n,m}$. Suppose P_m is the maximal total transmit power of mth secondary transmitter, then the set of $p_{n,m}$ should be subject to the following constraints

$$p_{n,m} \geq 0, \forall n \in \mathcal{N}, m \in \mathcal{M}, \tag{1}$$

and

$$\sum_{n=1}^{N} p_{n,m} \leq P_m, \forall m \in \mathcal{M}. \tag{2}$$

At the secondary base station, the technique of successive interference cancellation (SIC) is resorted to [3]. By assuming the channel gain between mth secondary transmitter and the secondary base station on channel n as $h_{n,m}$, the following throughput can be achieved for mth secondary transmitter on channel n,

$$C_{n,m}^{N} = \ln \left(1 + \frac{p_{n,m} h_{n,m}}{1 + \sum\limits_{m'=m+1}^{M} p_{n,m'} h_{n,m'}} \right). \tag{3}$$

Then the system throughput of the M mobile users can be written as

$$C_N = \sum_{n=1}^{N} \sum_{m=1}^{M} C_{n,m}^{N} = \sum_{n=1}^{N} \ln \left(1 + \sum_{m=1}^{M} p_{n,m} h_{n,m} \right). \tag{4}$$

When the secondary users are transmitting information, they will also generate interference to primary users. Denote the channel gain from mth secondary transmitter to the primary receiver on channel n as $g_{n,m}$. At the beginning of every fading block, mth mobile user will measure $h_{n,m}$ and then determine the transmit power $p_{n,m}$. On the other hand, due to the separation between primary user system and secondary user system, it is hard for the secondary transmitter m to measure $g_{n,m}$ in every fading block. In this paper, $g_{n,m}$ is assumed to be an identically and independently distributed random variable. In this case, to limit the interference to primary receiver, the interference outage probability should be limited, i.e.,

$$\Pr \left(\sum_{m=1}^{M} p_{n,m} \cdot g_{n,m} \geq I_n \right) \leq \varepsilon_n, \forall n \in \mathcal{N} \tag{5}$$

where I_n is the threshold of interference and $\varepsilon_n \in (0,1)$ is tolerable outage probability for channel n. Note that (5) is equivalent with the following constraint

$$\Pr\left(\sum_{m=1}^{M} p_{n,m} \cdot g_{n,m} \leq I_n\right) \geq 1 - \varepsilon_n, \forall n \in \mathcal{N}. \tag{6}$$

For the random variable $g_{n,m}$, only a limit number of samples of $g_{n,m}$ can be obtained by investigating the historical signaling signal. Suppose the number of samples is S, denote S samples of $g_{n,m}$ as $\hat{g}_{n,m}^1, \hat{g}_{n,m}^2, ..., \hat{g}_{n,m}^S$, and define $\mathcal{S} \triangleq \{1, 2, ..., S\}$. We wish to describe the distribution of $g_{n,m}$, i.e., $f(g_{n,m})$ according to $\hat{g}_{n,m}^s$ for $s \in \mathcal{S}$. Specifically, with $\hat{g}_{n,m}^s$ for $s \in \mathcal{S}$, we want to make sure that $f(g_{n,m})$ falls into the uncertain region of distribution functions $\mathcal{F}(g_{n,m})$ with probability at least $(1 - \alpha)$, which can be also written as $\mathcal{F}(g_{n,m}, \alpha)$ for the ease of presentation in the following. In this case, given α predefined, a robust form of constraint (6) should rewritten as

$$\inf_{\substack{f(g_{n,m}) \in \mathcal{F}(g_{n,m}, \alpha), \\ \forall m \in \mathcal{M}}} \Pr\left(\sum_{m=1}^{M} p_{n,m} \cdot g_{n,m} \leq I_n\right) \geq 1 - \varepsilon_n, \forall n \in \mathcal{N}. \tag{7}$$

In this paper, our target is to maximize C_N by optimizing $p_{n,m}$ for $m \in \mathcal{M}$ by conforming to the associated constraints. Specifically, the following optimization problem is to be solved

Problem 1

$$\max_{p_{n,m}, \forall n \in \mathcal{N}, m \in \mathcal{M}} \sum_{n=1}^{N} \ln\left(1 + \sum_{m=1}^{M} p_{n,m} h_{n,m}\right)$$

$$s.t. \quad p_{n,m} \geq 0, \forall n \in \mathcal{N}, m \in \mathcal{M},$$

$$\sum_{n=1}^{N} p_{n,m} \leq P_m, \forall m \in \mathcal{M},$$

$$Constraint\ (7).$$

3   Transformation and Optimal Solution

Looking into Problem 1, it can be seen that constraint (7) is not in closed-form, which leads to the hardness. For the ease of discussion, the following notational conventions are claimed first. Denote $\boldsymbol{g}_n = (g_{n,1}, g_{n,2}, \ldots, g_{n,M})^T$, $\boldsymbol{p}_n = (p_{n,1}, p_{n,2}, \ldots, p_{n,M})^T$, and $\hat{\boldsymbol{g}}_n^s$ as the sth group of sampling of the vector $\boldsymbol{g}_n$, for $n \in \mathcal{N}$ and $s \in \mathcal{S}$. Define $\hat{\boldsymbol{\mu}}_n$ as the sample mean of the vector $\boldsymbol{g}_n$ and $\hat{\boldsymbol{\Sigma}}_n$ as the sample covariance of the vector $\boldsymbol{g}_n$ over the S samples for $n \in \mathcal{N}$, respectively. Suppose $\mathbb{P}_{n,m}$ is the general probability measure for the random variable $g_{n,m}$ for $n \in \mathcal{N}$ and $m \in \mathcal{M}$. Let $\mathbb{P}_n$ and $\mathbb{P}_n^*$ indicate a general and the true (which is unknown in advance) probability measure for the random vector

g_n for $n \in \mathcal{N}$, respectively. Define $\mathbb{P}_{n,S}$ $(\mathbb{P}^*_{n,S})$ as the measure of the samples g^s_n for $s \in \mathcal{S}$, which is actually a S-fold product distribution of $\mathbb{P}_n$ $(\mathbb{P}^*_n)$. $\mathbb{E}^{\mathbb{P}}[\cdot]$ is the expectation of a random vector or matrix under the probability measure $\mathbb{P}$.

Define G_n as the value such that $||g_n||_2 \leq G_n$ almost surely for $n \in \mathcal{N}$, when $S > (2 + 2\ln(2/\alpha))$, it can be proved that with probability at least $(1 - \alpha)$, the probability measure $\mathbb{P}^*_n$ falls into the following region [4]

$$
\mathcal{F}^{\mathrm{I}}(g_n, \alpha) = \left\{ \mathbb{P}_n \middle| \mathbb{P}_n \in \Xi(G_n), ||\mathbb{E}^{\mathbb{P}_n}[g_n] - \hat{\mu}_n||_2 \leq \Lambda_1(\alpha, S, G_n), \right.
$$
$$
\left. ||\mathbb{E}^{\mathbb{P}_n}[g_n g_n^T] - \mathbb{E}^{\mathbb{P}_n}[g_n]\mathbb{E}^{\mathbb{P}_n}[g_n^T] - \hat{\Sigma}||_F \leq \Lambda_2(\alpha, S, G_n) \right\} \tag{8}
$$

where

$$
\Lambda_1(\alpha, S, G_n) = \frac{G_n}{S}\left(2 + \sqrt{2\ln(1/\alpha)}\right), \Lambda_2(\alpha, S, G_n) = \frac{2G_n^2}{S}\left(2 + \sqrt{2\ln(2/\alpha)}\right),
$$

$\Xi(G_n)$ means the set of probability measures such that the norm of the associated random vector g_n is no larger than G_n for $n \in \mathcal{N}$, and $||\cdot||_F$ is the Frobenius norm of a matrix.

With $\mathcal{F}^{\mathrm{I}}(g_n, \alpha)$ defined, the next step is to transform constraint (7) to be closed-form expression. The following lemma can be expected.

Lemma 1. *Given* $\mathcal{F}^I(g_n, \alpha)$, *constraint (7) will hold when the following constraint holds*

$$
\hat{\mu}_n^T p_n + \Lambda_1(\alpha, S, G_n) \cdot ||p_n||_2
$$
$$
+ \sqrt{\frac{1-\varepsilon_n}{\varepsilon_n}}\sqrt{p_n^T\left(\hat{\Sigma}_n + \Lambda_2(\alpha, S, G_n) \cdot I\right)p_n} \leq I_n, \forall n \in \mathcal{N} \tag{9}
$$

where I is the identity matrix.

Proof. Define

$$
\mathrm{VaR}^{\mathbb{P}_n}_{\varepsilon_n}(g_n^T p_n) = \inf\left\{t | \mathbb{P}_n(g_n^T p_n \leq t) \geq 1 - \varepsilon_n\right\}. \tag{10}
$$

According to [5], there is

$$
\sup_{\mathbb{P}_n \in \mathcal{F}^{\mathrm{I}}(g_n, \alpha)} \mathrm{VaR}^{\mathbb{P}_n}_{\varepsilon_n}(g_n^T p_n) = \hat{\mu}_n^T p_n + \Lambda_1(\alpha, S, G_n)||p_n||_2
$$
$$
+ \sqrt{\frac{1-\varepsilon_n}{\varepsilon_n}}\sqrt{p_n^T\left(\hat{\Sigma}_n + \Lambda_2(\alpha, S, G_n)I\right)p_n}. \tag{11}
$$

Thus when the right-hand side of (11) is smaller than I_n, which is the exact expression of (9), it is straightforward to see that constraint (7) will hold.

This completes the proof.

Remark: It can be seen that the transformed constraint in (9) is a second-order cone constraint with the vector p_n, which defines a convex region of p_n for $n \in \mathcal{N}$. Additionally, it can be checked that the objective of Problem 1 is concave, thus Problem 1 is a convex optimization problem, whose global optimal solution can be achieved by existing methods.

References

1. Zhou, Z., Jia, Y., Chen, F., Tsang, K., Liu, G., Han, Z.: Unlicensed spectrum sharing: from coexistence to convergence. IEEE Wirel. Commun. **24**(5), 94–101 (2017)
2. Kim, D., Le, L.B., Hossain, E.: Joint rate and power allocation for cognitive radios in dynamic spectrum access environment. IEEE Trans. Wirel. Commun. **7**(12), 5517–5527 (2008)
3. Ding, Z., et al.: A survey on non-orthogonal multiple access for 5G networks: research challenges and future trends. IEEE J. Sel. Areas Commun. **35**(10), 2181–2195 (2017)
4. Shawe-Taylor, J., Cristianini, N.: Estimating the moments of a random vector with applications (2003). http://eprints.soton.ac.uk/260372/1/EstimatingTheMomentsOfARandomVectorWithApplications.pdf
5. Calafiore, G., El Ghaoui, L.: On distributionally robust chance-constrained linear programs. J. Optim. Theory Appl. **130**(1), 1–22 (2006)

An Asynchronous Schwarz-IC Cascade MUD Detection Algorithm for Multiple Access Mobile Communication System

Weiting Gao[1($\boxtimes$)], Haifeng Zhu[1], Guobing Cheng[1], Fei Ma[2], and Weilun Liu[1]

[1] Institute of Information and Navigation, Airforce Engineering University, Xi'an, China
519105941@qq.com, kunta0089@sina.com,
Guobingcheng12@163.com, 1197853086@qq.com
[2] Xi'an Modern Control Technology Research Institute, Xi'an, China
365553071@qq.com

Abstract. The multi-user detection precision and interference suppression processing efficiency of single successive interference cancellation (SIC), parallel interference cancellation (PIC) algorithm and the cascade structure interference cancellation (IC) algorithm are always affected by single-stage and multi-stage detection error diffusion. This paper focuses on the detection error diffusion of multiple access communication system with strong multiple access interference (MAI) and inter symbol interference (ISI). In this paper, a Schwarz-IC (S-IC) cascade multi-user detection (MUD) algorithm based on the monotone convergence characteristics of the Schwarz algorithm, is proposed for the multiple access mobile asynchronous communication system. The Schwarz-IC algorithm can precisely control the sub-domain boundary value of the user load power in system, efficiently track the change wireless channel, avoid the single-stage or multi-stage detection error diffusion of single interference cancellation algorithm and improve the detection convergence. Simulation results show that the Schwarz-IC algorithm is of better BER performance, detection accuracy and dynamic tracking capability.

Keywords: Multiple access interference ·
Successive interference cancellation · Parallel interference cancellation ·
Schwarz

1 Introduction

In the modern wireless mobile communication, public land mobile network (PLMN), satellite communication (SC) and other fields, the multiple access systems such as direct sequence spread spectrum code division multiple access (DS-CDMA), frequency division multiple access (FDMA), time division multiple access (TDMA), space division multiple access (SDMA), packet division multiple access (PDMA) and pulse address multiple access (PAMA) are all popular wireless communication technology [1]. For example, in the DS-CDMA system, the multiple users can share the same frequency band to improve the spectrum utilization greatly. However, in the actual communication

B. Li et al. (Eds.): IoTaaS 2018, LNICST 271, pp. 431–447, 2019.
https://doi.org/10.1007/978-3-030-14657-3_45

process, there are some issues such as the unsatisfactory spreading code correlation characteristics, uncoordinated signal synchronization, transmission distortion, undesirable cross-correlation and the implementation of the spreading code with zero correlation value, which always cause the phenomenon such as incompletely orthogonal of spreading waveform, and further lead to multiple access interference (MAI), ultimately seriously restrict the capacity and performance of the multiple access communication systems [2].

The interference cancellation multi-user detection technology is an research focus of multiple access mobile communication system, which includes the successive interference cancellation (SIC) [3], the parallel interference cancellation (PIC) [4] and the cascade structure interference cancellation (IC) algorithms [5]. The basic principle of interference cancellation multi-user detection is firstly to estimate the MAI of each user at the receiving terminal, and then eliminate some or all MAI in the received signal [6]. This class detector is generally formed by a multi-level detection structure.

SIC detector gradually reduces the maximum power user-generated interference by the multi-user data decisions of all users, then estimates the user signal source of each detection level, and finally, restructures the load information of each received user signal at the receiving terminal by the signal source estimation results [7]. Because this processing must constantly order the new succession sort of multi-users to priority process the maximum power user, so any sort error would increase MAI and lead detection error diffusion [8]. PIC detector generates the MAI by the decision estimate value of the $m - 1$th detection level, then removes it completely in the received signal on the mth level. This processing is good of short processing delay and high detection efficiency [9]. But each detection error on any level would cause unacceptably high complexity, low convergence rate, poorer magnitude estimates and near-far problem (NFP). Because a reliable detection result depends on the accurate amplitude estimation, therefore, the low amplitude estimation accuracy may cause the reduction of system performance gain, and then limit the system performance. In summary, any detection error diffusion on any detection level would severely affect the precision of SIC and PIC detector in the subsequent levels. In addition, when the power control is not satisfactory, as in the multi-path channel, the performance of SIC is better than the performance of PIC. Conversely, the performance of PIC is superior to the performance of SIC. In general, while the SIC has a better detection performance on the weak power user signal, but the processing always reduce the detection performance of the maximum power user [10].

Cascade structure IC detector is a compromise approach between SIC and PIC in the aspects of system delay and performance. Generally, IC detector improves the multi-interference suppression ability by the multi-level cascade structure detection, which adapts phases iteration for all users severally by the same iterative calculation method [11]. Although this processing can effectively avoid detection error diffusion, but the excessive repeated computations always result in high complexity and instability convergence, which heavily limit the project implementation.

Based on the weak decomposition theory, the Schwarz algorithm can achieve the real-time acquisition of weighted norm error convergence rate, then assigns the users power sub-domain boundary value to internal boundary unit by information transfers, and finally, obtains the overall numerical solution of all users in the channel [12].

In this paper, we present a combination of Schwarz and cascade structure IC detection algorithm to achieve the power sub-domain boundary value precise control of user signals, then restrict the generation and diffusion of detection error, finally ensure detection precision and convergence through real-time estimation of channel.

2 MUD Model for Multiple Access Communication System

Assume an asynchronous multiple access communication system with $2P + 1$-length transmitted symbol, the bit interval is T, the user equivalent channel response maximum order is P, and the time sequence is $\{-P, -P + 1, \ldots, -1, 0, 1, \ldots, P - 1, P\}$. Supposing the 1-user is the expected user, let all K users send asynchronous signals in the sum noise channel and add sum noise for each user signal respectively [13]. Then adapt spread spectrum secondary on chaotic sequence and makes power sub-domain boundary value precise control for each user respectively. All of these send signals after been double spread spectrum processing by the asynchronous S hexadecimal transmission mode [13]. The S hexadecimal signal of the k-user is expressed as:

$$S = \{s_{k,0}, s_{k,1}, \ldots, s_{k,S-1}\}, (k = 1, \ldots, K) \tag{1}$$

for all serial number k and i, the sending data of signal transmitting symbol $\{b_k(i)\}$ is independent and identically distributed, the output signal model is expressed as:

$$\begin{cases} y_k = \int_0^T r(t)s_k(t)\mathrm{dt} = \Lambda_k b_k(i) + \sum_{i=1, i \neq k}^K A_i b_i \rho_{ik} + n_k(t), \ b_k(i) \in \{-1, +1\} \\ \rho_{ik} = T^{-1} \int_0^T s_i(t)s_k(t)\mathrm{dt}, \ n_k(t) = T^{-1} \int_0^T n(t)s_k(t)\mathrm{dt}, \ t \in [iT, (i+1)T] \end{cases} \tag{2}$$

where: A_k is the amplitude of the received signal; ρ_{ik} is the spread spectrum inter symbol correlation coefficient; $s_k(t)$ is the characteristic waveform; $n_k(t)$ is the noise-related output of AWGN.

Generally, the tradition SIC and PIC algorithm process MAI as the background noise, then detect the output of the matched filter (MF) detector directly as $\hat{b}_k = \mathrm{sgn}(y_k)$, which easily increase bit error rate (BER) and reduce the system capacity.

When $t \notin [0, T]$, then $s_k(t) = 0$, if the energy characteristic waveform is limited to $t \in [0, T_C]$, the characteristic waveform is:

$$\begin{cases} s_k(t) = \sum_{l=0, l \in \{-L, L\}}^{N-1} s_{k,l} P_{T_C}(t - lT_C) \Leftrightarrow \int_0^T s_k^2(t)\mathrm{dt} = 1 \\ \|s_k\|^2 = \int_0^T s_k(t)\mathrm{dt} = 1, \ P_{T_C}|_{t \in [0, T_C]} = 1/\sqrt{T_C}, \ T_C = T/N \end{cases} \tag{3}$$

where: N is the spread spectrum processing gain; $\{s_{k,0}, s_{k,1}, ., s_{k,N-1}\}$ is the normalized spectrum sequence $(\pm N^{-1/2})$; P_{T_C} is the T_C-cycle matrix code piece.

Because MAI can be equivalent to the pseudo-random (PN) sequence signal with a strong correlation structural, and the correlation function between each user is known [14]. Therefore, we can use the structural information and statistics information of these PN sequence to eliminate MAI and thus improve the system performance. The PN sequence generator is defined as:

$$\begin{cases} PN(z) = 1 + PN_1z + \ldots + PN_{K-1}z^{m-1} + z^m \\ b_k[m] = \sum_1^K b_k(m)PN_k = U(V^T) = [b_1, b_2, \ldots, b_K][PN_1, PN_2, \ldots, PN_m]^T \\ U = [b_1, b_2, \ldots, b_K], V = [PN_1, PN_2, \ldots, PN_m], L_{PNmax} = 2^m - 1, \end{cases} \quad (4)$$

where: L_{PNmax} is the maximum cycle of the PN sequence generator; U is the state vector of shift register; V is the connect vector; $b_k[m]$ is the feedback signal outputted by modulo-two adder.

The PN sequence generator and the corresponding results (one sample per symbol) of PN sequence generator are shown in Figs. 1 and 2.

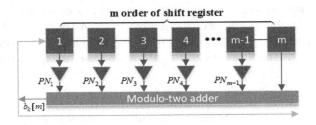

Fig. 1. PN sequence generator

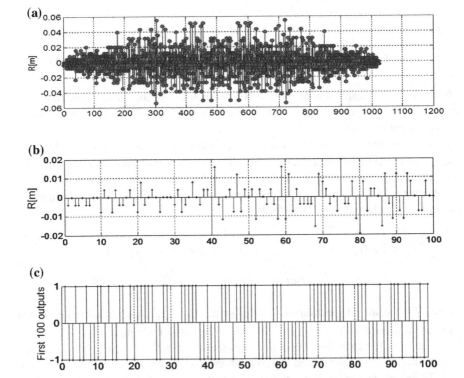

Fig. 2. (a) The single result of each PN sequence generator unit. (b) The corresponding results of PN sequence generator. (c) The first 100 outputs

Supposing the system noise $e_k(t)$ of the k-th user is equivalent to the sum of AWGN component $n_k(t)$ and zero mean colored noise component $\zeta_k(t)$ as:

$$\begin{cases} e_k(t) = n_k(t) + h_k\zeta_k(t) \\ e_k(n,i) = e(nL+i) \end{cases} \tag{5}$$

where: h_k is the colored noise intensity; $e(n,i)$ is the sum sequences noise component, and $e(n,i) = e(nL+i)$.

At the receiving terminal, the received signal after dealing with adaptive filter and binary phase shift keying (BPSK) modulate can be equivalent as follow:

$$r(t) = \sum_{k=1}^{K} \left\{ \sum_{i=-\infty}^{\infty} [A_k b_k(i)s_k(t - iT - \tau_k) + p_k(t - iT - \tau_k)]\cos(\omega_k t) + e_k(t) \right\} \tag{6}$$

where: p_k (± 1) is the waveform of the secondary spread spectrum; T is the bit interval; τ_k is the time delay; ω_{Ck} is the adaptive weight vector of filter unit.

Generally, short (cycle) spreading sequence is commonly used in direct sequence spread spectrum signal model systems, namely, the spreading sequence cycle of each user is equal to the corresponding symbol cycle.

However, in the actual mobile communication transmission conditions, especially for the multiple access communication systems, it often appears phenomenon as the spreading sequence cycle is greater than the symbol cycle (long spread). Replacing $s_k(t - iT - \tau_k)$ by $s_{k,(i)_Q}(t - iT - \tau_k)$, the received signal model on Eq. 3 is:

$$s_{k,(i)_Q}(t) = \sum_{l=0}^{N-1} s_{k,[i/Q]N+l} P_{T_c}(t - lT_c) \tag{7}$$

where: $(i)_Q$ is i mod Q operation; $l \in [0, L]$, $L > N$ and $[\bullet]$ is the end function; L is the long spread spectrum sequences cycle, satisfy as $L/N \triangleq Q > 1$.

The element of time-shift characteristic waveform cross-correlation matrix $R(l)$ for the i-th and k the user can defined as:

$$R_{i,k}(l)|_{R(-l)=R^T(l)} = \int_{-\infty}^{\infty} s_i(t - \tau_i)s_k(t + lT - \tau_k)dt \triangleq \rho_{ik}(l) \tag{8}$$

then $R(l)|_{l\notin\{-1,0,1\}} = 0$ and $R(-l) = R^T(l)$.

Set the received signal sampling rate is equal to chip rate, r is the output vector of L-dimensional match filter in symbol interval T. So the vector form of asynchronous multiple access communication system base band received signal model is formed as:

$$\begin{cases} r = \sum_{k=1}^{K} A_k b_k s_k p_k + e \\ p = [p_1, ., p_K]^T, p_k = L^{-1}[p_1^k, ., p_L^k]^T \end{cases} \tag{9}$$

$$\begin{cases} y = RAb + n = [y_1, y_2, \ldots, y_K]^T \\ R = E[pp^T], A = \text{diag}[A_1, \ldots, A_K]^T \end{cases} \tag{10}$$

where: $E\{nn^T\} = \sigma^2 R, e$ is the zero mean sum noise covariance matrix.

In any relevant transmission interval, when $\max\{\tau_k\} \leq T$, estimates the bit symbols of transmitted user signals in any relevant transmission interval, then the asynchronous multiple access communication system with K users could be equivalent to a synchronous user multiple access communication system with $2K - 1$ users [15]. If $2K - 1 \leq L$ and all spreading code of the $2K - 1$ users are linearly irrelevant, the calculation of asynchronous system is similar to synchronization system.

3   Schwarz-IC Algorithm

The computing structure of S-IC algorithm is different from the traditional serial pattern detector. Every SIC unit of S-IC detector is formed by a traditional serial processing unit and a Schwarz boundary value control unit, this structure can effectively reduce the exception errors of sorting process. The asynchronous Schwarz algorithm takes advantage of the parallel computing of PIC detector. Due to the introduction of Schwarz unit, the single-level PIC unit can determine whether to continue operation according to the latest information variables without waiting for any data input at any time. This combination treatment can significantly reduce the computational complexity caused by excessive duplicate detection.

3.1   Schwarz Rules and Implementation Conditions

Define sets: $L^2(0, 1) = \{q(x)|q(x), q(x) \in (0, 1)\}$ $X = \{q(x)|q(x), q'(x) \in L^2(0, 1)\}$, here $q(x)$ is the square-integrable observation function In the interval as $(0, 1)$, and $q(1) = 0$. Supposing $u(x)$ is the decreasing function. For the one-dimensional boundary value control, which can be equivalent to the power control process problem as:

$$\begin{cases} -\frac{d}{dx}[k(x)\frac{du}{dx}] = f(x) \\ 0 < x < 1, \, u(0) = 0, \, u'(1) = 0 \end{cases} \tag{11}$$

if $p = k(x)\frac{du}{dx}$, so $-\frac{dp}{dx} = f(x)$, $0 < x < 1$ so $p - k(x)\frac{du}{dx} = 0$, $u(0) = 0$, $p(1) = 0$, for an arbitrary function $v(x) \in L^2(0, 1)$, equivalent integral weak form of Eq. 11 is:

$$\int_0^1 (\frac{dp}{dx} + f(x))v dx = 0 \tag{12}$$

the equivalent integral weak form of Eq. 11 is:

$$\int_0^1 (k^{-1}(x)p - \frac{du}{dx})q dx = 0 \tag{13}$$

for segment integral boundary condition as $u(0) = u'(1) = 0$, there is:

$$\int_0^1 (k^{-1}(x)pq + u\frac{dp}{dx})dx = 0 \tag{14}$$

Equations 15 and 16 are conducive to the finite element approximation. Define a bounded open set: $\Omega \subset R^n$, $\exists B(x)$, $(B(x) \geq 0)$ for the continuous Schwarz algorithm. When $x \subset \Omega$, A_{ij} is consistent positive definite. Assume a boundary value problem: $Lu \mid_\Omega = f$, $u \mid_{\partial\Omega} = g$, $u \mid_{\Omega|_{min}} = f(x,y)$, $u \mid_{\Omega|_{min}} = g(x,y)$, then the differential operator can be expressed as:

$$Lu = -\sum_{i,j=1}^{K} \frac{\partial}{\partial x_j}(A_{ij}\frac{\partial u}{\partial x_i}) + B(x)u \tag{15}$$

make non-average decomposition for the bounded open set Ω, let $\Omega_i \cap \Omega_j \neq \phi$, $i \neq j$, and $H_g^1(\Omega)$ is the boundary maximum approximation matrix, ε is the error-limitation. Parallel compute the sub-domain boundary value under condition of $u^0 \in H_g^1(\Omega)$, where $k = 0$, then we have:

$$\begin{cases} Lu_i^{k+1}\mid_{\Omega_i} = f \\ u_i^{h+1}\mid_{\partial\Omega_i} = u^k \end{cases}, (i = 1, 2, \ldots, k), \overline{\Omega} = \overset{k \leq K}{\underset{i=1}{U}} \overline{\Omega}_i \tag{16}$$

extend the final solution u_i^{k+1} on Ω_i to Ω, so:

$$\bar{u}_i^{k+1} = \begin{cases} u_i^{k+1}, & (x,y) \in \Omega_i \\ u^k, & (x,y) \in \Omega/\Omega_i \end{cases} \tag{17}$$

calculate average value of all sub-domain solutions, so:

$$u_i^{k+1} = \frac{1}{K}\sum_{i=1}^{k \leq K} \bar{u}_i^{k+1} \tag{18}$$

when ε is the approximate solution on u^{k+1}, there is $\|u^{k+1} - u^k\| \leq \varepsilon$.

3.2 The Algorithm Implementation of Basic IC Cascade Multi-user Detection

In the multi-path fading channel, supposing $c_k(i)$ is the decision vector, c_{opt} is the optimal decision vector, ω_{Ck} is the adaptive weight vector of each filter unit is also the adaptive update component of $c_k(i)$, ω_{opt} is the weight vector of c_{opt}, $g_k(i)$ is the equivalent channel response, $\{c_k(i)\}_{l=0}^{L-1}$ is L-length spread spectrum code, and $e_{optk}(i)$ is sun noise vector. The process equation of expected user is formed as:

$$\omega_{\text{opt1}}(i+1) = \omega_{\text{opt1}}(i) \tag{19}$$

where: $\omega_{\text{pot1}}(i)$ is the new dynamic state vector.

The observation equation of expected user is formed as:

$$y_1(i) = d_k^{\text{T}}(i)\omega_{\text{opt1}}(i) + e_{\text{opt1}} \tag{20}$$

where: $d_k^{\text{T}}(i)$ is a $R - 1$-dimensional row vector; R is the user symbols coherence length ($R = [(L+P-1)/L]$).

The dynamic system process equation is formed as:

$$\begin{cases} x_k(i+1) = F_k(i+1,i)x(i) + e_{\text{opt1}} \\ e_{\text{opt1}}(i) \equiv 0 \end{cases} \tag{21}$$

where: $x_k(i)$ is the $L \times 1$ state vector in the i-th moment; $F_k(i+1,i)$ is a known $L \times L$ state transfer matrix.

The dynamic system observation equation is formed as:

$$\begin{cases} y_k(i) = C_k(i)x_k(i) + \varepsilon_{k+1}(i) \\ y_k(i) \Rightarrow \bar{y}_k(i), \varepsilon_2(i) \Rightarrow e_{\text{opt1}}(i), C_k(i) \Rightarrow d_k^{\text{T}}(i) \end{cases} \tag{22}$$

where: $y_k(i)$ is a $R \times 1$-dimensional state vector of system in the i-th moment; $C_k(i)$ is a $R \times L$-order measurement matrix; $\varepsilon_{k+1}(i)$ is the error of measurement matrix.

Because $F_k(i+1,i)$ is a $L \times L$-order unit matrix I, so after through the channel fading, the spread spectrum code signal can be converted as:

$$\begin{cases} d_k(i) = c_k(i) * g_k(i) = \sum_{p=0}^{P-1} g_k(p)c_k(i-p) \\ i = 0, \ldots, L+p+1; p \in [0, P] \end{cases} \tag{23}$$

the r-th sampling of the received base band signal in i-th symbol period is:

$$\begin{cases} x_k(r,i) = x_k(rL+i) = \sum_{k-1}^{K} \sum_{r=0}^{R-1} A_k d_k(r,i)b_k(i-r) \\ i = 0, \ldots, L-1; d_k(r,i) = d_k(rL+i) \end{cases} \tag{24}$$

the L samples in the i-th symbol periods can be represented as a $L \times 1$-order matrix:

$$\begin{cases} x_k(i) = Ad_1b_k(i) + D_{\text{int}}d_{\text{intk}}(i) + e_k(i) + \varepsilon_{k+1}(i) \\ E[e_k(i)e(i)^{\text{H}}] = \sigma_e^2 I_L, e_k(i) = [e_k(i,0),..,e_k(i,L-1)]^{\text{T}} \end{cases} \tag{25}$$

where: D_{int} is interference matrix contains ISI and MAI; d_{intk} is the interference symbol vector.

Set a L-dimensional decision vector $f(k)$ for expected user, the MUD model is:

$$\hat{b}_1(i) = \text{sgn}(<f_k(i), x_k(i)>) \tag{26}$$

supposing the iterative initial condition is $W_k(1,0) = I$, for the expected user, the iterative calculation of S-IC is:

$$\begin{cases} g_k(i) = W_k(i, i-1)d_k(i)\{d_k^H(i)W_k(i, i-1)d_k(i) + \xi_{\min}\}^{-1} \\ W_k(i+1, i) = W_k(i, i-1) - g_k(i)d_k^H(i)W_k(i, i-1) \\ \hat{\omega}_{\text{opt1}}(i) = \hat{\omega}_{\text{opt1}}(i-1) + g_k(i)\{y_k(i) - d_k^H(i)\hat{\omega}_{\text{opt1}}(i-1)\} \\ c_{\text{opt1}}(i) = S_1 - C_{1,i}\hat{\omega}_{\text{opt1}}(i) \\ \xi_{\min} = \text{cov}\{e_{\text{optk}}\} = F(e_{\text{opt1}}^2(i)) = A_1^2 + \varepsilon_{\min} \end{cases} \tag{27}$$

where: $\varepsilon_{\min}$ is the minimum mean square error of the optimal decision vector; $\xi_{\min}$ is minimum output energy; S_1 is the S scale received signal of expected user.

So, the dynamic system equation of expected user is:

$$\begin{cases} \tilde{x}_1 = F_1^H(i)\omega_1(i) + e_2(i) + \varepsilon_2(i) \\ \omega_1(i) = \omega_1(i-1) + \Delta\omega_1(i-1) \\ \tilde{x}_k(i) = s_k^H x_k(i) \\ F_1^H(i) = x_1^H(i)U_{null} \\ \varepsilon_2(i) = c_{\text{opt1}}^H x(i) \end{cases} \tag{28}$$

where: $\tilde{x}_k(i)$ is the observation vector; $F_1^H(i)$ is the system observation matrix of expected user; $e_2(i)$ is the observation noise matrix.

Because both the number of activity users and noise characteristics are time-varying, the traditional SIC, PIC and IC detector always lead detection divergence or abnormal filtering, which reduces the detection accuracy. So, it is necessary to improvement of iterative calculation processing for traditional interference cancellation algorithm. Estimate the adaptive update section $\omega_1(i)$ of $c_{\text{opt1}}(i)$ by Schwarz algorithms rules and let it to be the tap weight vector of expected user. Set m is the number of Schwarz iteration, let i and m unified, δ is the corresponding vector forgetting factor, so we have:

$$\begin{cases} \omega_1[m] = \omega_1[m|m-1] + K[m]\delta[m] \\ \omega_1[m|m-1] = \omega_1[m-1] + q[m-1] \\ \delta_k[m] = \tilde{x}_k[m] - F_k^H[m]\omega_1[m|m-1] - r[m-1] \\ r_k[m] = \{I - d_k[m-1]\}r_k[m-1] + d_{m-1}\{\tilde{x}_k[m] - F_k^H[m]\omega_1[m|m-1]\} \\ R_k[m] = (1 - d_{m-1})R_k[m-1] \\ \quad + d_{m-1}\{\delta_k[m]\delta_k^T[m] - F_k^H[m]P_k[m|m-1]F_k[m]\} \end{cases} \tag{29}$$

So the Schwarz iteration rule is as follows:

$$\begin{cases} \omega_k(i) = \omega_k(i|i-1) + W_k(i)\delta_k \\ \quad = \omega_1(i-1)[I - W_k(i)F_k^H(i)] + q_k(i-1)[I + W_k(i)F_k^H(i)] \\ \quad + W_k(i)[\hat{x}_k(i) - r_k(i-1)] \\ q_k(i) = (1 - \delta_{k-1})q_k(i-1) + \delta_{k-1}[\omega_1(i) - \omega_1(i-1)] \end{cases} \tag{30}$$

$$\begin{cases} r_k(i) = (I - d_{k-1})r_k(i-1) + \delta_{k-1}[\hat{x}(k) - F_k^H(i)\omega_1(i|i-1)] \\ W_k(i) = \dfrac{P_k(i|i-1)F_k(i)}{F_k^H(i)P_k(i|i-1)F_k(i) + R_k(i-1)} \\ R_k(i) = (1 - \delta_{k-1})R_k(i-1) + d_{k-1} - \delta_{k-1}F_k^H(i)P_k(i|i-1)F_k(i) \\ P_k(i) = [I_R - W_k(i)F_k^H(i)]P_k(i|i-1) \\ P_k(i|i-1) = P_k(i-1) + Q_k(i-1) \\ Q_k(i) = (1 - d_{k-1})Q_k(i-1) + d_{k-1} + d_{k-1}P_k(i)P_k(i-1) \end{cases} \qquad (31)$$

where: $\delta_{k-1} = (1-b)/(1-b^k)$, $0 < b < 1$, Q is the $K \times (2P+1)$-dimensional processing boundary conditions of square-integrable observation function, (k, i)-element in Q is $q_k(i)$.

3.3 The Successive and Parallel Detection Units of Schwarz-IC Cascade Process Implementation

The improved SIC and PIC unit single-stage structure of S-IC detector for the multiple access communication system are shown in Figs. 3 and 4.

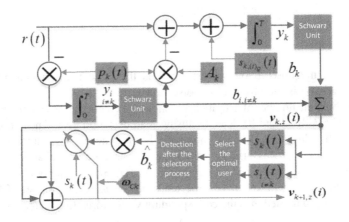

Fig. 3. Single-stage structure of S-IC processing

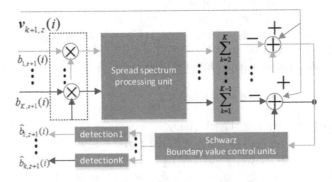

Fig. 4. Single-stage structure of S-IC processing

The improved structure of S-IC detection unit can use the relevant information of observational data to estimate the properties of time-varying unknown noise statistics real-time while conducting state filtering. Set y_k is the actual output vector of filter, $v_k(i)$ is the input vector, $u_k(i)$ is the energy input vector of S-IC unit, $g_k(i)$ is the expected response vector, and μ is the step parameter, so

$$\begin{cases} v_k(i) = d_k(i) + g_k(i) - \omega_{\text{optk}}^{\text{H}}(i)u_k(i) \\ \omega_k(i+1) = \omega_{\text{optk}}(i) + \mu u_k(i) * v_k(i) \end{cases} \tag{32}$$

after dealing with double spread spectrum, the signal output of k-user in the z-th (z is an integer greater than 1) detection level of traditional S-IC detector can be formed as:

$$b_{k,z}(i) = \text{sign}[S_k^{\text{T}} v_{k,z}(i)] \tag{33}$$

the maximum power user is expressed as:

$$r_{\max}(t) = \arg \max_{1 \le k \le K} \{|S_k^{\text{T}} v_{k,z}(i)|\} = \arg \max_{1 \le k \le K} \{|[s_{k,0}, .., s_{k,K-1}] v_{k,z}(i)|\} \tag{34}$$

where: S_k is S band asynchronous spread spectrum code after double spread spectrum. On the z-th detection level, $v_{1,z}(i) = r(t)$, detected signal $b_{k,z}(i)$ can be changed as:

$$v_{h+1,z}(i) = v_{h,z}(i) - \omega_{\text{opth}}(i)b_{h,z}(i)S_h \tag{35}$$

the weight for the detector after through adaptive filter detection unit is adjusted as:

$$\omega_k(i+1) = \omega_k(i) + 2\mu v_{k,z}(i)b_{k,z}(i)S_k \tag{36}$$

According to Eq. 9, the received PIC signal vector is expressed as:

$$\begin{cases} r(t) = S(t, Q)P(t, Q) + e \Rightarrow r = \sum_i \sum_{k=1}^{K} A_k b_k^{(i)} s_k p_k + e \\ S(t, Q) = \sum_i \sum_{k=1}^{K} s_{k,q(i)}(t - iT - \tau_k) \\ P(t, Q) = \sum_i \sum_{k=1}^{K} p_{k,q(i)}(t - iT - \tau_k) \end{cases} \tag{37}$$

the estimated value of b on the z-th PIC level is:

$$\hat{b}_{k,z}(i) = [\hat{b}_{1,z}(i), \hat{b}_{2,z}(i), .., \hat{b}_{K,z}(i)]^{\text{T}} \tag{38}$$

the information bit on the $z + 1$-th IC level is:

$$
\begin{cases}
\hat{b}_{k,z+1}(i) = \arg\{ \underset{\substack{q_k \in (-1,1) \\ q_i = b_k(i), \forall \neq k}}{\max} [2\boldsymbol{y}^{\mathrm{T}}\boldsymbol{b} - \boldsymbol{b}^{\mathrm{T}}\varphi\boldsymbol{b}]\} = \mathrm{sgn}[\boldsymbol{\phi}_{k,z}(i)] = \mathrm{sgn}[y_k - \sum_{i \neq k} \hat{b}_{k,z}(i)\varphi_{kl}] \\
\boldsymbol{\phi}_{k,z}(i) = [\phi_{1,z}(i), \phi_{2,z}(i), \ldots, \phi_{K,z}(i)]^{\mathrm{T}} \\
\varphi_{kl} = \int_0^T s_k(t)p_k(t)s_l(t)p_l(t)dt \\
y_k = \int_0^T r(t)s_k(t)p_l(t)dt
\end{cases}
$$

$$(39)$$

where: $\phi_{k,z}(i)$ is the z th statistical results; φ_{kl} is the element of correlation matrix φ.

The Schwarz-IC detector re-estimates MAI by the estimated value of $\boldsymbol{b}$ obtained on the m-th detection level, then cancels out the latest produced MAI from $\boldsymbol{y}$, finally obtains the estimated value of $\boldsymbol{b}$ on the $m + 1$-th detection level. When there is no new user information added in the channel, the IC cascade detection decision process is terminated. Because the signal power and the wireless communications parameters are time-varying, so the introduction Schwarz unit can precise track the channel variation and a single user signal power by the precise control of the user power sub-domain boundary value. Furthermore, it can ensure the accuracy of the interference suppression processing and improve the overall system performance.

4   Simulation Results and Performance Analysis

In a multiple access communication system (Multi-path number $P = 10$, K users), let each user send an information symbol in multi-path channel in each simulation step (1s), then use m-sequences (The number of sequences is K, $N = 31$) to make independent spread spectrum and add sum noise processing, while make adding processing in user's order respectively. The K users send asynchronous signal in S band transmission asynchronous after through double spread spectrum processing. Then, use the same K m-sequence to despread the information symbols. Finally, complete the symbol recovery processing of these K users (the symbol number is equal to the transmission time) by the integral decision at receiving terminal and sending terminal (Fig. 5).

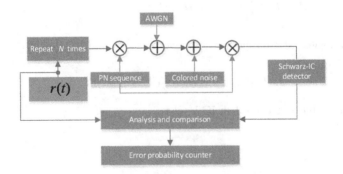

Fig. 5. Multiple access communication spread spectrum system model

set the k-user as the minimum power user, every bit energy is $A_k^2 T/2$. Use Schwarz-IC and SIC detector to detect the excess output energy (EOE) performance of Schwarz-IC, PIC and SIC algorithm. The k-th iterative output signal to interference ratio (SIR) performance on the m-th level of this system is defined as:

$$\text{SIR} = \frac{E^2\{c_{\text{optk}}^{\text{T}}(n)r\}}{\text{var}\{c_{\text{optk}}^{\text{T}}(n)r\}} = \frac{A_k^2(c_{\text{optk}}^{\text{T}}(n)p_k)^2}{\sum_{k=2}^{K} A_k^2(c_{\text{optk}}^{\text{T}}(n)p_k)^2 + \sigma^2 c_{\text{optk}}^{\text{T}}(n)c_{\text{optk}}(n)} \tag{40}$$

the output SIR is defined as:

$$\text{SIR} = \frac{E^2\{c_{\text{optk}}^{\text{T}}(m)r\}}{\text{var}\{c_{\text{optk}}^{\text{T}}(m)r\}} = \frac{A_k^2(c_{\text{optk}}^{\text{T}}(m)p_k)^2}{\sigma^2 c_{\text{optk}}^{\text{T}}(m)c_{\text{optk}}(m) + \sum_{k=2}^{K} A_k^2(c_{\text{optk}}^{\text{T}}(m)p_k)^2} \tag{41}$$

the BER is defined as:

$$\begin{cases} \text{BER} = \phi(\sqrt{e_k(\sigma)\sigma^{-2}}) = \frac{1}{\sqrt{2\pi}} \int_{\sqrt{e_k(\sigma)\sigma^{-2}}}^{\infty} \ln(-\frac{u^2}{2})du \\ \phi(j) = \frac{1}{\sqrt{2\pi}} \int_{j}^{\infty} \ln(-\frac{u^2}{2})du \end{cases} \tag{42}$$

where: $e_k(\sigma)$ is the equivalent effective noise energy of the k-th user; ϕ is the conversion function.

EOE is defined as the excess energy of transmitted user signal in order to achieve single-user error performance for MUD algorithm in the mobile communication system, namely the more stable and rapidly for the EOE decay, the more stable the system transmission performance is. The EOE is defined as:

$$\text{EOE} = \xi(m) - \xi_{\min} = \text{MOE}[x_k(m)] - (A_1^2 + \varepsilon_{\min}) \tag{43}$$

Use spreading sequence adapts GOLD sequence, source adapts BPSK signal, step size is $\mu = 0.0005$, the sampling rate is equal to the chip rate. The difference power value between the maximum user and the minimum user is 8 dB, the BER is defined as:

$$P_k(\sigma) = Q\sqrt{e_k(\sigma)\sigma^{-2}} \tag{44}$$

where: $e_k(\sigma)$ is the equivalent energy of the kth user.

4.1   SIR Performance Comparison Analysis

Set K users with different power values. Assume there are no addition and no withdrawal of existing users in the process of the whole communication. Add a set of users with lager power when the iteration number is 600, then withdraw these users and a part of original user when the iteration number is 1200. This program simulate the actual dynamic mobile communication environment.

As shown in Fig. 6: In the multi-user static static conditions, when the iteration number is greater than 400, the SIR performance of S-IC algorithm is significantly better than SIC and PIC algorithm. So S-IC algorithm is of faster convergence rate and

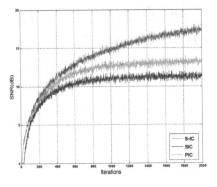

Fig. 6. The   static   SIR   performance analysis

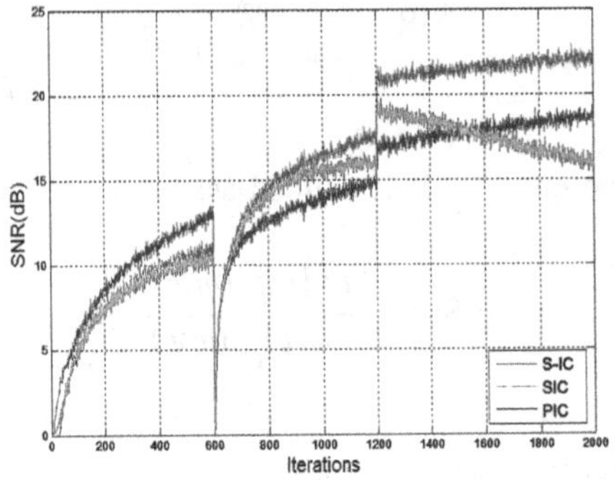

Fig. 7. The   dynamic   SIR   performance analysis

stronger multi-user interference inhibition ability in static condition.

As shown in Fig. 7: In the multi-user dynamic conditions, when there is new interference added in the system (the iteration number is 600), the SIR curve of S-IC and SIC algorithm just appear little bit down peak and recover fast at a high speed before the interference been receded, then basically restore stability convergence after the interference been receded (the iteration number is greater than 1200), while the SIR performance of S-IC algorithm is significantly better than SIC algorithm in the whole process. But the SIR curve of PIC algorithm appears a great attenuation volatility and even becomes unstable convergence after the interference is withdrawn. So S-IC is of better dynamic tracking performance than SIC and PIC algorithm.

4.2   BER Performance Analysis

As shown in the Fig. 8: In the multi-user static conditions, the blind adaptive Schwarz-IC is of better BER performance, namely the new detector can effectively inhibit the interference of strong NFP and improve the detection precision. So Schwarz-IC is of better MAI rejection ability, and MUD ability.

As shown in the Fig. 9: In the dynamic conditions, the S-SIC algorithm is of better BER performance, namely the new detector can effectively inhibit the interference of strong NFP and improve the detection precision. So S-SIC is of better MAI rejection ability, and MUD ability.

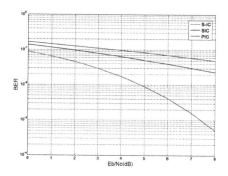

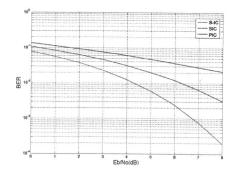

Fig. 8. The static BER performance analysis

Fig. 9. The dynamic BER performance analysis

4.3 EOE Performance Analysis

As shown in Fig. 10: In the multi-user static conditions, the EOE curve of SIC algorithm appears a great instability fluctuations when the interference brought into the system and recover very slow at a low speed before the interference users been receded while basically higher than 0.1 dB. The EOE of PIC appears serious divergence after the interference brought into the system and ultimately failed to converge. The EOE of S-IC start remarkable convergence before the interference been receded and basically lower than 0.1 dB, final attenuate close to 0 dB value theory. So S-IC is of better interference rejection capability, convergence stability and MUD ability.

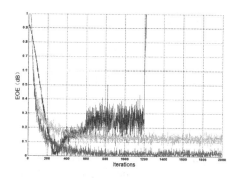

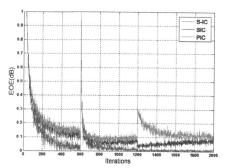

Fig. 10. The static EOE performance analysis

Fig. 11. The dynamic EOE performance analysis

As shown in Fig. 11: In the multi-user dynamic conditions, the EOE curve of SIC algorithm appears a great instability fluctuations when the interference brought into system and recover very slow at a low speed before the interference users been receded while basically higher than 0.1 dB. The EOE of PIC appears serious divergence after the interference brought into system and ultimately failed to converge. The EOE of S-IC start remarkable convergence before the interference been receded and nearly lower than 0.1 dB, final attenuate close to 0 dB value theory. So S-IC algorithm is of better interference rejection capability, convergence stability and MUD ability.

4.4   Error Probability Performance Analysis

As shown in Fig. 12: In the multi-user static conditions, the error probability of SIC algorithm is always higher than 10^{-4}, the error probability of PIC algorithm is always higher than 10^{-5}, furthermore, the decay rate of SIC and PIC algorithm is slower than S-IC algorithm. In contrast, the error probability of S-IC algorithm showed a rapid decay state in the whole detection process and closest to the single-user case. So S-IC algorithm is of better static detection accuracy than SIC and PIC algorithm.

As shown in Fig. 13: In the multi-user dynamic conditions, the error probability of PIC algorithm is always higher than 10^{-2}, the error probability of SIC algorithm is always higher than 10^{-3}, furthermore, like situation in the multi-user static conditions, the decay rate of SIC and PIC algorithm is significantly slower than S-IC algorithm. Although the error probability of S-IC algorithm is slightly worse than the same situation in the multi-user static conditions, there has been no noticeable performance degradation, which still showed a rapid decay and closest to the single-user case. So S-IC algorithm is of better dynamic detection accuracy than others.

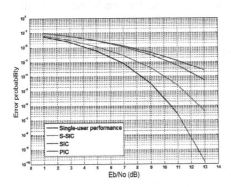   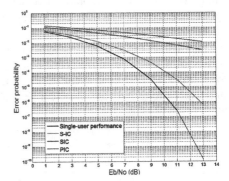

Fig. 12. The static EP performance analysis   **Fig. 13.** The dynamic EP performance analysis

5   Conclusion

The Schwarz-IC algorithm can adaptively determine whether to continue operation according to the latest information variables without waiting for data input at any time, while it is helpful for implementation of project. Because there is no system background noise adding in the MAI cancellation processing, the new algorithm can fully track the time-varying channel in complex condition and completely eliminate the co-channel interference. Because it is no need to sort the order of different power user in the MUD processing, so it can effective avoid detection error diffusion caused by the intermediate link detection error of traditional SIC algorithm. Simulation results show that, the Schwarz-IC algorithm outperforms the Schwarz algorithm, PIC algorithm and SIC algorithm in term of BER performance, dynamic tracking capability, convergence and precision control capability. Therefore, the blind adaptive Schwarz-IC MUD algorithm is an efficient MUD scheme.

Acknowledgment. This work was supported in part by:

(1) The National Science Foundation of China (61701521).

(2) The Certificate of China Postdoctoral Science Foundation Grant (2016M603044).

(3) The National Science Foundation of ShannXi (2018JQ6074).

References

1. Yang, K., Hu, J.: Wireless data and energy integrated communication networks. ZTE Commun. **16**(01), 1 (2018)
2. Zou, J.: Low-complexity interference cancellation receiver for sparse code multiple access. In: Proceedings of 2015 IEEE 6th International Symposium on Microwave, Antenna, Propagation, and EMC Technologies (MAPE 2015), p. 6 (2015)
3. Tian, T.: Dichotomous coordinate descent based successive interference cancellation algorithm for MIMO systems. In: Proceedings of 2017 IEEE 9th International Conference on Communication Software and Networks (ICCSN 2017), p. 4 (2017)
4. Xu, G.: New parallel interference cancellation for coded MC-DS-CDMA. In: Proceedings of 2016 International Conference on Communications, Information Management and Network Security (CIMNS2016), p. 4 (2016)
5. Lu, H., Huang, C., Shao, S., Tang, Y.: Novel multi-tap analog self-interference cancellation architecture with shared phase-shifter for full-duplex communications. Sci. China (Inf. Sci.) **60**(10), 139–154 (2017)
6. Zhai, X.: Compressive sensing multi-user detection for MC-CDMA system in machine to machine communication. In: Proceedings of 2017 2nd International Conference on Wireless Communication and Network Engineering (WCNE 2017), p. 6 (2017)
7. Li, X., Shi, Y., Wang, X., Xu, C., Sheng, M.: Efficient link scheduling with joint power control and successive interference cancellation in wireless networks. Sci. China (Inf. Sci.) **59**(12), 23–37 (2016)
8. Wei, L.: Performance of successive interference cancellation scheme using VW-OOC in OCDMA systems. In: Proceedings of 2015 IEEE International Conference on Communication Problem-Solving (ICCP), p. 4 (2015)
9. Liang, W.: Performance of parallel interference cancellation scheme In: Proceedings of SCIEI 2015 Rome Conference (SCIEI), VW-OCDMA Systems, p. 4 (2015)
10. Tian, Y.-H.: Parallel interference cancellation for MIMO radar receiver. In: Proceedings of 2015 4th International Conference on Mechatronics, Materials, Chemistry and Computer Engineering (ICMMCCE 2015), p. 5 (2015)
11. Chuang, G.: A new adaptive noise cancellation method of non-continuous communication signal submerged in multi-interference noise. In: Proceedings of the Seventh Asia-Pacific Conference on Environmental Electromagnetics (CEEM 2015), p. 4 (2015)
12. Liu, Y., Chen, Z., Pan, Y.: A boundary schwarz lemma for holomorphic mappings on the polydisc. Chin. Ann. Math. **39**(01), 9–16 (2018)
13. Ma, X.: Wireless transmission space coupling characteristics and new coupling mode proposed. In: Proceedings of 2018 5th International Conference on Key Engineering Materials and Computer Science (KEMCS 2018), p. 6 (2018)
14. Tian, L., Zhang, L., Li, J.: Pseudo-random coding side-lobe suppression method based on CLEAN algorithm. J. Beijing Inst. Technol. **26**(03), 375–380 (2017)
15. Chen, S.L., Ponnusamy, S., Rasila, A., Wang, X.T.: Linear connectivity, Schwarz–Pick lemma and univalency criteria for planar harmonic mapping. Acta Mathematica Sinica **32**(03), 297–308 (2016)

A Novel DAG Spectrum Sensing Algorithm with Reducing Computational Complexity

Weiting Gao[1(✉)], Fuwei Jiang[1], Fei Ma[2], and Weilun Liu[1]

[1] Institute of Information and Navigation, Airforce Engineering University,
Xi'an, China
519105941@qq.com, 1197853086@qq.com,
wanyuyubei@hotmail.com
[2] Xi'an Modern Control Technology Research Institute, Xi'an, China
365553071@qq.com

Abstract. Aiming at problems that the eigenvalue based spectrum sensing algorithms don't perform well in the situation of low SNR, small sample and need high computational complexity with eigenvalue decomposition, based on the difference value between maximum and minimum eigenvalue spectrum sensing algorithm (DMM), a difference value between the arithmetic mean and geometric mean eigenvalue spectrum sensing algorithm (DAG) with low computational complexity and dynamic threshold was proposed, which via the power method. Simulation results show that the DAG can improve performance over the classical algorithms in situation of low SNR, small samples and increased second users without reduction of computational complexity.

Keywords: Eigenvalue · Spectrum sensing · Arithmetic mean ·
Geometric mean · Computational complexity · DMM

1 Introduction

As the key link of Cognitive Radio, spectrum sensing can be used by the cognitive users (the Second User, SU) through detecting the real-time frequency "spectrum holes" (White Space), real-time and accurately judge the authorized frequency resources of primary user (PU) is idle or not, which is acknowledged as an important technique to solve the spectrum tense problem [1].

The existing spectrum sensing algorithms mainly include: matching filtering detection, cyclic stable feature detection, energy detection, multi-user collaborative detection and so on [2]. The matching filter algorithm requires the primary user prior signal to design the filter mechanism, which does not conform to actual application. The performance of cyclic stable feature detection is better, but computation complexity is high and the real-time capability is poor [3]. The energy detection is simple and easy to implement, and is not a priori to the primary user signal, but its performance is greatly influenced by the noise uncertainty, which has the threshold requirement of SNR. The great advantage of the eigenvalue spectrum sensing algorithm is that it can overcome the influence of noise uncertainty. But the performance of the classical eigenvalue spectrum sensing algorithm is poor in low SNR and low sampling

© ICST Institute for Computer Sciences, Social Informatics and Telecommunications Engineering 2019
Published by Springer Nature Switzerland AG 2019. All Rights Reserved
B. Li et al. (Eds.): IoTaaS 2018, LNICST 271, pp. 448–458, 2019.
https://doi.org/10.1007/978-3-030-14657-3_46

condition, and the computational complexity of sampling covariance eigenvalue decomposition is high. The MME algorithm based on the eigenvalue of the maximum and the minimum structured statistic, which use the Tracy-Widom distribution of Wishart matrix minimum limit value and maximum eigenvalue to overcame the influence of noise uncertainty and obtained a good detection result, but the constant detection threshold value is not in conformity with practical application of the scene [5]. The approximate value of the maximum eigenvalue is obtained by the method of eigenvalue approximation, which can reduce the operation complexity, but the threshold theoretical expression cannot be obtained. The "cumulative method" of MMS algorithm could reduce the computational complexity by the iterative computation, but the threshold value is also unable to change dynamically according to the noise [6].

The DMM algorithm based on the difference value of the maximum and minimum eigenvalues, and its dynamically changeable threshold value is further expand the development of this field [7]. Based on the dynamic detection threshold of DMM algorithm, this paper proposed an improved DAG algorithm, which improve the detection quantity by arithmetic mean and geometric mean. The DAG algorithm approximate the maximum eigenvalue by "cumulative method" iterative to reduce the computational complexity.

2   The System Model

2.1   Actual Spectrum Sensing Scene

In the cognitive radio wireless network as Fig. 1, primary user (PU1, PU2, PU3,...) communicate through the primary base station (PBS) [8]. The cognitive users (SU1, SU2, SU3) collaboratively detect the PU signal, send detection data to the secondary station (Second Base Station, SBS) for data processing, then make decision whether there is any white space can be used by SU in PU authorized frequency [9].

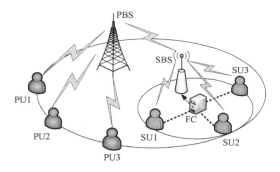

Fig. 1. The real spectrum sensing scene

2.2 Spectrum Sensing Model and DMM Algorithm

Take the classical binary hypothesis math model of cognitive radio, assume there is only one primary user (PU) in the narrow band cognitive radio network [10], the cognitive users detect and judg the signal as follows:

$$x_i(n) = \begin{cases} \omega_i(n) & H_0 \\ h_i(n)s_i(n) + \omega_i(n) & H_1 \end{cases} \tag{1}$$

where:

H_0 and H_1 represent the primary user signal exists or not;
$x_i(n)$ is the i th sampling signal received by cognitive user;
$s_i(n)$ is the detect signal of primary user;
$\omega_i(n)$ is the interference noise;
$h_i(n)$ is the channel fading factor.

for generality, assume as follows:

(1) Interference noise is the white gaussian noise, which need obey $\omega_i(N) \sim N(0, \sigma^2)$;
(2) The amplitude of PU signal $s_i(n)$ obey the gaussian distribution of the mean variance μ and variance σ^2, which is independent with the noise;
(3) M cognitive users detect the same frequency band of a primary user;
(4) During the detection period, the channel characteristics are stable and $h_i(n)$ is unchanged;

Sample the M cognitive users N times, set X is the $M \times N$ matrix, the multi-user collaborative spectral perception model can be summarized as $X = Hs + \omega$, each element $x_i(n)$ represents a sample value of the i-th cognitive user at time n, ($i = 1,2,3,\ldots M; k = 1,2,3,\ldots N$). When the cognitive user sampling signal covariance is large, it can be approximately expressed as:

$$R_x(N) = \frac{1}{N}\sum_{n=0}^{N-1} x(n)x(n)^T = \frac{1}{N}XX^T = R_{Hs}(N) + \sigma^2 I_M \tag{2}$$

set $\lambda_i = \rho_i + \sigma^2$ is the eigenvalues of $R_x(N)$, ρ_i is the eigenvalues of ($i = 1,2,3,\ldots M$) of $R_{Hs}(N)$, $\bar{\lambda}$ is the mean value of λ_i, ρ_{max} is the maximum eigenvalues of $R_{Hs}(N)$, ρ_{min} is the minimum eigenvalues of $R_{Hs}(N)$. The eigenvalue statistic mean value $\bar{\lambda}$ of $R_x(N)$ can provide the feasibility of detection in presence differences of primary user signals.

When H_0 set up, there is only the Gauss white noise, and obey:

$$\bar{\lambda} = \lambda_{max} = \lambda_{min} = \sigma^2 \Leftrightarrow R_x(N) = \sigma^2 I_M \tag{3}$$

When H_1 set up, because the different self sampling times of $s(n)$ has correlation, and ρ_i make λ_i do not equal, so:

$$\overline{\lambda} = \overline{\rho} + \sigma^2 \neq \sigma^2, (\lambda_{max} = \rho_{max} + \sigma^2, \lambda_{min} = \rho_{min} + \sigma^2) \tag{4}$$

The DMM algorithm based on the difference value of the maximum and minimum eigenvalue of receiving signal in the presence of the primary user [8], therefore, the detection statistics of DMM algorithm can be represented as:

$$T_{DMM} = \lambda_{max} - \lambda_{min} \underset{H_0}{\overset{H_1}{\underset{<}{\gtrless}}} \gamma_{DMM} \tag{5}$$

where:

γ_{DMM} is the decision threshold, compare with the threshold value to determine whether the primary user signal exists or not.

Although DMM algorithm can dynamically estimate noise information and set detection threshold to overcomes the influence of noise uncertainty [9], but exist following problems as:

(1) DMM algorithm need the eigenvalue decomposition of covariance matrix to calculate λ_{max} and λ_{min}, which resulted in greater computational complexity;
(2) The detection statistics do not make full use of eigenvalues, but only use the maximum and minimum eigenvalues, which only reflect part characteristics of the matrix;

2.3   The Optimization of Mean Eigenvalues

For the problem of DMM algorithm, the arithmetic mean and geometric mean eigenvalues are used to improve the detection statistics. The expressions of arithmetic mean and geometric mean are as follows:

$$\begin{cases} \lambda = \frac{1}{M} \sum_{i=1}^{M} \lambda_i = \frac{tr(R_x(N))}{M} \\ \tilde{\lambda} = (\prod_{i=1}^{M} \lambda_i)^{1/M} = (\det(R_x(N)))^{1/M} \end{cases} \tag{6}$$

The sum of is equal to the sum of the square matrix diagonal elements, and the product of all the eigenvalues is equal to the matrix determinant value. Respectively take $\overline{\lambda}$ and $\tilde{\lambda}$ in to replace λ_{max} and λ_{min} in the DMM algorithm, which can avoid the reducing of computational complexity caused by the eigenvalue decomposition of sampling covariance. When H_0 set up, $R_x(N)$ is the diagonal matrix, $\overline{\lambda} = \tilde{\lambda} = \sigma^2$, when H_1 set up, $\overline{\lambda}$ and $\tilde{\lambda}$ are not equal. Take the difference value between $\overline{\lambda}$ and $\tilde{\lambda}$ to measure the difference of the primary user signal, and construct the detection statistic by $\overline{\lambda} - \tilde{\lambda}$ to optimize the detection amount of the DMM algorithm.

Usually, it could set the maximum detection probability P_d $(P(H_1|H_1))$ as the basis for the algorithm detection performance in case of fixed virtual alarm probability $P_{fa}(P(H_1|H_0))$. In the same P_{fa} condition, let P_d increase, the algorithm is considered been optimized.

3 DAG Spectral Sensing Algorithm

3.1 Algorithm Detection Threshold Structure

In the ideal situation, the value of $\overline{\lambda} - \tilde{\lambda}$ is 0 when the primary user signal is absent. In case of virtual alarm probability, structure the detection statistics as:

$$T_{DAG} = \overline{\lambda} - \tilde{\lambda} \underset{H_0}{\overset{H_1}{\underset{<}{\gtrless}}} \gamma_{DAG} \tag{7}$$

where:

γ_{DAG} is the detection threshold, which affects the detection performance.
The virtual alarm probability of DAG algorithm is expressed as:

$$P_{fa} = P(T_{DAG} > \gamma_{DAG}|H_0) = P(\overline{\lambda} - \tilde{\lambda} > \gamma_{DAG} \mid H_0) \tag{8}$$

when N is large enough, $\overline{\lambda}$ obey normal distribution $N(\sigma^2, 2\sigma^4/MN)$, so:

$$\begin{cases} P_{fa} = P(\overline{\lambda} > \gamma_{DAG} + \tilde{\lambda} \mid H_0) = P(\frac{\overline{\lambda} - \sigma^2}{\sigma^2\sqrt{2/MN}} > \frac{\gamma_{DAG}}{\sigma^2\sqrt{2/MN}}) = Q(\frac{\gamma_{DAG}}{\sigma^2\sqrt{2/MN}}) \\ Q(x) = \int\limits_{x}^{\infty} \frac{1}{\sqrt{2\pi}}\exp(-t^2/2)dt \\ \gamma_{DAG} = \sigma^2 Q^{-1}(P_{fa})\sqrt{2/MN} \end{cases} \tag{9}$$

where:

$Q(x)$ is the probability integral function, which satisfies the expression.
Because the threshold value is related to the noise energy, so use the eigenvalue noise estimation method of DMM algorithm to estimate the noise in real time as:

$$\sigma^2 \approx (Tr(R_x(N)) - \lambda_{\max})/(N - 1) \tag{10}$$

By using the trace and maximum eigenvalue of the sampling covariance matrix to update the noise in real time, obtain the detection threshold $\gamma_{DAG}(\sigma^2)$ in different dynamic change condition with different SNR.

3.2 Maximum Eigenvalue Calculation of DAG Algorithm

Because the way of using eigenvalue decomposition to calculate $\lambda_{\max}$ always increase the algorithm complexity, so another way is using "cumulative method" to calculate $\lambda_{\max}$, which through continuous iterative to gradually approach the maximum eigenvalue of positive definite symmetric matrix [10].

For the sample covariance matrix, the eigenvalues and the eigenvectors are corresponding one by one as $\{v_m, \lambda_m\}_m^M \, (0 < m \leq M)$. Because the covariance matrix is symmetric matrix, and the eigenvector set $\{v_m\}_m^M$ can form an orthonormal basis for M dimension field R^M [11], therefore, assume the initial iterative eigenvector is $v(v \in R^M)$, v can be represented as a linear combination of $\{v_m\}_m^M$ as:

$$v = \sum_{m=1}^{M} \alpha_m v_m \tag{11}$$

If λ_1 is the maximum eigenvalue, the corresponding eigenvector is v_1, and $\alpha_1 \neq 0$, the sampling covariance matrix of k iteration is expressed as $R^k_{x\,M \times M}$, and the corresponding eigenvalue is λ^k_m, which satisfies:

$$R^k_x v = \sum_{m=1}^{M} \alpha_m R^k_x v_m = \sum_{m=1}^{M} \alpha_m \lambda^k_m v_m \tag{12}$$

the left and right sides divided into λ^k_1 as:

$$\frac{R^k_x v}{\lambda^k_1} = \alpha_1 v_1 + \sum_{m=2}^{M} \alpha_m \left(\frac{\lambda_m}{\lambda_1}\right)^k v_m \tag{13}$$

Because $\lambda_1 > \lambda_m\,(2 \leq m \leq M)$, as k increases, $(\lambda_m/\lambda_1)^k$ will approaches to 0, therefore, $R^k_x v \approx \alpha_1 v_1 \lambda^k_1$. The algorithm can concluded as follows (Table 1):

Table 1. The basic step of DAG algorithm.

algorithm:
Input: $R_x(N)$; $v_0 (\|v_0\|_2 = 1)$
For $k = 1, 2, 3, \ldots, do$
$w \leftarrow R_x v_{k-1};$
$v_k \leftarrow w / \|w\|_2;$
$\lambda_k \leftarrow v_k^T R_x v_k;$
End for
Output: λ_k (λ_k is the approximation of λ_{max} after k iteration)

For the $M \times M$ matrix, if only take multiplication into consideration, the computational complexity of "cumulative method" to calculate λ_{max} is $O(kM)$, while the computational complexity of eigenvalue decomposition to calculate λ_{max} is $O(M^3)$. Because the computational complexity of the eigenvalue class spectrum sensing algorithm is mainly resulted from the eigenvalue decomposition of sampling covariance matrix [12], the computational complexity of DAG algorithm is much lower than DMM algorithm. The steps of DAG algorithm could be summarized as follows:

(1) Sampling the detection signal and calculate the covariance matrix $R_x(N)$ of the detected signal;
(2) Calculate the matrix trace $Tr(R_x(N))$ and the determinant $\det(R_x(N))$;
(3) Use $Tr(R_x(N))$ and $\det(R_x(N))$ to calculate $\bar{\lambda}$ and $\tilde{\lambda}$, and construct detection $\bar{\lambda} - \tilde{\lambda}$;
(4) Use "cumulative method" to calculate the maximum eigenvalue λ_{max};
(5) According to $\sigma^2 \approx (Tr(R_x(N)) - \lambda_{max})/(N-1)$, use the eigenvalue and trace to estimate the noise, and calculate the threshold γ_{DAG} under the specific virtual alarm probability P_{fa} according to the estimated noise;
(6) Carry out judgment, cumulative number of times, and calculate detection probability.

4   Simulation Results and Performance Analysis

4.1   Validation of Threshold Validity

If there is no primary user signals, and only input additive Gaussian white noise to simulate the detection scene of H_0. In case that the cognitive users are $M = 5$, the virtual alarm probability is $P_{fa} = 0.1$, the relationship between the detection statistic $\bar{\lambda} - \tilde{\lambda}$ and the detection threshold is obtained under different sampling points N, which is shown in Fig. 2:

Figure 2 expressed the vast majority of actual detection quantities are lower than the threshold theoretical value, because the allowed virtual alarm probability threshold setting is 0.1, very few sample points value are higher than the detection threshold value, which proved the validity of threshold value setting. At the same time, "virtual alarm points" is increasing with the increase of sampling points number, by reason that the sampling points number increase and the threshold value decrease. So for the fixed detection structure, the lower the threshold is, the higher the detection probability is, and the virtual alarm probability will also increase.

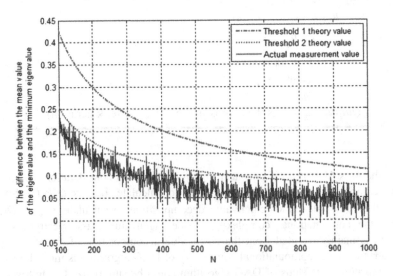

Fig. 2. The threshold validity of DAM algorithm

4.2   Analysis of Algorithm Performance

Assuming the signal of the primary user is QPSK signal, set the virtual alarm probability as P_{fa}, after 2000 Monte- Carlo simulation experiment, take the statistical detection probability P_d as the index, and then compared with classic ED, MME and DMM algorithm.

Figure 3 expressed the relationship between detection probability and SNR, when virtual alarm probability is P_{fa}=0.05, cognitive users M = 5, and the number of sampling points is 1000. set $ED - x$dB to express the ED algorithm with the noise uncertainty is x., the detection probability curves of the algorithms increase rapidly with the increase of SNR, and the optimal ED algorithm is best. However, the performance decrease sharply when there exists 1 dB noise uncertainty, and there appears "SNR wall" in -10 dB. In low SNR condition, the performance of DAG algorithm is better than DMM and MME algorithm, by reason that the detection threshold value of MME algorithm is not related to noise, so MME algorithm cannot use noise estimation to make dynamic adjustment of threshold value like DMM and DAG algorithm. The DAG algorithm use the arithmetic mean value of eigenvalue to approximate the average signal energy, and its "energy" characteristic makes it maintain a certain detection performance in low SNR condition.

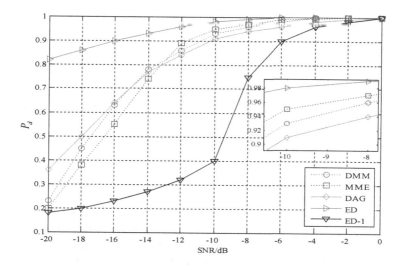

Fig. 3. The curve of detection probability with SNR

Figure 4 expressed the relation between detection probability and signal-to-noise ratio when virtual alarm probability $P_{fa} = 0.05$, cognitive users M = 5, and the SNR is -20 dB, which aims to further verify the influence of the sampling points on the detection performance under the low SNR condition of Fig. 3. Beside the ED algorithm in 1 dB noise uncertainty condition, the detection probability under low SNR is compensated by increasing the number of sampling points. Compared with MME and DMM algorithm, Before the number of sampling points is 4000, the DAG algorithm always maintains the advantage of detection probability under low SNR. With the number of sampling points continue to rise, its advantages gradually disappear and close to MME and DMM algorithm, by reason that the DAG algorithm based on normal distribution is not accurate enough to describe the threshold value in large sample compared with MME and DMM algorithm, which based on the Tracy-Widom distribution, and the eigenvalue algorithm is very sensitive to threshold value under the low SNR condition.

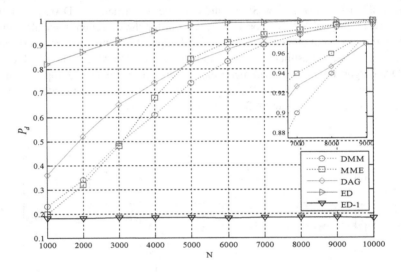

Fig. 4. The change curve of detection probability with number of sampling points

Figure 5 expressed the relation between detection probability and cognitive users, when virtual alarm probability $P_{fa} = 0.05$, SNR is -20 dB, and the number of sampling points is 1000, which aims to test the influence on detection probability by increasing the number of cognitive users participating in the collaboration.

As shown in Fig. 5, the increase of cognitive users number and sampling points number cannot eliminate the influence of noise uncertainty on ED algurithm. With the congitive users number increasing, the sampling covariance dimension and the eigenvalues number both increased, the advantage of eigenvalue mean value is highlighted, so the detection probability of DAG, MME and DMM algorithm are all increased, and situation of DAG algorithm shows best, by reason that the eigenvalue mean value use all eigenvalue, which can more reflect the "features" of matrix.

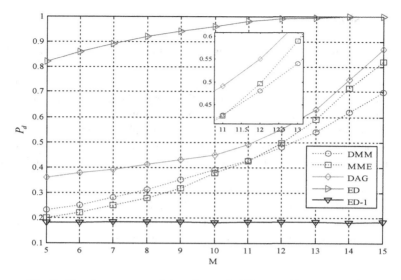

Fig. 5. The change curve of detection probability cognitive user number

Figure 6 expressed the receiver operating characteristics (ROC) in different virtual alarm probability when the sampling points is 1000, SNR is −20 dB, cognitive users M = 5. In low virtual alarm probability and high virtual alarm probability conditions, although the detection probability of DAG algorithm is lower than ideal ED algorithm, but the situation is higher than that of MME and DMM algorithm. The ROC curve comprehensive indicates that in low SNR and the relatively low sampling condition, the DAG algorithm has the higher average detection probability.

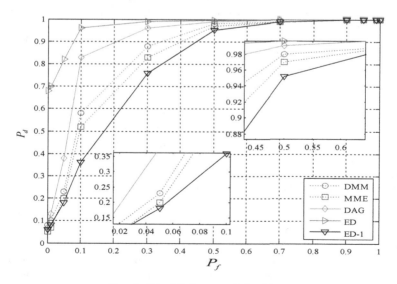

Fig. 6. The ROC curve of algorithm

5 Conclusion

This paper proposed a spectrum sensing optimization DAG algorithm based on the difference value between the geometric mean eigenvalues and arithmetic mean eigenvalues, which on theory of classical DMM algorithm. The DAG algorithm calculate the maximum eigenvalue by 'cumulative method' to obtain the dynamic detection threshold, which avoid the eigenvalue decomposition of sampling covariance matrix. Simulation results show that, the DAG algorithm has the higher detection probability and lower computing complexity than MME and DMM algorithm, with situations under the condition of low SNR and relatively low sampling and collaboration of cognitive users.

Acknowledgment. This work was supported in part by:
(1) The National Science Foundation of China (61701521).
(2) The Certificate of China Postdoctoral Science Foundation Grant (2016M603044).
(3) The National Science Foundation of ShannXi (2018JQ6074).

References

1. Ramani, V., Sharma, S.K.: Cognitive radios: a survey on spectrum sensing, security and spectrum handoff. China commun. **14**(11), 185–208 (2017)
2. Daniel, M., Vieira, J., Tomé, A.: Adaptive threshold spectrum sensing based on expectation maximization algorithm. Phys. Commun. **21**, 60–69 (2016)
3. Yu, C., Wan, P., Wang, Y.: Spectrum sensing algorithm based on improved MME-cyclic stationary fearture. In: International Conference on Natural Computation, Fuzzy Systems and Knowledge Discovery, pp. 857–861 (2016)
4. Pillay, N., Xu, H.J.: Blind eigenvalue-based spectrum sensing for cognitive radio networks. IET Commun. **6**(11), 1388–1396 (2012)
5. Dikmese, S., Wong, J.L., Gokceoglu, A.: Reducing computational complexity of eigenvalue based spectrum sensing for cognitive radio. In: International Conference on Cognitive Radio Oriented Wireless Networks, pp. 61–67 (2013)
6. Ren, X., Chen, C.: Spectrum sensing algorithm based on sample variance in multi-antenna cognitive radio systems. AEUE - Int. J. Electron. Commun. **70**(12), 1601–1609 (2016)
7. Kapoor, G., Rajawat, K.: Outlier-aware cooperative spectrum sensing in cognitive radio networks. Phys. Commun. **17**, 118–127 (2015)
8. Huang, X., Shi, L., Zhang, C., Zhang, D., Chen, Q.: Distributed resource allocation with imperfect spectrum sensing information and channel uncertainty in cognitive femtocell networks. EURASIP J. Wirel. Commun. Netw. **2017**(1), 201 (2017)
9. Fu, C., Li, Y., He, Y., Jin, M., Wang, G., Lei, P.: An inter-frame dynamic double-threshold energy detection for spectrum sensing in cognitive radios. EURASIP J. Wirel. Commun. Netw. **2017**(1), 118 (2017)
10. Han, Y., Li, Y., Chen, Q., Wang, J.: A cooperative spectrum sensing algorithm based on fuzzy set and D-S theory. Energy Proc. **13** (2011)
11. Corral-Ruiz, A.L.E., Cruz-Perez, F.A., Castellanos-Lopez, S.L., Hernandez-Valdez, G., Rivero-Angeles, M.E.: Modeling and performance analysis for mobile cognitive radio cellular networks. EURASIP J. Wirel. Commun. Netw. **2017**(1), 159 (2017)
12. Fang, H., Xu, L., Choo, K.K.R.: Stackelberg game based relay selection for physical layer security and energy efficiency enhancement in cognitive radio networks. Appl. Math. Comput. **296**, 153–167 (2017)

Workshop on Space-Based Internet of Things

Robust Lossless Data Hiding Using Chaotic Sequence and Statistical Quantity Histogram

Xiaobo Li[(⊠)], Jianhua Zhang, and Quan Zhou

China Academy of Space Technology, Xi'an, People's Republic of China
lxb619@126.com

Abstract. In order to solve the security problem of robust lossless data hiding, A method to cover up detectable traces is proposed based on chaotic sequence and statistical quantity histogram. The proposed method divides a original image into non-overlapping blocks and selects a smaller block in each block by using chaotic sequence. Then calculates the statistical quantity of each selected block as a robust parameter and shifts it by appropriate thresholds. Finally, secret information can be hidden into blocks by modifying robust parameter values. With our proposed method, the embedding trace of secret information is concealed. Experimental results show that the proposed method can achieve high performances in security, robustness and visual quality.

Keywords: Robust lossless data hiding · Statistical quantity · Chaotic sequence

1   Introduction

Recently, many information hiding methods have been proposed [1–5]. However, most lossless data hiding methods are exposed to an open environment to deliver the hidden information, and thus they are not safety for practical applications [6–11]. To solve the problems, we propose a novel robust information hiding method. The proposed method divides a carrier image into a number of non-overlapping blocks and selects a smaller block of each block by utilizing chaotic sequence. Then calculates the robust parameter of each selected block called statistical quantity. Secret information are hidden into blocks by shifting the robust parameter histogram. Without private keys, the secret information can't be extracted, even if the third party detects the existence hidden information under carrier image. Experimental results show that the proposed method has higher visual quality and better robustness than that of other robust lossless data hiding methods being mentioned in the paper.

The rest of this paper is organized as follows, the proposed method is presented in Sect. 2. Experimental results are discussed in Sect. 3. Finally, the conclusions of paper are given in Sect. 4.

B. Li et al. (Eds.): IoTaaS 2018, LNICST 271, pp. 461–468, 2019.
https://doi.org/10.1007/978-3-030-14657-3_47

2 Proposed Method

2.1 Robust Parameter

First of all, an 8-bit original image with size $M \times N$, denoted by I, is divided into a number of un-overlapping blocks each of size $m \times n$. And divide the $m \times n$ block into un-overlapping smaller blocks with size $h \times w$ consecutively. Then, we can establish a one-to-one mapping between $m \times n$ blocks and random integers of chaotic sequence. Based on the above results, we introduce a $h \times w$ matrix E, the matrix E is shown below:

$$E(i,j) = \begin{cases} -1 & if \quad \mathrm{mod}(i,2) \neq \mathrm{mod}(j,2) \\ 1 & if \quad \mathrm{mod}(i,2) = \mathrm{mod}(j,2) \end{cases}$$

Where $i \in [1,h], j \in [1,w]$ and mod(i, 2) or mod(j, 2) is a mod-2 function. The robust parameter of selected bock is given by

$$\alpha^{(Z)} = \sum_{i}^{h} \sum_{j}^{w} (C_Z(i,j) \times E(i,j))$$

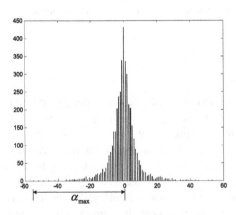

Fig. 1. The distribution of α of image "Lena" with $m \times n = 8 \times 8$ and $h \times w = 2 \times 2$

Where $C_z = \{C_z(i,j) | C_z(i,j) \in \{0, 1, \ldots, 255\}, i = 1, 2, \ldots, h, j = 1, 2, \ldots, w\}$, and the subscript (Z) means the Z-th selected block with size $h \times w$, and $\alpha^{(Z)}$ means the robust parameter value of the Z-th selected block. Because small changes to the blocks caused by attacks such as JPEG2000 compression will not cause the robust parameter α to change much, and robustness is achieved. Thus, the statistical quantity of α can be used to hide information. The distribution of α is shown in Fig. 1, where the x-axis being the value of α, while the y-axis is the number of α. Supposing that α_{max} is the maximum absolute value between the values of α. From the Fig. 1 it is seen that the α values are close to zero, that is because pixels in a local area have a good relativity. This distribution of α can be used to produce extra space for embedding information.

2.2 Information Hiding Process

In this part, we will use the statistical quantity histogram formed by α as the hiding carrier for information embedding. First, we introduce four thresholds, denoted by T_1, T_2, R_1, and R_2, both of which are respectively positive integers. We let $0 < T_1 \leq \alpha_{max}$, $T_2 = \alpha_{max} - T_1$, and R_1, R_2 are used to separate the different zones as robust thresholds.

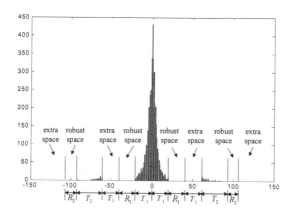

Fig. 2. Distribution of α after shifting

Next, for generating the robust space, we shift statistical quantity by modifying the pixels of selected blocks C_Z as follows:

$$D_Z(i,j) = \begin{cases} C_Z(i,j) + \beta_1 & \text{if } \alpha > T_1 \& E(i,j) = 1 \\ C_Z(i,j) - \beta_1 & \text{if } \alpha > T_1 \& E(i,j) \neq 1 \\ C_Z(i,j) - \beta_1 & \text{if } \alpha < -T_1 \& E(i,j) = 1 \\ C_Z(i,j) + \beta_1 & \text{if } \alpha < -T_1 \& E(i,j) \neq 1 \\ C_Z(i,j) & \text{otherwise} \end{cases}$$

Where $i \in [1,h]$, $j \in [1,w]$, $\beta_1 = \lceil (T_1 + R_1)/(h \times w) \rceil$.

From the Fig. 2, the resulting distribution of shifted α is shown, and it is seen that four extra spaces are obtained for information embedding, i.e., $(-\infty, -2T_1 - T_2 - R_1 - R_2]$, $[-2T_1 - R_1, -T_1 - R_1]$, $[T_1 + R_1, 2T_1 + R_1]$ and $[2T_1 + T_2 + R_1 + R_2, \infty)$. Besides, the range of $(-2T_1 - T_2 - R_1 - R_2, -2T_1 - T_2 - R_1)$, $(-T_1 - R_1, -T_1)$, $(T_1, T_1 + R_1)$ and $(2T_1 + T_2 + R_1, 2T_1 + T_2 + R_1 + R_2)$ are four robust spaces. Then, we scan each block and check up the corresponding α, the secret bits can be hidden into these blocks. The hiding process is as follows:

(1) While $-T_1 \leq \alpha \leq T_1$, the hiding regulation is given by
If $B = 0$, let $G_Z(i,j) = D_Z(i,j)$.
If $B = 1$, let

$$G_Z(i,j) = \begin{cases} D_Z(i,j) + \beta_1 & \text{if } 0 \leq \alpha \leq T_1 \& E(i,j) = 1 \\ D_Z(i,j) - \beta_1 & \text{if } 0 \leq \alpha \leq T_1 \& E(i,j) \neq 1 \\ D_Z(i,j) - \beta_1 & \text{if } -T_1 \leq \alpha < 0 \& E(i,j) = 1 \\ D_Z(i,j) + \beta_1 & \text{if } -T_1 \leq \alpha < 0 \& E(i,j) \neq 1 \\ D_Z(i,j) & \text{otherwise} \end{cases}$$

(2) While $\alpha > T_1$ or $\alpha < -T_1$, the hiding regulation is given by
If $B = 1$, let $G_Z(i,j) = D_Z(i,j)$.
If $B = 0$, let

$$G_Z(i,j) = \begin{cases} D_Z(i,j) + \beta_2 & \text{if } \alpha > T_1 \& E(i,j) = 1 \\ D_Z(i,j) - \beta_2 & \text{if } \alpha > T_1 \& E(i,j) \neq 1 \\ D_Z(i,j) - \beta_2 & \text{if } \alpha < -T_1 \& E(i,j) = 1 \\ D_Z(i,j) + \beta_2 & \text{if } \alpha < -T_1 \& E(i,j) \neq 1 \\ D_Z(i,j) & \text{otherwise} \end{cases}$$

Where $i \in [1,h]$, $j \in [1,w]$, $\beta_2 = \lceil (T_2 + R_2)/(h \times w) \rceil$. When bits are hid into the image blocks, the pixel values of blocks need to be modified by β_1 or β_2. The values of β_1 and β_2 is known as hiding level.

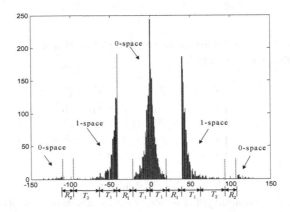

Fig. 3. Distribution of α after hiding data

The distribution of α after hiding data is shown in Fig. 3. When 0s are embedded, the ranges of α are kept within $[-T_1, T_1]$, $[-\infty, -(R_2 + T_2 + R_1 + 2T_1)]$ and $[2T_1 + R_1 + T_2 + R_2, \infty]$, and these ranges are called the 0-space; When 1 s are embedded, the ranges of α are kept within $[T_1 + R_1, 2T_1 + R_1 + T_2]$ and $[-(2T_1 + R_1 + T_2), -(T_1 + R_1)]$, these ranges are called the 1-space. Figure 3 shows the 0-space and the 1-space are separated by a distance R_1 or R_2, attacks applied to the stego-image will not cause the 0-space and the 1-space to overlap, i.e., thus, the hidden information are robust to attacks, the hidden bits can be extracted correctly. Clearly, the robust capability of hidden bits are corresponding to distance R_1 or R_2. That is, the larger the distance R_1 or R_2 is, the more robust hidden bits can be extracted correctly. Besides, the embedding capacity of our method is $Cap = \lfloor M/m \rfloor \times \lfloor N/n \rfloor$ bits.

2.3   Information Extraction

The information extraction of stego-image has two cases, one is stego-image remains intact in a lossless environment, the extracting procedure is the same as the hiding procedure exception the processing image is stego-image. For another case, once the stego-image has been attacked, for instance, JPEG2000 compression, the distribution of statistical quantity is changed and many α step into the wrong zone, as shown in Fig. 4. With the different, in addition to record the amounts of 0s and 1s in the hidden bits, denoted by N0 and N1, we must record the numbers of 0s embedded in the range $[-T_1, T_1]$ of statistical quantity α, denoted by N2. As shown in Fig. 4, we can obtain a bits_x such that the number of α in the range of $[-x\_bits, x\_bits]$ is equal to N2, a bits_y such that the number of α in the range of $[-y\_bits, y\_bits]$ is equal to (N2 + N1), and a bits_z such that the number of α in the range of $[-z\_bits, z\_bits]$ is equal to (N0 + N1), respectively. Thus, the hidden bits can be extracted by

$$B = \begin{cases} 0 & \begin{aligned} &if -x\_bits \leq \alpha \leq x\_bits \ or \\ &\alpha > y\_bits \ or \\ &\alpha < -y\_bits \end{aligned} \\ 1 & \begin{aligned} &if -y\_bits \leq \alpha < -x\_bits \ or \\ &x\_bits < \alpha \leq y\_bits \end{aligned} \end{cases}$$

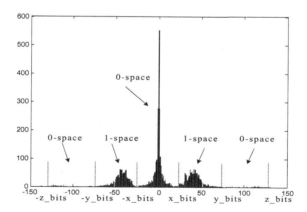

Fig. 4. Distribution of α of a stego-image has been attacked

3 Experiments

In order to verify the effectiveness of the proposed method in this paper, we test the proposed algorithm on six 8-bits grayscale images with 512 × 512 resolution, "Lena", "Baboon", "Boat", "Airplane", "Pepper", and "GoldHill" [12]. The secret bits used in our tests were generated by a random number generator. Six stego-images were compressed by JPEG2000 under the different image compression ratio in robustness testing. Robustness against compression was measured by bpp (the survival of bit rate) at BER (a bit error rate <1%). The survival of bit rate is used to adjust image quality under the different image compression ratio, and BER means the percentage of errors bits in total hidden bits. As a general rule, the lower is the bpp and BER, the better is the robustness. We use the general visual evaluation function of PSNR to evaluate the image quality.

Table 1. The test results of the proposed method with block size 8 × 8

Images (512 × 512)	PSNR (dB)	Capacity (bits)	Threshold (T_1, T_2)	Threshold (R_1, R_2)	EL* (β_1, β_2)	Rb* (bpp)	Key (k, x_0)	BER (%)	BERW* (%)
Lena	42.29	4096	(20, 35)	(20, 5)	(10, 10)	0.62	(5, 0.5)	0.85	50.9
Airplane	42.20	4096	(20, 37)	(20, 3)	(10, 10)	0.67	(6, 0.5)	0.85	48.6
Boat	41.70	4096	(20, 32)	(20, 8)	(10, 10)	0.94	(4, 0.6)	0.88	50.2
GoldHill	42.12	4096	(20, 36)	(20, 4)	(10, 10)	0.94	(4, 0.6)	0.93	49.1
Peppers	39.79	4096	(20, 36)	(20, 4)	(10, 10)	1.23	(8, 0.3)	0.66	49.3
Baboon	32.56	4096	(60, 58)	(60, 2)	(30, 15)	1.60	(3, 0.2)	0.24	49.2

*EL = Embedding level; Rb = Robustness; BERW = BER with wrong keys.

First, the test images was divided into blocks of size 8 × 8, and select blocks with size 2 × 2 in each 8 × 8 block by using chaotic sequence with the parameter k and initial value x_0 as private keys. Then calculate statistical quantity value α of each selected block, and choose appropriate threshold T_1, T_2, R_1, and R_2 to shift the statistical quantity histogram. Finally, 4096 secret bits can be hid into six test images. Then, all the stego-images under the different image compression ratio were compressed by JPEG2000 after embedding secret bits. In the end, the hidden bits were extracted from the attacked stego-images, and the test results are shown in Table 1. In the Table 1, it is seen that the PSNR range of stego-images show that from 42.29 dB to 32.56 dB, and robustness from 0.62 to 1.60 bpp when BER < 1%. To show the security of our method, the embedded bits were extracted from the stego-images with wrong keys, Table 1 shows that the BER values are all near to 50%. This implies that hidden data extracted under stego image are completely wrong. The distribution of α after embedding data with wrong keys is show in Fig. 5. In contrast with Fig. 3, our method cover up the bits embedding trace. In general, high security and robustness while keeping the distortion low is proved by the experiment result of our method.

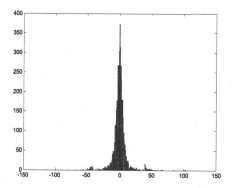

Fig. 5. The distribution of α after embedding data with wrong keys

Finally, comparing the results of our method to the Zeng et al.'s method [8], we list some experimental results of this methods on six test images. In this experiments, we used a block size of 8 × 8. A performance comparison results is shown in Table 2. Experimental results show that our proposed method can achieve higher performances in visual quality of stego-image, robustness, and data embedding capacity than other methods.

Table 2. Performance comparison results

Images (512 × 512)	Zeng's scheme				Our proposed scheme			
	PSNR (dB)	Capacity (bits)	Rb* (bpp)	BER (%)	PSNR (dB)	Capacity (bits)	Rb* (bpp)	BER (%)
Lena	38.60	4096	1.04	0.69	42.29	4096	0.62	0.85
Airplane	38.60	4096	1.05	0.80	42.20	4096	0.67	0.85
Boat	38.59	4096	1.56	0.77	41.70	4096	0.94	0.88
GoldHill	38.58	4096	1.72	0.90	42.12	4096	0.94	0.93
Peppers	37.26	4096	0.81	0.74	39.79	4096	1.23	0.66
Baboon	31.87	4096	1.70	0.94	32.56	4096	1.60	0.24

*Rb = Robustness.

4 Conclusion

The proposed method embeds secret information into carrier image by modifying the statistical quantity parameters of blocks selected by chaotic sequence. Without private keys, the secret information can't be extracted, even if the third party detects the existence hidden information under carrier image. Experimental results demonstrate that our method provides a security approach to embed information. Performance comparisons with other methods are provided to show that the proposed method has obtained an high performances in visual quality of images, information embedding capacity and robustness. It is expected that the proposed method can be applied in information safety fields.

References

1. Dragoi, I.C., Coltuc, D.: Reversible data hiding in encrypted images based on reserving room after encryption and multiple predictors. In: 2018 IEEE International Conference on Acoustics, Speech and Signal Processing, ICASSP, Calgary, AB, Canada, pp. 2102–2105 (2018)
2. Lee, C.-F., Shen, J.J., Lai, Y.H.: Data hiding using multi-pixel difference expansion. In: 2018 3rd International Conference on Computer and Communication Systems, ICCCS, Nagoya, Japan, pp. 56–60 (2018)
3. Wang, W., Ye, J., Wang, T., Wang, W.: Reversible data hiding scheme based on significant-bit-difference expansion. IET Image Process. **11**, 1002–1014 (2017)
4. Yi, S., Zhou, Y.: Adaptive code embedding for reversible data hiding in encrypted images. In: 2017 IEEE International Conference on Image Processing, ICIP, Beijing, China, pp. 4322–4326 (2017)
5. Shi, Y.-Q., Li, X., Zhang, X., Wu, H.-T., Ma, B.: Reversible data hiding: advances in the past two decades. IEEE Access **4**, 3210–3237 (2016)
6. Singh, A., Dutta, M.K.: Lossless and robust digital watermarking scheme for retinal images. In: 2018 4th International Conference on Computational Intelligence & Communication Technology, CICT, Ghaziabad, India, pp. 1–5 (2018)
7. Choi, K.-C., Pun, C.-M.: Difference expansion based robust reversible watermarking with region filtering. In: 2016 13th International Conference on Computer Graphics, Imaging and Visualization, CGiV, Beni Mellal, Morocco, pp. 278–282 (2016)
8. Zeng, X.-T., Ping, L.-D., Pan, X.-Z.: A lossless robust data hiding scheme. Pattern Recogn. **43**, 1656–1667 (2010)
9. Zeng, X.-T., Pan, X.-Z., Ping, L.-D., et al.: Robust lossless data hiding scheme. J. Zhejiang Univ.-SCIENCE C (Comput. Electron.) **11**(2), 101–110 (2010)
10. Yang, Q.T., Gao, T.G., Fan, L.: Lossless robust data hiding scheme based on histogram shifting. In: Hu, W. (ed.) Electronics and Signal Processing. LNEE, vol. 97, pp. 937–944. Springer, Heidelberg (2011). https://doi.org/10.1007/978-3-642-21697-8_120
11. An, L., Gao, X., Yuan, Y.: Robust lossless data hiding using clustering and statistical quantity histogram. Neurocomputing **77**, 1–11 (2012)
12. CVG-UGR Image Database. http://decsai.ugr.es/cvg/dbimagenes/index.php

Gradable Cloud Detection in Four-Band Remote Sensing Images

Shuwei Hou[1,2(✉)], Wenfang Sun[1], Baolong Guo[1], Xiaobo Li[2], and Huachao Xiao[2]

[1] School of Aerospace Science and Technology, Xidian University, Xi'an, People's Republic of China
hsw521@sina.com
[2] China Academy of Space Technology, Xi'an, People's Republic of China

Abstract. Cloud detection is one of the major techniques in remote sensing image processing. Many cloud detection algorithms have been developed recently. According to the type of remote sensing images that are used to detect cloud, they can be divided into two major categories: visible image-based methods and multispectral image-based methods. The first category mainly uses structure and texture characteristics for thick cloud detection, while the second category often uses the specific spectral bands for good results. In general, the existing methods above deal with cloud detection as a binary classification problem, cloud or non-cloud. However, as cloud has various forms and types, it is inappropriate to simply classify detection results into cloud or non-cloud. In this paper, we present a novel cloud detection method using orthogonal subspace projection (OSP), which can yield gradable cloud detection results. This detailed detection result not only conforms to the characteristics of cloud, but also brings more valuable guidance to subsequent interpretation of remote sensing images. Additionally, the proposed method only uses four universal bands including red, green, blue and near-infrared bands for detection, and has no requirement for special spectral bands, which make it more practical. Experiment results indicate that the proposed method has excellent results with high speed and accuracy.

Keywords: Remote sensing image · Cloud detection · Gradable

1 Introduction

Space imaging systems have the ability to collect digital images with high resolution and wide coverage of land surface. However, the captured rich information also brings great challenges to information storage, transmission, extraction and application. According to statistics, more than 50% of the optical remote sensing images are covered by clouds of different thickness, which consumes most transmitted time and link bandwidth, therefore affects the downlink priority of important information. Furthermore, cloud increases the difficulty in identifying important targets such as aircrafts and ships. Accordingly, the study of real-time cloud detection technology is of great significance for reducing data on-board and improving the intelligent processing ability of remote sensing images.

B. Li et al. (Eds.): IoTaaS 2018, LNICST 271, pp. 469–476, 2019.
https://doi.org/10.1007/978-3-030-14657-3_48

Cloud detection technology of remote sensing image has been one of the major techniques in remote sensing image processing. It can be summarized in two categories: visible image-based methods and multispectral image-based methods. The visible image-based methods include linear dimension compression for feature space [1] and multi-attribute fusion algorithm [2], etc. These methods mainly focus on how to effectively extract cloud features to achieve better cloud detection results.

Recently, carrying multispectral or hyperspectral imaging spectrometer [3] gradually become a development trend of satellite surveying approach, basically because it can provide remote sensing image processing with favorable conditions. Under this circumstance, multispectral image-based methods often use the specific spectral characteristics to detect cloud, of which the threshold, pattern classification and multi-dimensional space analysis methods are the most notable. The threshold method takes full advantage of the spectral characteristics of cloud, simple and easy to FPGA implementation, but relying on the specific spectrum to ensure the performance of the algorithm. The HCC algorithm [4] used on EO-1 satellite and the ACCA algorithm [5] used on Landsat-7 satellite, both select representative spectral bands to complete cloud detection. To improve the cloud detection result, some researchers use both the spectral and statistical features of cloud, and then use a classifier to separate clear and cloudy pixels. Reference [6] combined the texture features and spectral features of cloud with MODIS remote sensing data, followed by a neural network to complete detection. Generally the pattern classification method has better cloud detection results than the threshold method. However, it is more complex and more difficult to implement. Finally, the multi-dimensional space analysis method uses the signal processing principle to detect cloud. In [7], a cloud detection based on ICA is proposed, which is processed in higher space and has relatively large computation amount.

It turns out that many different methods generally deal with cloud detection as a binary classification problem. Cloud detection results are either cloud or non-cloud. However, as the form and type of cloud is various, it is not appropriate to simply classify the cloud detection results into cloud or non-cloud. Therefore how to further subdivide the cloud detection results to different levels of cloud products, is not only more accordant with the features of cloud, but also brings more valuable guidance to subsequent interpretation of remote sensing images. Although in [8], a multilevel cloud detection algorithm is proposed based on deep learning, it can only detect two levels of cloud, thick cloud and thin cloud, and also need more cloud samples corresponding to the two different levels, accordingly it increases the complexity of the algorithm.

In addition, most multispectral image-based methods are based on specific sensors, MODIS [11], Landsat or AVHRR [9, 10], etc., which cannot be widely used on different satellites. By contrast, four-band (red, green, blue, and near-infrared) remote sensing images are generally accessible and universal. Nearly all optical sensors such as those equipped on HJ-1a/1b, ZY-3 and GF series satellites can provide four-band remote sensing images. Although cloud detection in four-band remote sensing images is more general and applicable, it is too difficult to implement with its little spectral information. This paper focuses on orthogonal subspace projection for cloud detection in four-band remote sensing images. Our proposed algorithm not only can yield refined levels of cloud products, but also has low complexity.

In summary, the presented method has the following two main contributions.

(1) A gradable cloud detection approach is proposed. The new approach applied orthogonal subspace projection (OSP) to cloud detection and yields more detailed information about cloud thickness.
(2) The novel approach generalized the OSP algorithm from hyperspectral unmixing to cloud detection in four-band remote sensing images, which is more universal and practical.

The remainder of this paper is organized as follows. Section 2 introduces four-band cloud detection. Section 3 describes the experimental data, results and discussions. Section 4 analyzes the computation complexity and Sect. 5 draws the conclusion.

2 Four-Band Cloud Detection Based on OSP

Generally, input pixel can be seen as a mixed pixel with cloud contamination. Inspired by the OSP unmixing algorithm, we first introduced the concept of unmixing to cloud detection and tested its validity through experiments.

Different with hyperspectral unmixing problem, four-band cloud detection based on OSP has its difficulties and particularity. First, cloud has various forms and types, leading to the variability of the target subspace. Second, the spectral information of four-band (red, green, blue, and near-infrared) is much less than those in hyperspectral image. To solve these problems, our proposed algorithm is improved from two aspects: optimizing background subspace estimation and expanding cloud subspace. By using the improved OSP methods, graded cloud detection results can be generated for four-band remote sensing images.

2.1 Estimation of Background Subspace

In hyperspectral image processing, ATGP algorithm proposed by REN [13] can automatically generate the background with no required a priori knowledge. However, it is not ideal to directly apply the method to our four-band cloud detection because of the little spectral information. Furthermore, there is no guarantee that each generated signature is completely different from the desired target in the spectral characteristics, so leakage often happens from the cloud subspace to the background subspace.

As to the background generation of multispectral image, we proposed a multi-band automatic target generation process (MATGP) based on the ATGP algorithm. Followed by the background generation (ATGP), a false background removing processing is designed. The detailed implementation is given as follows.

Let the undesired background U generated by ATGP is

$$U = [u_1 \, u_2 \cdots u_i \cdots u_m]$$

where U is an L * m matrix and $u_i(1 \leq i \leq m)$ is the ith background signature.

Let t be an L * 1 column vector and denote the desired target. Then the spectral similarity between t and u_i is given by

$$specoff_i(t, u_i) = \frac{t^T u_i}{\sqrt{t^T t} * \sqrt{u_i^T u_i}}$$

Let *Th* be a spectral similarity threshold. The removing processing is given as follow:

If *specoff_i* is greater than *Th* then the corresponding background u_i is removed from <U>. Otherwise, the corresponding background u_i is preserved in <U>. Finally, the updated background is the true background with qualified conditions.

Figure 1 shows the four GF-1 bands labeled by (a)–(d). For the purpose of comparison, the maximum number of target required to search was all set to 11 to terminate the background generation. The ATGP result is shown in Fig. 2(a), where target labeled by 1–3 is of cloud subspace (cloud subspace is not shown here), and target labeled by 4–11 is of background subspace. It is clear that target 5 and target 8 are cloud-like targets and will result in the loss of cloud.

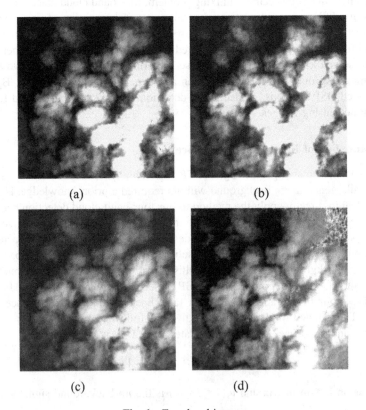

(a)   (b)

(c)   (d)

Fig. 1. Four-band images

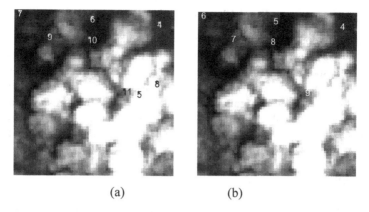

Fig. 2. Search of background (a) ATGP background (b) our MATGP background

The MATGP result is shown in Fig. 2(b). It can be seen that cloud-like targets are all removed and the remaining is a good estimation of the background.

Figure 3 gives the corresponding cloud detection results with the above two different background estimation algorithm. Figure 3(a) used ATGP background search algorithm, which resulted in the loss of most cloud. Figure 3(b) used MATGP background search algorithm and yielded better results.

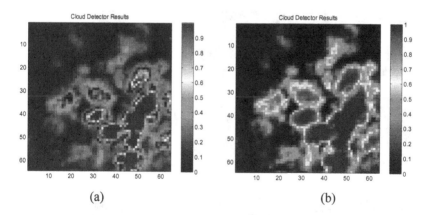

Fig. 3. Cloud detection result (a) using ATGP background (b) using MATGP background

2.2   Estimation of Target Subspace

In the OSP algorithm, target t is generally known as a priori information. Usually, the target is modeled by extracting a single spectral signal. However, not only the type and forms of cloud are varied, but also the thickness and the height of cloud are different. Therefore, the spectral features of cloud exhibit significant variability. In order to improve the modeling of spectral characteristics of cloud, we expand cloud subspace from a single vector to P-dimension subspace (P > 1).

2.3   Adaptation of Four-Band Cloud Detection Based on OSP

In our four-band cloud detection, we only have 4 bands, whereas the total objects in remote sensing images are mostly larger than 4. Therefore we cannot distinguish all the objects by their spectral information. In other words, directly using OSP for our four-band cloud detection is not ideal. Under such circumstances, the number of bands must be expanded to meet the classification conditions [12]. In remote sensing images, less than ten objects are usually adequate to model any given areas of a given spatial resolution [14]. Here we extended the four-band remote sensing images to 18 bands.

3   Experiments

The four-band remote sensing images collected by GF-1 were used for experiments, which were blue band (0.45 μm–0.52 μm), green band (0.52 μm–0.59 μm), red band (0.63 μm–0.69 μm) and near-infrared band (0.77 μm–0.89 μm), with a spatial resolution of 8 m.

The following gives the experimental results with a wide range of GF-1 area. Figures 4, 5 and 6 show the original four-band images, the OTSU results and the gradable cloud detection results respectively. It can be found that our method not only detects the cloud regions accurately, but also provides more detailed information about

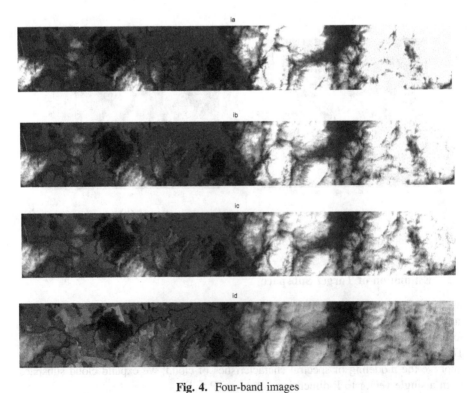

Fig. 4. Four-band images

cloud thickness. As is seen in Fig. 6, different color in the color bar illustrates different categories of cloud. For example, red areas represent thick clouds, yellow areas represent the edge of cloud, and light blue areas represent thin cloud, etc. At the same time, the scale values from high to low, in turn, indicate that the detected area is cloud, very likely to be cloud, may be cloud or non-cloud. The gradable cloud products can be more helpful for subsequent intelligent processing tasks (for example, cloud removal task [15] and ROI compression [16]).

Fig. 5. OTSU results

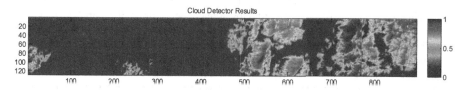

Fig. 6. Gradable cloud detection results (Color figure online)

4   Computation Complexity

In this paper, an OSP-based method is applied to detect cloud. In the algorithm, background U and cloud target t only need to be computed once. When real-time cloud detection is used on satellite, only multiplication and addition operations are implemented in the cloud detector. If the number of bands is L, our detector only needs L multiplications and L-1 additions for each pixel. As cloud detection in different pixel is independent of each other, multiple detectors can be used in parallel to speed up processing. Therefore, our method is a rapid cloud detection method with high degree of parallelism and very low computational complexity.

5   Conclusion

In this paper, a gradable cloud detection method in four-band remote sensing images is proposed. The method not only can yield gradable cloud detection results, but also has no requirement for special spectral bands. Moreover, it even has no strict limit on the number of bands. Therefore our method is more practical to satellite cloud detection. Our method has been tested on real remote sensing images with different scene, and the experimental results have proved its efficiency.

References

1. Bian, C., Hou, Q.: Cloud detection in remote sensing image based on linear dimension compression. J. Harbin Inst. Technol. **46**(1), 29–33 (2014)
2. Zhao, X., Hou, Q.: A method for cloud detection in high-resolution remote sensing image based on multi-attribute fusion. Opt. Tech. **40**(2), 145–150 (2014)
3. Feng, S., Zhang, N.: Method of cloud detection with hyperspectral remote sensing image based on the reflective characteristics. China Opt. **8**(2), 198–204 (2015)
4. Ricciardelli, E., Romano, F.: Physical and statistical approaches for cloud identification using meteosat second generation-spinning enhanced vision and infrared imager data. Remote Sens. Environ. **112**, 2741–2760 (2008). Science Direct
5. Williams, J.A., Dawood, A.S.: FPGA-based cloud detection for real-time onboard remote sensing. In: IELCONF (2002)
6. Song, X., Yingshi, Z.: Cloud detection and analysis of MODIS Image. Journal of Image and Graphics **8A**(9), 1079–1083 (2003)
7. Du, H., Wang, Y.: Studies on cloud detection of atmospheric remote sensing image using ICA algorithm. In: Image and Signal Processing Conference (2009)
8. Xie, F., Shi, M.: Multilevel cloud detection in remote sensing images based on deep learning. IEEE J. Sel. Top. Appl. Earth Obs. Remote Sens. **10**(8), 3631–3639 (2017)
9. Saunders, R., Kriebel, K.T.: An improved method for detecting clear sky and cloudy radiances from AVHRR data. Int. J. Remote Sens. **9**, 123–150 (1987)
10. Bankert, R.L.: Cloud classification of AVHRR imagery in maritime regions using a probabilistic neural network. J. Appl. Meteorol. **33**, 909–918 (1994)
11. Ackerman, S., Strabala, K.: Discriminating clear-sky from clouds with MODIS. J. Geophys. Res. **103**, 141–157 (1998)
12. Ren, H., Chang, C.-I.: A generalized orthogonal subspace projection approach to unsupervised multispectral image classification. IEEE Trans. Geosci. Remote Sens. **38**(6), 2515–2528 (2000)
13. Ren, H., Chang, C.-I.: Automatic spectral target recognition in hyperspectral imagery. IEEE Trans. Aerosp. Electron. Syst. **39**(4), 1232–1249 (2003)
14. Adams, J.B., Smith, M.O.: Imaging Spectroscopy: interpretation based on spectral mixture analysis. In: Pieters, C.M., Englert, P.A. (eds.) Remote Geochemical Analysis: Elemental and Mineralogical Composition, pp. 145–166. Cambridge University Press, Cambridge (1993)
15. Gomez-Chova, L., Amoros-Lopez, J.: Cloud masking and removal in remote sensing image time series. J. Appl. Remote Sens. **11**(1), 015005 (2017)
16. Christine, M., Samuel, R.: A feasibility study of on-board cloud detection and compression. In: Aerospace Conference. IEEE (2010)

A Novel Algorithm of Congestion Control Based on Satellite Switching

Yi Zhang[1,2(✉)], Jingling Li[1,2], Jun Li[1,2], Tao Cui[1,2], Wei Liang[1,2], and Gong Li[3]

[1] China Academy of Space Technology, Xi'an 71000, China
lbzytogether@163.com
[2] National Key Laboratory of Science and Technology on Space Microwave, Xi'an 71000, China
[3] Beijing Institute of Space Mechanics and Electricity, Beijing, China

Abstract. One of the characteristics of satellite switching network is limited link resources. When the satellite switching network is congested, there are not enough resources to send the congestion notification message to the ground terminal. In order to solve this problem, a novel algorithm of congestion control based on satellite switching is proposed in this paper. The congestion notification messages are sent while transmitting the switching data according to the specific hiding approach, which uses the redundant information in the switching data to embed the congestion control information into the transmission data. The method can send congestion control messages to the ground terminal without any additional physical resources, and also increase the amount of information that can be transmitted by each port of the onboard switch. The performance analysis and simulation result show that the proposed algorithm can not only avoid congestion in the onboard switch, but also save link resources and improve the performance of the satellite switching network.

Keywords: Congestion control · Information hiding · Onboard switch · Data embedding

1 Introduction

Onboard satellite switching is one of the development trends of satellite communications network. With the increasing demand for satellite communication, satellite communication network needs to have the ability of large-capacity information exchange and forwarding. Onboard switching is one of the key technologies to meet the development needs of satellite communication. The satellite uses switching technology to carry out multi-beam exchange, which facilitates communication among multiple ground stations and constitutes a wireless communication network integrating space and earth. Therefore, onboard switching technology will greatly promote the development of satellite communication system [1]. One characteristic of this system is limited link resources. If the total load in the switching system exceeds the maximum throughput that the system can bear in a local time, a large amount of data will be discarded, and the system performance is greatly reduced, which means congestion.

© ICST Institute for Computer Sciences, Social Informatics and Telecommunications Engineering 2019
Published by Springer Nature Switzerland AG 2019. All Rights Reserved
B. Li et al. (Eds.): IoTaaS 2018, LNICST 271, pp. 477–486, 2019.
https://doi.org/10.1007/978-3-030-14657-3_49

Increasing the internal buffer of onboard switch can only delay congestion, and the congestion problem can't be resolved. Therefore, the effective algorithm of congestion control must be taken in the satellite switching system to improve the switching performance of satellite network [1–4].

This paper adopts different congestion control strategies based on the current state of onboard buffer. In this method, the onboard switch must response its own congestion state to the ground terminal. However, when the congestion occurs, the load on the onboard switch is saturated and the congestion notification message cannot be sent to the ground terminal. In order to improve this problem, this paper proposes a novel congestion control algorithm based on specific information hiding method for onboard processing and switching system. This specific algorithm embeds congestion control messages into the switching data, and the congestion control message is sent while transmitting data in order to avoid the congestion.

This paper is arranged as follows: the composition of the satellite processing switching system is described in Sect. 1, and then introduces the novel algorithm of congestion control based on information hiding. Section 3 describes the information hiding and de-hiding algorithm proposed in this paper in detail. Finally, the new algorithm is simulated and verified with specific examples.

2   The Satellite Processing and Switching System

The satellite processing and switching system is composed of an onboard switch, ground terminals, and a ground network control center. In this system, N ground source terminals transmit data to N ground destinations through an onboard switch, and the ground network control center performs basic satellite network management functions. The satellite processing and switching system is shown below (Fig. 1):

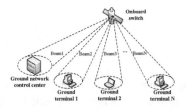

Fig. 1. The satellite processing and switching system

- Onboard switch: The onboard switch in this paper mainly refers to the packet switch, which is responsible for exchanging packet service data (including ATM data and IP data) and performing congestion control according to the onboard status.
- Ground terminal: To complete the sending and receiving of data. Each ground terminal has the function of sending and receiving cells, which can be used as a ground source and a ground destination.

- Ground network control center: Complete basic satellite network management functions (include network authentication, onboard device status supervision, network registration, etc.).

The onboard packet switching system mainly includes ATM switching and IP switching.

In the onboard ATM switching system, a connection establishment message is sent to the onboard switching by ground source before sending data. The onboard switch decides whether to agree the request according to the resources and link state of the satellite. If this request is accepted, the onboard switch sends the connection setup success message to the ground source terminal. Then the ground source terminal can send cells by the reservation rate. When the data rate sent by the ground source exceeds the reserved resources, or when the burst strength of the service is too high, the total load of the onboard switch will exceed the maximum throughput that can be supported during the local time [5].

Because the onboard IP switching system doesn't establish connection in advance, the total load of the onboard switch in the local time will exceed the maximum throughput that can be supported [6, 7]. In order to improve this problem, the satellite processing and switching system must take appropriate measures for congestion control to ensure the performance and normal operation of the system [8–11].

3 A Novel Algorithm of Congestion Control Based on Satellite Switching

In this paper, a new congestion control algorithm is proposed by the characteristics of the satellite link. This algorithm adopts different measures to control the congestion according to the current state of onboard switch. Because the onboard switch sends control messages to the ground terminal, it also occupies some link resources. Therefore, this paper uses the information hiding method to embed the congestion control message into the switching data. The switching data and the congestion control information are sent to the ground terminal together.

3.1 The Type of Congestion Control Information

The onboard switch determines whether to send an alarm message to the ground source or discards the entire cell according to the congestion level. Therefore, the congestion control information of the onboard switch mainly includes the following types:

(1) Alarm message

When the onboard switch is about to be congested, the onboard switch sends a feedback message to the corresponding ground source according to the number of connections and cells in the buffer. This process is shown in Fig. 2.

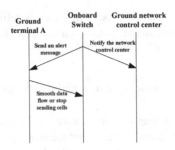

Fig. 2. Send an alert message to the ground terminal by onboard switch

When the buffer reaches the alarm threshold, the onboard switch sends an alarm message to the ground source, which has the most serious impact on the congestion of onboard switch, then send the notified message to the ground control center. The alarm message includes the onboard switching cache usage, service parameters of the connection, service type, and QoS, and the onboard switch has obtained the service parameters, such as service type, and QoS requirements of the connection while establishing the connection.

After receiving the alarm message, the ground source determines whether to smooth the data stream or stop sending the cell according to the current cache usage of the onboard switch. After this state is maintained for a period of time, the ground terminal restarts sending data.

(2) Drop notification message

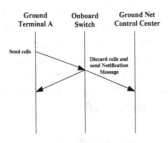

Fig. 3. After congestion, the onboard switch discards cell and sends notification cell to the ground terminal

If the onboard switch has sent the alarm message, but the congestion problem has not been resolved and the onboard ATM switch reaches the set discard threshold, the switch must choose to discard some cells. While discarding a connection, the onboard ATM switch should send a drop notification message to the connected ground source and the ground control center. This process is shown in Fig. 3. The drop notification message includes some parameters, such as service type, QoS, and so on.

3.2   The Format of Congestion Control Information

According to the Sect. 3.1, the onboard switch performs congestion control by sending different control messages to the ground terminal. The format of control information includes information type, cache status, service parameters, service type, QoS, and so on, as shown in Table 1:

Table 1. The format of control information

Information type (2bits)	Cache status (2bits)	Service type (2bits)
Service parameters (10bits)		
QoS (3bits)	Reserved (5bits)	

- Information Type: Indicates the type of the control information.
  "01" indicates an alarm message, and "10" indicates a drop notification message;
- Cache Status: Indicates the cache status of the current onboard switch. If the alarm message is sent, "01" means slowing down the sending cell rate, "10" means stopping sending the cell; if sending a drop notification message, the state is "11";
- Service Type: Indicates the type of the service.
  "00" means an unicast data; "01" means a multicast data;
  "10" means a broadcast datagram; "11" means reserved data;
- Service parameter: Indicates the connection parameter of the service;
- QoS: Indicates the QoS level of the service;
- Reserved: reserved domain.

4   Information Hiding and De-hiding Algorithm

4.1   Information Hiding Algorithm

The new congestion control algorithm proposed by this paper mainly uses the redundant information in the traffic payload to hide the control information. The redundant information here is the number that does not appear in the traffic load.

In this chapter, the onboard ATM switch is taken as an example for specific description. Since the ATM payload has a total of 48 bytes, and each byte can be represented as a number between 0 and 255, the payload of each cell includes 48 numbers between 0 and 255, that means, at least 207 numbers which between 0–255 do not appear in the cell payload.

Assuming that the number "X" (between 0 and 255) does not appear in the cell payload of 48 bytes, then you can choose number "X" to hide information. The specific hiding principle is:

- If the binary number to be hidden is "1", then all "X − 1" digits in the payload become "X";

- If the binary number to be hidden is "0", then all "X − 1" digits in the payload are unchanged.

There will be at least 207 different "X" in the payload of each cell. If the total number of "X − 1" appears the most in a 48-byte payload, it means that more digits can be used for hiding information, so the hiding algorithm proposed in this paper use the number "X" as the hiding location. This information hiding location is stored in the cell header.

In extreme cases, if all "X − 1" occurrences are less than 8, that is, if any "X" is selected, the hidden information in the payload is less than 8 bits. In this case, the information to be hidden can be directly stored in the cell header.

Figure 4 shows the workflow of the information hiding algorithm.

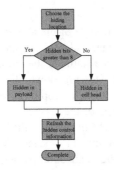

Fig. 4. The workflow of the information hiding algorithm

4.2 Information De-hiding Algorithm

The information de-hiding algorithm is the opposite process of hidden method. The workflow is shown in Fig. 5. After receiving the cell, first determine whether the cell hides the congestion information, and then find the hiding location. If the information is hidden in the header, the 8-bit hidden information is extracted directly from the hiding location field of the cell header; if the information is hidden in the payload, the hidden information is extracted in the payload by the information de-hiding algorithm. Assuming that the hidden position of the cell head is "M", then the information de-hiding algorithm is:

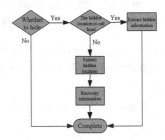

Fig. 5. The workflow of the information de-hiding algorithm.

- The byte of "M − 1" appears in the payload, and the hidden information "0" is restored. The byte remains unchanged.
- The byte of "M" appears in the payload, and the hidden information "1" is restored. The byte number becomes "M − 1".

4.3 Hidden Control Information

In order to complete the hiding and de-hiding of control information, this paper designs a format of hidden control information, which is stored in the header of the switching cell, including identification, type and hidden location.

- ID: "1" indicates that there is hidden information in the cell, and "0" indicates that there is no hidden information in the cell;
- Type: When the flag is "1", indicates that the information is hidden in the cell header, and "0" means that the information is hidden in the cell payload;
- Hidden position: When the flag ID and the flag Type are both "1", means this 8 bits of hidden position are the hidden information; when the ID is "1" and the type bit is "0", means this 8 bits are the hidden position in the payload.

In summary, the information hiding and de-hiding algorithm in this paper can hide the N-bit control information in the switching data. Taking ATM cells as an example, the algorithm can guarantee at least 8 bits of control information in each cell, that is, the worst case transmission of 3 ATM cells can transmit 24-bit congestion control information to ground terminal.

5 Algorithm Verification

This section takes the onboard ATM switching system as an example to verify the new congestion control algorithm proposed in this paper. Firstly, the feasibility of the proposed information hiding and de-hiding algorithm is verified by the enumerated method, which analyses whether the algorithm can transmit the congestion control information correctly. Then the cell loss ratios are simulated and analyzed between the algorithm of congestion control based on satellite switching and the other algorithm without congestion control strategy.

5.1 Algorithm Analysis

The congestion control algorithm proposed in this paper embeds the congestion control information into the switching data while sending cells to the ground terminal. This section mainly verifies the correctness and feasibility of the proposed information hiding algorithm by analyzing.

The implementation process of this algorithm mainly includes several steps: control information generation, information hiding, data reception and information extraction.

Suppose there is congestion in the onboard ATM switch, and an alarm message needs to be sent according to the cache condition.

Step 1. Generate alarm control message.

Because an alarm message needs to be sent, the information type is set as "01". Assuming that the current cache state only requires the ground terminal to slow down the transmission rate to resolve the congestion, the bit of buffer status is set as "01". If the cell is unicast, the connection reference value is "5", and the QoS level is 1, then the binary control message is:

"01010000000000010100100000".

Step 2. The generated alarm message is hidden in the switching data.

Suppose the first ATM cell to be sent is (hexadecimal):

"0241800022,0000101033334444101055556666101077771010000010103334444411010555566661010777710100000101033334444".

Analysis of the cell load shows that:

The decimal number 17 does not appear in the payload and the account of decimal number 16 (hexadecimal is 10) appears the most (18 times) in the payload, so the number 17 is selected as the hidden bit, there is 18 digits which can be hidden in the cell.

Using the hiding algorithm introduced in 3.2, the data including control information becomes:

"2241811089,000010113333444410115555666610107777101000001010333444441010555566661010777711100000111033334444".

In the same way, the remaining control message content is hidden into the following data.

Step 3. The switching data is sent to the corresponding ground terminal by the output address.

Step 4. After receiving the data, the ground terminal extracts the congestion control information and restores the original service data.

The received cells are:

"2241811089,0001011333344441011555566661010777710100000101033334444101055556666101077771110000011103334444".

According to the de-hiding algorithm introduced in 3.3, the hidden information finally extracted from the cell is "01010000000000001010".

The other hidden information is then extracted from the subsequent data in the same way.

Step 5. The ground terminal slows down the transmission rate according to the requirement by the alarm control information.

The congestion control process based on information hiding is completed.

The above analysis proves that the novel algorithm proposed in this paper can correctly transmit congestion control information to the terminal without any additional physical resources.

5.2 Simulation

In this chapter, the processes including sending, buffer and switching, are simulated by Matlab. Then the performance between the proposed algorithm of congestion control and the other algorithm without congestion control strategy are compared. Because the satellite has multi-beams and every beam has one corresponding port of switch. The onboard switch we choose to simulate has eight ports.

The simulation time in this section has 100000 slots, and the cells arrive by the process of burst. Then the cell loss ratio is simulated and analyzed between the proposed algorithm of congestion control and the other algorithm without congestion control strategy. The performance of the proposed congestion control algorithm can be verified, and it will be found that whether the congestion can be controlled. If there are not any violation cells, the onboard switch doesn't drop the cells with the congestion control algorithm proposed in this paper. Consequently the cell loss ratios of the two strategies are simulated and compared when there are some violation cells.

Figure 6 compares cell loss ratio between the algorithm of congestion control based on satellite switching and the other algorithm without congestion control strategy, for which the payload is from 0.55 to 0.95 and the burst length is 16. In this figure, the cell loss ratios of some different probabilities for violation cells are also compared.

The figure shows that the proposed algorithm has the smaller cell loss ratio than the other algorithm. The conclusion is found from the simulation that the proposed algorithm in this paper, which combines the open-loop and close-loop based on the current state of onboard buffer, can reduce cell loss ratio of onboard switching obviously.

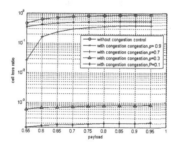

Fig. 6. Cell loss ratio between the algorithm of congestion control based on satellite switching (different probabilities for violation cell) and the other algorithm without congestion control strategy

6 Conclusion

Because satellite network has the character of limited resources, congestion control is an important issue which needs to be solved in satellite processing and switching system. If the onboard switch is congested, it will greatly reduce the performance of the system. The congestion control algorithm based on information hiding is proposed to embed congestion control messages into the switching data, and the congestion control message is sent while transmitting data in order to achieve the purpose of congestion control. This algorithm not only saves link resources, but also improves the performance of the onboard processing and switching system. Consequently, the novel algorithm can be considered to adopt in the onboard switch for which the resource are limited.

Acknowledgment. This work is supported by the Sustainedly Supported Foundation by National Key Lab. of Science and Technology on Space microwave(2018SSFNKLSMT-12).

References

1. Qin, H., Zeng, Z., Wang, B.: The satellite network application based on onboard switching. In: The Ninth Communication Conference, Beijing, 1 March 2013
2. Han, B., Zong, L., Liang, S.: Research satellite communication network congestion control protocol based on OPNET. Electron. Qual. **5**, 52–55 (2015)
3. Yan, H., Zhang, Q., Sun, Y.: On congestion control strategy for space delay/disruption tolerant networks. J. Commun. **37**(1), 142–150 (2016)
4. Tian, B., Cai, S.: Multi-rate multicast congestion control mechanism for GEO satellite network. Acta Electronica Sinica **44**(7), 1599–1604 (2016)
5. Zhang, Y., Li, J., Li, J.L., et al.: The algorithm of congestion control based on satellite switching. Int. J. Future Comput. Commun. **4**(6), 399–402 (2015)
6. Li, J.L., Zhang, Y., Li, J., Hu, Y.L., et al.: A simulation of congestion control algorithm based on satellite IP switching. In: The International Conference on Electronics, Communications and Control (ICECC2011), Ningbo, China, 9–11 September 2011
7. Ma, J.: Research of Rate Mechanism on Internet Congestion Control Protocol. Ningbo University, Ningbo (2015)
8. Zhou, Y., Xu, C.: Hopf bifurcation in a fluid-flow model of congestion control in wireless networks with state feedback control. Chin. J. Comput. Phys. **32**(3), 352–360 (2015)
9. Lin, C., Dong, Y.: Research on space internetworking service based on DTN. J. Comput. Res. Dev. **51**(5), 931–943 (2014)
10. Cao, Y., Sun, Z.: Routing in delay/disruption tolerant networks a taxonomy, survey and challenges. IEEE Commun. Surv. Tutor. **15**(2), 654–677 (2013)
11. Sun, Y., Ji, Z., Wang, H.: An improved congestion control algorithm for satellite broadband network. J. Astronaut. **30**(6), 2366–2371 (2009)

A Load-Balanced and Heterogeneous Switch Path (LHSP) Algorithm in Space Optical-Electrical Hybrid Switching System

Jingling Li[1(✉)], Yi Zhang[1], Wei Liang[1], Tao Cui[1], Jun Li[1],
and Gong Li[2]

[1] National Key Laboratory of Science and Technology on Space Microwave,
CAST, Xi'an 710100, China
ljlspirit@foxmail.com
[2] Beijing Institute of Space Mechanics and Electricity, Beijing, China
13651058529@189.cn

Abstract. Currently, with both optical and packet heterogeneous paths existing in space optical-electrical hybrid switching system, the method of independently calculating the optical/packet switching path cannot comprehensively consider the availability of the system's path resources, which result in the possibility of selecting irrational switching path and making the path load of the switching system unbalanced, even may cause high path blocking probability. In this paper, the homogeneous path influence factor and the heterogeneous path influence factor are defined to quantitatively measure path weights, and based on the optical/packet integrated signaling, a load-balanced and heterogeneous switch path algorithm (LHSP) is proposed to calculate the internal heterogeneous switching path. Moreover, the load balancing degree and the path blocking probability are simulated based on Switching Structure Simulation Model in OPNET. Extensive simulation results show that the LHSP algorithm can reduce average load balancing degree by about 32% and average path blocking probability by about 51% with previous algorithm.

Keywords: Optical-electrical hybrid switching · Integrated signaling ·
Switching path · Load balancing

1 Introduction

The space information network needs the large capacity and strong anti-jamming capability of optical communication, and also needs the wide beam and easy access characteristics of microwave communication. The future spatial information network will be a form of optical/microwave hybrid network in a certain period of time. The switching system in this network should have the capability of optical-electrical hybrid switching, which may exist on a certain satellite node, or a number of tightly coupled spatial nodes to complete the aggregation or distribution switching function of optical-electrical services. This paper mainly focuses on the optical-electrical hybrid switching on satellite node.

B. Li et al. (Eds.): IoTaaS 2018, LNICST 271, pp. 487–495, 2019.
https://doi.org/10.1007/978-3-030-14657-3_50

At present, there are many forms of optical networks combining with electricity. Their common feature is the integration of optical and electrical data communication at the data plane, and the unified control at the control plane. For example, POTN [1] networks based on OTN, has the unified control plane, leading PTN Ethernet and MPLS-TP packet switching and processing functions to realize the organic convergence of optical and electric. And there is no optical-electrical hybrid situation in POTN switching system. In terms of the optical protocol, the GMPLS technology based on the IP backbone network has been proposed to integrate the IP layer with the optical layer [2], the literature [3, 4] propose the RSVP-TE signaling in GMPLS protocol which can be used in the satellite networks and based on a unified control, but there is no relevant description of the optical-electrical hybrid switching system. In the research of space information network, the literature [5] proposes an optical-electrical hybrid on-board switching technology scheme, with the edge nodes packing the packet data into optical data and the core nodes switching the optical data, which doesn't have implemented an integral optical-electrical hybrid switching system in one satellite node.

In the satellite switching system, optical switching mainly includes resources such as wavelength/band, matrix switch, and time slice. Packet switching mainly includes resources such as routing table/forwarding table, bandwidth, and time slot. When the optical service needs to be switched to packet output port, or the packet service needs to be switched to the optical output port, such services need to be switched through heterogeneous switching system, which may pass through different combination paths of optical and electrical. If irrational switching paths are selected, some of the path loads may be oversaturated which may lead to path blocking of the paths. At the same time, other available paths in the switching system are idle and are not fully utilized. This situation may result in unbalance path loads and lower resource utilization. Currently, most of the load balancing-based path algorithms in satellite consider hops control [6], the load balancing of homogeneous path [7], or the effect of other constraints, for instance control link constraints [8] are introduced in the SDN network routing algorithm and energy constraints are combined [9] to calculate the paths.

In this paper, the load-balanced and heterogeneous switch path algorithm (LHSP) is proposed to build heterogeneous path source weight function, which is based on the space integrated signaling. And according to the homogeneous path influence factor and the heterogeneous path influence factor etc., the path load capacity can be quantitatively measured by the path source weight function, then the optimal hybrid switching path satisfied the load balancing could be selected with the maximum path resource weight. The simulation results based on switching model with OPNET shows that the LHSP algorithm not only can effectively distribute the optical-electrical load, but also can improve the utilization of path resources and reduce the blocking probability of the switching system.

2 Integrated Signaling Design in Optical-Electrical Hybrid Switching

The switching structure of the satellite optical-electrical hybrid switching system includes two types of heterogeneous switching structure: the optical switching matrix and the packet (electrical) switching matrix. Figure 1 shows an example of satellites optical-electrical heterogeneous switching structure [10], which is composed of optical switching matrix and the electrical switching matrix, with optical switching capabilities, packet switching capabilities, and optical-electrical heterogeneous switching capabilities.

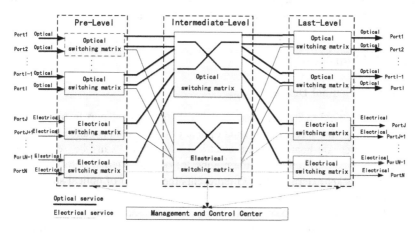

Fig. 1. An example of optical-electrical hybrid switching structure

The optical switch path must be configured before the service transmission, for the optical data is hard to cache, and the packet data can be stored and forwarded through the table. For this situation, an optical-electrical integration signaling mechanism [11] is proposed.

The switching paths inside the satellite switch are considered as a set of the combination of the optical switching path and the packet switching path, which can be expressed as: $\{P_o, P_e\}$, where P_o is the subset of optical switching paths, $P_o = \{g_1, \cdots, g_y\}$, $1 \leq y \leq G$, G is the number of optical internal ports, and P_e is the subset packet switching paths, $P_e = \{l_1, \cdots, l_x\}$, $1 \leq x \leq K$, K is the number of the packet internal port. When the input and output ports are heterogeneous, an integrated signaling mechanism is used for joint reservation of the optical-electrical resources.

In the integrated signaling mechanism, signaling bundling is used as shown in Fig. 2. The input service (input port) type is used as the signaling message 1, the output service (port) is used as the signaling message 2, and two types of signaling messages are used. It is nested in IP datagrams in order of priority. The signaling header must be properly modified.

Fig. 2. The format of optical-electrical integration signaling message

After receiving the optical-electrical integrated signaling, the satellite switch determines whether to accept the signaling request according to the message request, the available path resources and network resource status of the optical/packet. And the optimal switching path is generated according to LHSP algorithm.

3   LHSP Algorithm

3.1   Network Symbol Definition

In an optical-electrical switching system, the paths from input port to output port are called switching paths, where the passing optical or packet path is called sub-path. The topology of the on-board optical-electrical switching can be expressed as $S(M, N, G, K)$, where M represents the number of optical external ports, N represents the number of packet external ports, G represents the number of optical sub-paths, and K represents the number of packet sub-paths. For the given input and output ports, the number of sub-paths included in a switching path is the same, but the number of path conversion (optical sub-path converse to packet sub-path or packet sub-path converse to optical sub-path) times in a switching path is not necessarily the same. The switch service requests obey uniform distribution, and only one service request arrives at the same time. To facilitate the algorithm description, the following related symbols are defined:

$T(l, w)$: The optical-electrical switching structure resources status information table, which labels currently available resources for all sub-paths;

$R_{(x,y)}$: The signaling request from input port X to output port Y;

$P_{(x,y)}$: The set of all switching path combinations from input port X to output port Y in the optical-electrical switching system;

$H_{(x,y)}$: The number of path conversion (optical sub-path converse to packet sub-path or packet sub-path converse to optical sub-path) times in a switch path. If the number of path conversion is zero, $H_{(x,y)}$ equal to 0.01, that is, the switching path without path conversion is preferentially selected.

Example: Take an optical-electrical hybrid switching path $\{O_{11}, E_{12}E_{13}E_{14}E_{15}, E_{21}E_{32}E_{43}E_{54}\}$ of Fig. 1 as an example: where an optical service needs to be transmitted to four packet paths, and the switching path takes a total of nine sub-paths and one path conversion.

3.2   Algorithm Description

If the optical-electrical integrated signaling message containing the input and output port information are accepted by the optical-electrical switching system, the optical-electrical heterogeneous switching path based on load balancing needs to be calculated. The algorithm steps proposed in this paper are as follows:

(1) Firstly, initialize the optical-electrical heterogeneous switching system to record the resources usage of each sub-path in each switching structure, and establish a resource status information table $T(l, w)$. When the new switching path request is received, or the existing switch path is removed, the corresponding path resource information of the table $T(l, w)$ will be updated;

(2) When $R_{(x,y)}$ arrived, calculate the set $P_{(x,y)}$ which combining all the sub-paths in the switching system according to the service input port x and the output port y;

(3) Calculate the number of path conversions $H_{(x,y)}$ of each switching path in the set $P_{(x,y)}$, which should be avoided as much as possible, because it would bring extra resource overhead and complexity of the system;

(4) Measure the impact of choosing different homogeneous sub-paths (such as optical path set or packet path set) on the load balancing of the switching system through the homogeneous path influence factor ω;

(5) Measure the impact of different heterogeneous sub-paths on the load balancing of the switching system through the heterogeneous path influence factor υ_{α};

(6) Construct the optical-electrical heterogeneous path weight function $Cp(x, y) = \frac{1}{H_{(x,y)}} \cdot \sum_{k=1}^{K} \sum_{l=1}^{L} Lw_l \cdot \omega_l \cdot \upsilon_{\alpha_k}$ to calculate the path weight $Cp(x, y)$ of all switching paths in the set $P_{(x,y)}$, where $H_{(x,y)}$ is the number of path conversion; Lw_l is the available resource value of any one of the sub-paths; ω_l is homogeneous path influence factor and υ_{α_k} is heterogeneous path influence factor;

(7) Arrange the path weights $Cp(x, y)$ of the switching paths in the set $P_{(x,y)}$ in ascending order, and select the switching path with the largest $Cp(x, y)$ value;

(8) Determine whether the selected switching path for all sub-path available resource value Lw_l is not less than the service request resource minimum Lw_{min}, if not less than Lw_{min}, go to the next step, if less than the value, go to step (10);

(9) Select the switching path of the previous step, at the same time, reserve corresponding optical switching and packet switching resources according to the signaling request message, and update the path resource status information in $T(l, w)$;

(10) Delete the path of the maximum resource weight obtained in step (8) from the set $P_{(x,y)}$, and then determine whether the set $P_{(x,y)}$ is empty, if it is not empty, go to step (8), if it is empty, the service request fails this time.

3.3   Key Factors Generation Method

The homogeneous path influence factor ω and the heterogeneous path influence factor υ_{α} are the key factors of the path weight function, and the calculation method is as follows:

1. Method for calculating the homogeneous path influence factor ω.
   (1) Query the available resources Lw included in each sub-path are through table $T(l, w)$;
   (2) Calculate the percentage of the available resource value Lw of each sub-path in packet/optical path set to the total number of available resource values of the same type of sub-paths in each switching path, and sort the percentage in ascending order;
   (3) Allocate the percentage of the previous step to each sub-path in reverse order, that is, allocate the maximum percentage to the sub-path with the smallest resource, and allocate the smallest percentage to the sub-path with the largest resource, and obtain the homogeneous path influence factor ω, which satisfies the constraint condition $\sum_{l=1}^{L} \omega_l = 1$, where L is the number of sub-paths in the path set of this type.
   By this method, the sub-paths can be balanced as much as possible to avoid causing hot sub-paths.
2. Method for calculating the heterogeneous path influence factor υ_α.
   (1) Calculate the greatest common divisor of the two types (optical path bandwidth and packet path bandwidth) of maximum path bandwidth values, and define the divisor to be the minimum granularity bandwidth α, which is the minimum unit to measure the bandwidth of the heterogeneous path;
   (2) Calculate the quotient of two types path bandwidth values and the minimum granularity bandwidth α respectively, and divide the quotient of optical/packet path bandwidth by the sum of them to obtain the ratio of the path bandwidth of the two types path;
   (3) Multiply the total available resources of the optical path with its corresponding path bandwidth ratio to obtain the optical path granularity ratio $Tw_{\alpha o}$; and multiply the total available resources of the packet path with its corresponding path bandwidth ratio to obtain the packet path granularity ratio $Tw_{\alpha e}$;
   (4) Calculate the percentage of ratio $Tw_{\alpha o}/Tw_{\alpha e}$, respectively, to their total;
   (5) Assign the percentage obtained from the previous step to the optical path and the packet path in reverse order, that is, the larger percentage is allocated to the path type with a smaller granularity ratio, and a smaller percentage is allocated to the path type with a larger granularity ratio. In this case, the percentage value assigned to the path is the heterogeneous path influence factor υ_α, which satisfies the constraint condition $\sum_{k=1}^{K} \upsilon_{\alpha k} = 1$, where K is the number of paths type, and here $K = 2$.

By calculating the heterogeneous path influence factor, it can be said that the impact of different types of paths on the load balancing of the overall switching system is converted into specific values related to the available resources of the heterogeneous paths, which is beneficial to the trade-off comparison between different switching paths.

4   Performance Evaluation

The optical-electrical switching structure shown in Fig. 1 has be simulated in OPNET. The front stage and the rear stage each contain two 2 × 2 optical switching matrices and two 8 × 8 packet switching matrices, and the intermediate stage contains one 4 × 4 optical switching matric and one 16 × 16 packet switching matric, as shown in Fig. 3.

In the incremental service mode, that is, when a service connection is established, it will occupy the path resources and will not be released. In this paper, the load balancing degree (LBD) of switching system based on LHSP algorithm and traditional load balance path selection algorithm [12] are compared.

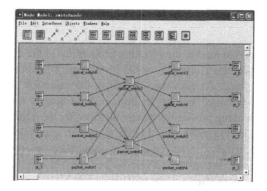

Fig. 3. Switching structure simulation model in OPNET

The LBD can indicate the balance degree of the path load through the statistical variance of the path resources, which is defined as LBD $= \sum_{l \in L}(w(l) - w)^2/(|L| - 1)$, Where L is the path set of the switching system, $|L|$ is the total number of paths in switching system, $w(l)$ is the number of available resources in the current path, and $w = \sum_{l \in L} w(l)/|L|$ is the average number of used resources in each path. A smaller LBD indicates better network load balancing, which means to distribute service more effectively on all paths. In order to compare the LBD of optical-electrical switching networks under different service load conditions, a total of 0-120 service requests are sent. The simulation results are shown in Fig. 4.

The traditional algorithm is based on the input port to select the optimal optical switching path and the corresponding packet switching path, which does not consider the impact of the optical switching path on the selection of the packet switching path. Therefore, as shown in Fig. 4(a), after ten calculations, the LBD obtained by the traditional algorithm is random and not the optimal result. The LBD values obtained by the traditional algorithm are averaged as shown in Fig. 4(b), it can be seen that the LHSP algorithm can effectively reduce the LBD, which can be reduced by about 32% When the network load reaches 120. When the network load is larger, the LHSP algorithm can more effectively enhance the network load balancing capability and improve the utilization of switching path resources.

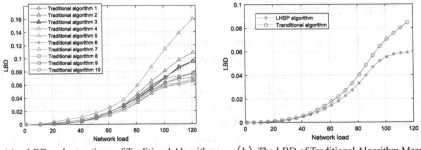

(a)   LBD under ten times of Traditional Algorithms   (b) The LBD of Traditional Algorithm Means
and LHSP Algorithm

Fig. 4. LBD under different network loads

In the dynamic service model, the blocking probability of the LHSP algorithm and the traditional load balancing algorithm [12] in the case of different service request arrival rates are compared. As shown in Fig. 5(a), the same for ten calculations, the blocking probability obtained by the traditional algorithm is random and not the optimal result. The blocking probability values obtained by the traditional algorithm are averaged as shown in Fig. 5(b), the LHSP algorithm can effectively reduce the Blocking probability, which can be reduced by about 51% when the service request arrival rate reaches 240. When the service request arrival rate is larger, the LHSP algorithm can more effective reduce the blocking probability.

As shown in Fig. 5, when the blocking probability starts to be greater than zero, the corresponding service request arrival rate of LHSP algorithm is larger than that of traditional algorithm, which can be explained by the fact that the LHSP algorithm comprehensively considers the available resources of both the optical path and the packet path. So, the path is balanced to avoid bottleneck paths and delay path congestion. The traditional load balancing algorithm only considers the resources of the homogeneous path, which cannot comprehensively balance the overall path resources of the optical-electrical hybrid switching system.

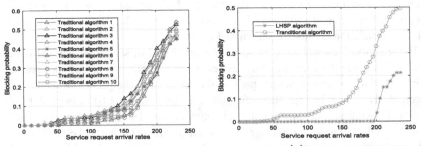

(a)   Blocking probability under ten times of Traditional Algorithms (b) The Blocking probability of
Traditional Algorithm Means and LHSP Algorithm

Fig. 5. Blocking probability under Different Service request arrival rates

5   Conclusion

In this paper, a load-balanced path calculate method for optical-electrical heterogeneous switch is proposed, which can be applied to the space optical-electrical hybrid switching system. Optical switching resources and packet-switching resources are compared quantitatively, and a weight function of the switching path resources is built to comprehensively consider the current optical/electrical available switching resources and generate a switching path that satisfies the load balancing of the switching network.

In the mode of adding services, with simulation model of OPNET, the network load balancing degree and blocking probability of the LHSP algorithm and the traditional path algorithm are compared through multiple simulations, and the simulation results show that the LHSP algorithm proposed in the paper can not only improve the path load capacity, but also reduce the path blocking probability of the switching system, and ultimately improve the overall switching performance of the optical-electrical hybrid switching system. In the following work, the LHSP algorithm is further optimized by adding characterization of optical and packet heterogeneous resources.

Acknowledgment. This work is supported by the sustainedly supported foundation by national key lab. of science and technology on space microwave (2018SSFNKLSMT-12).

References

1. Lee, J., Shim, B.: Network dimensioning methodology in packet-optical transport network. In: 2014 16th Asia-Pacific Network Operations and Management Symposium (APNOMS), pp. 1–4 (2014)
2. Yamanaka, N., Shiomoto, K., Oki, E.: GMPLS Technologies: Broadband Backbone Networks and Systems. Taylor & Francis Group, LLC, USA (2006)
3. Lu, L.: Research and Prototype Implementation of the Protocol in Hybrid Satellite Communication Networks. Xidian University, Xi'an (2014)
4. Wang, H., Hang, J.: The research on traffic engineering in GMPLS optical networks. Microelectron. Comput. **31**(5), 67–70 (2014)
5. Li, R., Zhao, S., Yao, Z., et al.: Research of on-board mixed optical/electric switching of GEO broadband multimedia satellite. Opt. Commun. Technol. **35**(6), 51–53 (2011)
6. Hu, J., Li, T., Wu, S.: Routing of a LEO&MEO double layer mobile satellite communication system. Acta Electronica Sinica **28**(4), 1–5 (2000)
7. Wang, J., Guo, Y., Sun, L.: Load balancing algorithm for multi-traffic in double layered satellite networks. Syst. Eng. Electron. **38**(9), 2156–2161 (2016)
8. Wang, P., Xu, H.: Control link load balancing and low delay route deployment for software defined networks. IEEE J. Sel. Areas Commun. **35**(11), 2446–2455 (2017)
9. Yin, J., Yang, X.: ELQS: an energy-efficient and load-balanced queue scheduling algorithm for mobile ad hoc networks. In: 2009 International Conference on Communications and Mobile Computing, pp. 121–126 (2009)
10. Cui, T., Li, J.: A distributed optical-electrical hybrid switching structure, China, 201710571631.2 (2017)
11. Li, J., Li, J.: Design and processing method of satellite signaling with a mixed optical/electric switching capability. Microelectron. Comput. **35**(6), 112–116 (2018)
12. Wang, L., Huang, S., Yang, X., et al.: A dynamic routing and wavelength assignment algorithm with traffic-balance in WDM network. Commun. Technol. **40**(11), 255–257 (2007)

Author Index

Printed in the United States
By Bookmasters